Abū Bakr Ibn al-ʿArabī (d. 543/1148)

Handbook of Oriental Studies

Handbuch der Orientalistik

SECTION ONE

The Near and Middle East

Edited by

Maribel Fierro (*Madrid*)
M. Şükrü Hanioğlu (*Princeton*)
D. Fairchild Ruggles (*University of Illinois*)
Florian Schwarz (*Vienna*)

VOLUME 184

The titles published in this series are listed at *brill.com/ho1*

Abū Bakr Ibn al-ʿArabī (d. 543/1148)

Legacy and Impact on the Transmission of al-Ghazālī's (d. 505/1111) Thought in al-Andalus

Edited by

Ilyass Amharar

BRILL

LEIDEN | BOSTON

Cover illustration: Abū ʿAbd Allāh al-Sāḥilī al-Muʿammam al-Mālaqī (d. 753/1352), *Bughyat al-Sālik fī ashraf al-masālik*. Private manuscript collection. Photograph taken by ʿAbd al-ʿAzīz al-Sāwurī.

The Library of Congress Cataloging-in-Publication Data is available online at https://catalog.loc.gov
LC record available at https://lccn.loc.gov/2024039102

Typeface for the Latin, Greek, and Cyrillic scripts: "Brill". See and download: brill.com/brill-typeface.

ISSN 0169-9423
ISBN 978-90-04-70684-2 (hardback)
ISBN 978-90-04-70685-9 (e-book)
DOI 10.1163/9789004706859

This book is printed on acid-free paper and produced in a sustainable manner.

Contents

Preface VII
List of Figures and Tables VIII
Notes on Style, Transliteration and Dates IX
Notes on Contributors X

Introduction: More Than One Hundred and Fifty Years of Research on Ibn al-ʿArabī: Assessment and Perspectives 1
Ilyass Amharar

1 The Writings of Ibn al-ʿArabī
A Quantitative Approach 20
Maribel Fierro

2 Law, Theology, and Language in Ibn al-ʿArabī's Thought
Steps in a Personal Approach (2002–2022) 36
Delfina Serrano-Ruano

3 مَفهومُ الإلحادِ في فِكرِ أَبي بَكرٍ ابنِ العَرَبِيّ
جَعفَر ابن الحاج السُّلَميّ

The Concept of *ilḥād* in the Thought of Ibn al-ʿArabī 61
Jaafar Ben El Haj Soulami

4 Ibn al-ʿArabī and al-Ṭurṭūshī
Are Their al-Ghazālī Critiques Related? 90
Kenneth Garden

5 القاضي ابن العربي أصوليًا
عبد الغني إدعيكل

al-Qāḍī Ibn al-ʿArabī as an *uṣūlī* 108
Abdelghani Daikal

6 الاستشفا من كتاب الشِّفا:
بيانٌ للنظر المعافريّ في مَشْرَع أبي حامد الأُصُولي
عبد اللّٰه التَّوْرَاتِي

al-Istishfā min Kitāb al-Shifā
Ibn al-ʿArabī's Analysis of al-Ghazālī's uṣūlī *Project* 131
Abdallah Taourati

7 De Bagdad à Séville : Étude d'une généalogie du *kalām*, d'al-Ashʿarī à Ibn al-ʿArabī 146
Ilyass Amharar

8 جهود الإمام أبي بكر بن العربي في النحو واللغة
قراءةٌ فيما كتب عنه
فريد بن عبد العزيز الزامل السُّلَيْم

Ibn al-ʿArabī's Work on Grammar and Lexicography
A Recent Bibliography 177
Fareed Abdulaziz Alzamil Alsulaim

9 Ibn al-ʿArabī, *adīb* et transmetteur de l'*adab* et de la poésie en al-Andalus, d'après le témoignage de quelques contemporains 184
Bruna Soravia

10 أبو بكر ابن العربي المعافري
رائد التأليف في « فن المسلسلات الحديثية » بالغرب الإسلامي
نورالدين الحميدي

Ibn al-ʿArabī, the Pioneer of Writings on *musalsalāt ḥadīthiyya* in the Islamic West 203

Nourredine Elhmidy

11 مخطوطات القاضي أبي بكر ابن العربي المعافري بخزائن سوس
محمد علوان

Manuscripts of Ibn al-ʿArabī's Texts Preserved in the Private Libraries of Souss (Morocco) 221
Mohammed Aalouane

Index of Terms 241
Index of Works 243
General Index 245

Preface

This volume is the product of the international symposium “Le Qāḍī Abū Bakr Ibn al-ʿArabī (m. 543/1148): parcours, héritage et transmission” which was hosted by the Centre Jacques Berque (Rabat) on 28–29 January 2022.

The colloquium was part of the activities I carried out during my post-doc at the CJB, in line with the director’s vision of revitalising Islam Studies within the institution.

I have received precious help and support from a great number of colleagues and friends. I received additional support for the arduous task of tidying up the language and unifying the referencing style of pieces written by different authors.

Our warmest thanks go to the Centre Jacques Berque for hosting the symposium, to Brill for publishing this work and to the anonymous evaluators. Special thanks to Elies Roumili, Ilias Lahmidi, and, last but not least, Abdurraouf Oueslati for the technical assistance.

The last word here is for my beloved family, to whom I remain indebted.

Ilyass Amharar
Rabat, Morocco
April 2024

Figures and Tables

Figures

0.1 Some of Ibn al-ʿArabī's *asānīd* from the Mashriq 3
7.1 Transmission de l'ashʿarisme, du fondateur à Ibn al-ʿArabī 170
11.1 لوحة من كتاب مناجاة الأبرار لابن القطان، خزانة خاصة بأكادير 223
11.2 لوحة من كتاب شرح أبيات الكندي لابن فضالة الغرناطي، خزانة خاصة بضواحي أكادير 224
11.3 لوحة من آخر كتاب شرح أبيات الكندي لابن فضالة الغرناطي، خزانة خاصة بضواحي أكادير 225
11.4 لوحة من أول كتاب إصلاح الغلط لابن العطار، خزانة تدسي 226
11.5 لوحة من آخر الجزء الثاني من كتاب إصلاح الغلط لابن العطار، خزانة تدسي 227
11.6 لوحة من أول كتاب شرح الجمل لأبي عبد اللّٰه المرسي، خزانة خاصة بسوس 228
11.7 لوحة من آخر كتاب شرح الجمل لأبي عبد اللّٰه المرسي، خزانة خاصة بسوس 229
11.8 لوحة من كتاب غريب الموطأ لأبي علي الأشيري، خزانة خاصة بسوس 230
11.9 لوحة من كتاب جنى الأزاهر النضيرة وسنى الزواهر المنيرة في صلة المطمح والذخيرة مما ولّدته الخواطر من المحاسن في هذه المدّة الأخيرة؛ لأبي الحسن علي ابن الفخار الرّعيني (ت666هـ)؛ محفوظة في خزانة أزاريف 231
11.10 لوحة من كتاب رسالة في الكلام على مشكل حديث السُّبحات والحجاب لابن العربي المعافري؛ محفوظ بخزانة خاصة بأكادير 233
11.11 لوحة من كتاب أحكام القرآن لابن العربي المعافري؛ محفوظة بخزانة أزاريف 234
11.12 لوحة من كتاب أحكام القرآن لابن العربي المعافري؛ محفوظة بخزانة أزاريف 235
11.13 لوحة من كتاب واضح السبيل إلى معرفة قانون التأويل وفوائد التنزيل؛ محفوظة بخزانة خاصة بضواحي تارودانت 235
11.14 لوحة من نسخة أخرى من كتاب واضح السبيل إلى معرفة قانون التأويل وفوائد التنزيل؛ محفوظة بخزانة خاصة بضواحي أكادير 236
11.15 لوحة من أول كتاب الناسخ والمنسوخ؛ محفوظة بخزانة الحاج علي بأكلو 237
11.16 لوحة من أول كتاب أوجز السير لخير البشر لابن فارس؛ محفوظة بخزانة تدسي 237

Tables

1.1 The most prolific Andalusī scholars 21
1.2 Number of works by renowned Andalusī scholars 22
1.3 Number of works by Ibn al-ʿArabī's contemporaries 23
1.4 Number of works by various scholars from the generation of Ibn al-ʿArabī's pupils 23
1.5 Number of Ibn al-ʿArabī's works and transmissions according to the fields of knowledge in *HATA* 25
1.6 The fields of knowledge of the works written by Ibn al-ʿArabī and Ibn Ḥazm 33

Notes on Style, Transliteration and Dates

For referencing, this volume follows the *Chicago Manual of Style* author-date in-text citation system.

Arabic words and names are transliterated according to the system used in Brill's *Encyclopaedia of Islam Three*:

> Consonants: ʾ, b, t, th, j, ḥ, kh, d, dh, r, z, s, sh, ṣ, ḍ, ṭ, ẓ, ʿ, gh, f, q, k, l, m, n, h, w, y.
> Short vowels: a, i, u.
> Long vowels: ā, ī, ū.
> Diphtongs: aw, ay.
> *Tāʾ marbūṭa*: -a, -at (construct state).

While classical proper names are fully transliterated (e.g., al-Ghazālī), for modern names, i.e., since 1900, also the official or common spellings are adopted (e.g., Mohammed Ali al-Bar). The "l" of the definite article "al-" is always retained, regardless of whether it is assimilated in pronunciation to the initial consonant of the word to which it is attached (*idghām*).

If not otherwise specified, the dates given are common era (CE) dates. If two dates are provided (e.g., 505/1111), the first one is the year according to the Islamic hijrī calendar (AH) and the second the CE date. For dates after 1900 only the CE date is provided.

Notes on Contributors

Mohammed Aalouane (محمد علوان)
is independent researcher in Islamic Law and Islamic Manuscripts. He is a specialist of private libraries in Souss (South of Morocco).

Fareed Abdulaziz Alzamil Alsulaim (فريد بن عبد العزيز الزامل السليم)
is an Associate Professor of Arabic Syntax and Morphology in the Department of Arabic Language and Literature at Qassim University, Kingdom of Saudi Arabia.

Ilyass Amharar (إلياس أمحرار)
is post-doc researcher in the Institut de Recherches sur les Mondes Arabe et Musulman (IREMAM). He has written monographs, translations and many articles on the intellectual history of the Maghrib (with a focus on ashʿarism, malikism and sufism). His last monograph is *Langage et théologie chez Abū Bakr Ibn al-ʿArabī (543/1148)* (Gorgias Press, 2023).

Jaafar Ben El Haj Soulami (جعفر بن الحاج السلمي)
is Full Professor in Moroccan and Andalusian heritage in general, Arabic philology, Arabic mythology, history of Arabic thought, sufism and translation, in the Faculty of Letters and Human Sciences at 'Abd al-Malik al-Sa'di University, Tetouan.

Abdelghani Daikal (عبد الغني ادعيكل)
is Associate Professor in Islamic Law and Jurisprudence at Mohammed I University of Oujda (Morocco).

Noureddine Elhmidy (نور الدين الحميدي)
is Associate professor in Hadith sciences at Qadi Iyad University of Marrakech.

Maribel Fierro
is Research Professor at Institute of Languages and Cultures of the Mediterranean, CSIC (Madrid).

Kenneth Garden
is an Associate Professor in the Department of Religion at Tufts University. His research is on the life and thought of Abu Hamid al-Ghazali and his reception in the Islamic East and the Islamic West.

Delfina Serrano-Ruano
is PhD Tenured Researcher at the Spanish National Council for Scientific Research (CSIC). She specializes in the history of Islamic law and theology. Her research work focuses on the study of Islamic jurisprudence (*fiqh*), and its relationship with other Islamic religious sciences such as *kalām* or discursive theology, and Sufism. Her publications address these issues in their doctrinal and practical aspects, and in specific socio-historical contexts, including the contemporary period.

Bruna Soravia
is independent scholar. She had taught history of al-Andalus and Medieval Islam in Italy and France.

Abdellah Taourati (عبد الله التوراتي)
is an Associate Professor in Hadith studies at Mohammed I University.

INTRODUCTION

More Than One Hundred and Fifty Years of Research on Ibn al-ʿArabī

Assessment and Perspectives

Ilyass Amharar

Born in 468/1076 in Seville, then capital of the Abbadid state, Muḥammad ibn ʿAbd Allāh ibn Muḥammad al-Maʿāfirī, known as Abū Bakr Ibn al-ʿArabī or simply Ibn al-ʿArabī, was the son of a vizier of the eponymous dynasty, and grew up in an affluent Andalusī noble family.[1] While still just a teenager, he was forced to leave the peninsula with his father for the East. He describes this journey in great detail in his travel account, *Tartīb al-riḥla li-l-targhīb fī l-milla*, which inaugurated an original literary genre: the *riḥla fahrasa*.[2] This journey brought him to Iraq—more precisely, to Baghdad, the Abbasid capital—where he met with the scholar Abū Ḥāmid al-Ghazālī, who remained his master and a major influence for him throughout his life. Returning to Seville in 495/1101–2, Ibn al-ʿArabī taught not only the sciences he had acquired in the East—Ḥadīth, *tafsīr* (exegesis), Ashʿarī theology, and *uṣūl al-fiqh* (legal theory)—but also *adab* (belles-lettres). He was ambitious enough to plan a reform of the teaching of Mālikī law, but this met with much opposition on the side of the Almoravid regime which had controlled Seville since 1091, making him feel like a foreigner in his own country.

Ibn al-ʿArabī's career was not always glorious. First elected to the high judiciary of Seville, he was criticised for his toughness,[3] and people went so far as to ransack his house, burn some of his books, and force him to flee to Cordoba. He did not return to Seville until the end of his life. His reputation as a great Ḥadīth specialist was tainted by rumours accusing him of fabricating *isnād*s (chains of transmission).[4] His link with the Almohads—who at first gave him a

1 Garden, Kenneth, "The *riḥla* and self-reinvention of Abū Bakr Ibn al-ʿArabī", in *Journal of the American Oriental Society* 135 (2015), 1–17 at p. 3.

2 Touati, Houari, *Islam et voyage au moyen âge: Histoire et anthropologie d'une pratique lettrée*, Paris: Le Seuil, 2000, 287. Ibn al-ʿArabī undertook this journey at the age of sixteen, accompanied by his father. The original travelogue having disappeared, one can nevertheless find extracts from this *riḥla* in various texts by Ibn al-ʿArabī, such as *Qānūn al-taʾwīl*.

3 Lagardère, Vincent, *Histoire et société en occident musulman au moyen âge: Analyse du Miʿyār d'al- Wanšarīšī*, Madrid: Casa de Velazquez, 1995, 70 (fatwa 259).

4 Robson, J., "Ibn al-ʿArabī", in *EI2*.

 | DOI:10.1163/9789004706859_002

warm welcome when he came to Marrakesh, before imprisoning him there for a year—raises many questions, as does his death from poisoning in 543/1148.[5]

The influence of Ibn al-ʿArabī is significant: he is credited with having been one of the great actors in the "Ashʿarisation" of al-Andalus by introducing books that quickly became major references in North African teaching, such as al-Juwaynī's *al-Irshād ilā qawātiʿ al-adilla fī uṣūl al-iʿtiqād*.[6] His influence was not limited to theology, but extended to all aspects of religious education—Ḥadīth studies, *uṣūl al-fiqh, taṣawwuf* (Sufism), and so forth—as well as *adab*, as evidenced by the Andalusī catalogues (*fahāris*),[7] in which many *isnād*s of books converge on Ibn al-ʿArabī. He was thus able to transmit into the West numerous *isnād*s from the East and was able to connect, either directly or through a minimum number of intermediaries, with famous scholars who were authoritative figures in their respective disciplines, such as al-Khaṭīb al-Baghdādī in Ḥadīth studies, al-Khaṭīb al-Tibrīzī in *adab*, and, of course, al-Ghazālī in *uṣūl al-fiqh*. In the various *isnād*s of later Maghribī scholars,[8] not only is the name of Qadi Ibn al-ʿArabī very often found, but it is also almost systematically the last (or first, depending on the reading direction) Maghribī name. This confirms the pivotal position that Ibn al-ʿArabī may have had in the flow of knowledge from East to West, a kind of channeler of Eastern *isnād*s, to which a large number of Maghribī and Andalusī scholars flocked. This centrality earned him the nickname of "the storehouse of knowledge" (*khizānat al-ʿilm*) and "the pole of the Maghrib" (*quṭb al-Maghrib*).[9]

He is also renowned for having introduced al-Ghazālī's work to al-Andalus. His relationship with al-Ghazālī, which he discussed in detail in his *riḥla* (now lost, but partially restored from his *Qānūn al-taʾwīl*[10] and his *al-ʿAwāṣim min al-qawāṣim*), allowed scholars to learn more about al-Ghazālī himself. Despite the hierarchy of the master/student relationship, Ibn al-ʿArabī allowed himself

5 Cano Ávila, P., A. García Sanjuán, and A. Tawfiq, "Ibn al-ʿArabī", in *Biblioteca de al-Andalus*, Almería: Fundación Ibn Ṭufayl de Estudios Árabes, 2009, 129–158.

6 On the works from the East introduced into al-Andalus by the Ibn al-ʿArabī, see Amharar, Ilyass, *Langage et théologie chez Ibn al-ʿArabī: Les informations subtiles de la somme de théorie légale* (*Nukat al-Maḥṣūl fī ʿilm al-uṣūl*), Piscataway NJ: Gorgias Press, 2023, xli–xliv.

7 Such as *Fahrasat Ibn Khayr*, in which Ibn al-ʿArabī's name appears 400 times.

8 Such as *Barnāmaj al-Wādī Āshī*, in which Ibn al-ʿArabī's name appears 306 times.

9 This appellation caused problems for some later Maghribī Sufis, who disputed Ibn al-ʿArabī's holiness and thus pointed to the possible confusion with *quṭb* status, reserved in Sufi hierarchies for an elite of saints. See al-Samlālī, ʿAbbās b. Ibrāhīm, *al-Iʿlām bi-man ḥalla Marrākush wa-Aghmāt min al-aʿlām*, Rabat: Imprimerie Royale, 1993, 102–103.

10 ʿAbbās, Iḥsān, "Riḥlat Ibn al-ʿArabī ilā l-Mashriq kamā ṣawwarahā Qānūn al-taʾwīl", in *al-Abḥāth* 21 (1968), 59–91.

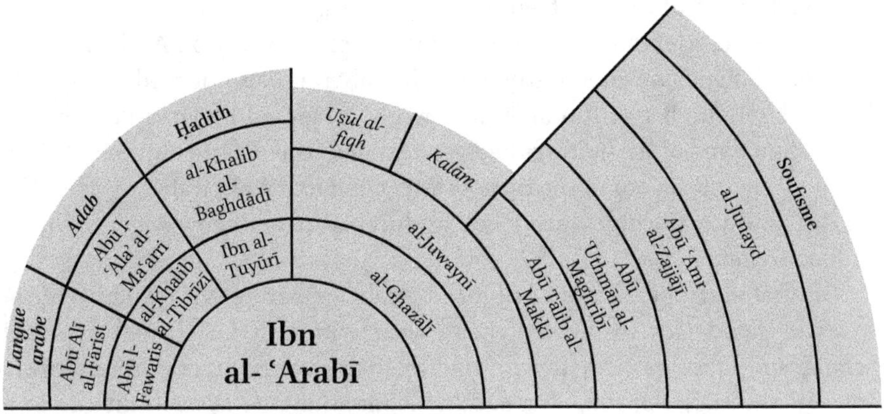

FIGURE 0.1 Some of Ibn al-ʿArabī's *asānīd* from the Mashriq

to criticise certain aspects of al-Ghazālī's thought, in particular the influence of *falsafa*.[11]

However, it would be simplistic to limit the scientific contribution of Ibn al-ʿArabī to that of a simple transmitter of Eastern texts. His abundant work testifies to an original thought, the influence of which went beyond the borders of al-Andalus and the Maghrib. Thus, his *Aḥkām al-Qurʾān* or his *Kitāb al-Nāsikh wa-l-mansūkh* remain staple works among scholars throughout the Muslim world. For some commentators, Ibn al-ʿArabī is one of the rare *mujtahid* scholars, that is to say, those who are capable of extracting legal statutes directly from the Qurʾān and the Sunna.[12]

Ibn al-ʿArabī is a symbolic figure in many respects: of Andalusī origin, but travelling as far as Iraq out of a thirst for knowledge, he illustrates the cultural encounter between East and West. A theologian, jurist, and specialist in Ḥadīth, but also a linguist and an exegete, he embodies the intersection of many domains of knowledge.

For nearly one hundred and fifty years, publications on his work have followed one another at a rapid pace, reflecting a strong interest of researchers in this complex figure of the 5th/11th century. In recent years, increased interest in this research has been encouraged by the discovery of several previously unpublished manuscripts, such as his *al-Amad al-aqṣā fī sharḥ asmāʾ Allāh al-ḥusnā* and *Sirāj al-murīdīn fī sabīl al-dīn*.

11 Griffel, Frank, *Al-Ghazālī's philosophical theology*, New York: Oxford University Press, 2009, 61.

12 Robson, "Ibn al-ʿArabī".

From the end of the nineteenth century onwards, the figure of Ibn al-ʿArabī has interested orientalists. In his bio-bibliographical study on Andalusī scholars, Pons Boigues devoted entry 217 to Ibn al-ʿArabī and offered a very brief biography.[13] The list of Ibn al-ʿArabī's works, prepared by Pons Boigues at a time when none of his writings had yet been published, is unsurprisingly lacking: only eight texts out of the 115 now attributed to Ibn al-ʿArabī are cited. This note is, however, the oldest academic study made, to our knowledge, on the life of Ibn al-ʿArabī.

The first biographical notice of Ibn al-ʿArabī appearing in the *Encyclopaedia of Islam* dates back to the sixties and was written by J. Robson,[14] a British scholar specialising in the study of Ḥadīth. In this very short and summary text (only 364 words), Robson, after succinctly mentioning the scientific career of Ibn al-ʿArabī and his accession to the post of qadi, essentially portrays him as a specialist in Ḥadīth: the only book which grasps his attention is Ibn al-ʿArabī's commentary on the Ḥadīth collection of al-Tirmidhī, and the only lines devoted to his reputation concern his transmission of Ḥadīth. For the rest of his article, Robson refers to a very late source, which nevertheless seems to be his main source, namely the book *Nafḥ al-ṭīb min ghuṣn al-Andalus al-raṭīb* by al-Maqqarī (d. 1041/1631).[15] The notice is also incomplete: none of Ibn al-ʿArabī's works seem to have been directly consulted. Similarly, nothing is said about the intellectual influence that al-Ghazālī could have had on Andalusī scholars through Ibn al-ʿArabī as intermediary, a phenomenon already pointed out by Cabanelas.[16] This can be partly justified[17] by the fact that in the first half of the twentieth century, the work of Ibn al-ʿArabī was still mostly in manuscript format and partially unavailable.

In the Arab world, Iḥsān ʿAbbās published in 1963 in the journal *al-Abḥāth* a study of a previously unpublished passage from Ibn al-ʿArabī's *Tartīb al-riḥla*,[18] which had been quoted in a book entitled *Kitāb Ansāb al-barbar wa-mulūkihim*

13 Pons Boigues, Fransisco, *Ensayo bio-bibliográfico sobre los historiadores y arábigo-españoles geógrafos*, Madrid: Establecimiento Tipográfico de San Fransisco de Sales, 1898, 216–217.

14 Robson, "Ibn al-ʿArabī".

15 See al-Maqqarī, Aḥmad, *Analectes sur l'histoire et la littérature des Arabes d'Espagne*, trans. Reinhart Dozy, Amsterdam: Oriental Press, 1967, i, 477–489.

16 Cabanelas, Darío, "Notas para la historia de Algazel en España", in *al-Andalus* 17 (1952), 223–232, at pp. 223–224.

17 We specify "in part" because *Aḥkām al-Qur'ān*, one of the major works of Ibn al-ʿArabī, was already published in Cairo by the Maktaba Saʿāda in 1914, and Ibn Badīs had published the *ʿAwāṣim* in 1928.

18 ʿAbbās, Iḥsān, "al-Jānib al-siyāsī min riḥlat Ibn al-ʿArabī ilā l-Mashriq", in *al-Abḥāth* 16 (1963), 217–236.

written by an unknown scholar.[19] We also find this passage in the anonymous work *Kitāb Mafākhir al-barbar* published by Lévi-Provençal in 1934.[20] This article contains elements of Ibn al-ʿArabī's *riḥla*, including the content of official documents that Ibn al-ʿArabī would have brought back from the East. This is, to our knowledge, one of the first studies covering the political role of this journey. In a second article five years later, ʿAbbās reconstructed more of the *riḥla* from elements found in the manuscript of another text of Ibn al-ʿArabī, *Qānūn al-taʾwīl*.[21]

In 1975 ʿAmmār Ṭālbī published his thesis in Algiers in which he offered the first serious critical edition of one of Ibn al-ʿArabī's theological works: *al-ʿAwāṣim min al-qawāṣim*[22] and highlighted a still little-known aspect of the author: his *mutakallim* status. The edition is in two volumes: the first, devoted to the study of the text, is entitled *Ārāʾ Abī Bakr ibn al-ʿArabī al-kalāmiyya wa-naqduhu li-l-falsafa al-yūnāniyya*, while the second is an edition of the book *al-ʿAwāṣim*. We discover there that Ibn al-ʿArabī, far from being a mere transmitter of the theses of his Ashʿarī predecessors, had his own opinions on theological questions and that he did not hesitate to express them, even if this meant criticising—sometimes in a virulent way—the positions of scholars considered among the greatest names of the school, such as al-Juwaynī, or even his own master, al-Ghazālī. Ṭālbī's work has the advantage of having shed light on this previously unobserved facet of Ibn al-ʿArabī.

Eleven years later, Muḥammad Slīmānī, professor at the University of Algiers, published with Dār al-Qibla an edition of another book by Ibn al-ʿArabī: *Qānūn al-taʾwīl*.[23] It abounds in factual biographical elements and confirms Ibn al-ʿArabī's status as *mutakallim*, as already argued by Ṭālbī. Disappointed with

---

19 It is a certain Muḥammad b. Abī al-Maḥd al-Maghīlī. See Viguera, María, "Las cartas de al-Gazālī y al-Ṭurṭūšī al soberano almorávid Yūsuf b. Tāšufīn", in *al-Andalus* 42 (1977), 340–374, at p. 341.

20 Lévi-Provençal, E., *Nubadh taʾrīkhiyya fī akhbār al-barbar fī l-qurūn al-wusṭā, muntakhaba min al-majmūʿ al-musammā bi-Kitāb Mafākhir al-barbar*, Rabat: al-Maṭbaʿa al-Jadīda, 1934.

21 ʿAbbās, "Riḥlat Ibn al-ʿArabī" 59–91.

22 In 1951, Muḥibb al-Dīn al-Khaṭīb partially republished the *ʿAwāṣim* as printed by Ibn Bādīs in 1928: he retained only the second part, which concerned exclusively the Companions (pp. 98–193), and which he republished under the title *al-ʿAwāṣim min al-qawāṣim fī taḥqīq mawāqif al-ṣaḥāba baʿda wafāt al-Nabī*. For many years, it was this version of the text that received re-editions, among them that of Maḥmūd Mahdī al-Istambūlī, who preserved al-Khaṭīb's work by adding a few notes. It is thus with ʿAmmār Ṭālibī that we finally have the first truly complete edition of the *ʿAwāṣim*.

23 See Ibn al-ʿArabī, Abū Bakr, *Qānūn al-taʾwīl*, ed. Muḥammad Slīmānī, Tunis: Dār al-Gharb al-Islāmī, 1990, 20.

the first edition, of which he deplored the many typographic errors, Slīmānī decided to republish the book, this time with Dār al-Gharb al-Islāmī publishing house. The section initially devoted to the study of the Ibn al-ʿArabī's life and work was substantial enough to be published in a separate volumed titled *Abū Bakr ibn al-ʿArabī: ḥayātuhu wa-turāthuhu al-fikrī.*[24]

Meanwhile, in Western academia, Dominique Urvoy devoted much of his research to the Andalusī scholarly milieu, including that of Ibn al-ʿArabī.[25] In 1977, in an article published in the journal *al-Andalus*, María Viguera offered a Spanish translation of letters written[26] by al-Ghazālī and al-Ṭurṭūshī to the Almoravid ruler Yūsuf b. Tāshfīn.[27] By directly consulting the manuscript K1275, which includes *Kitāb Ansāb al-barbar* and is kept at the National Library of Rabat, Viguera discovered a text attributed to Ibn al-ʿArabī in which the latter reproduced passages from his *Tartīb al-riḥla* for an unidentified recipient. Interspersed between these passages are the said letters, which are mainly advice on good governance or arguments for the merits of Ibn Tāshfīn's policy. This text highlights two things: the Abbasid legitimisation of the Almoravid empire, already pointed out by Lévi-Provençal,[28] and also the major role that Ibn al-ʿArabī's father played in this support. It was at his request that al-Ghazālī issued a fatwa (also translated in the article) justifying the intervention of Ibn Tāshafīn in al-Andalus.[29] All this confirms the hypothesis formulated fifteen years earlier by ʿAbbās that this *Tartīb al-riḥla* was not simply for scientific purposes, but also had a political motivation, a fact yet absent from most of the biographical notices of Ibn al-ʿArabī.

In 1985, in an issue of the *Revue des mondes musulmans et de la Méditerranée* devoted to al-Andalus, Vincent Lagardère, a historian of the Muslim West, published "Abū Bakr b. ʿArabī, grand cadi of Seville". For this researcher, working on the high judicature among the Almoravids, Ibn al-ʿArabī was "the best illustration of the nature of the Grand Cadi, of his training, his appointment, and his

24 This section is absent from both the Dār al-Qibla and the Dār al-Gharb al-Islāmī editions, and we have not been able to locate it, nor establish whether it has been since republished.

25 Urvoy, Dominique, *Le monde des ulémas andalous du* v*ème*/xi*ème au* vii*ème*/xiii*ème siècle*, Geneva: Droz, 1978, 125–126.

26 For fragments of their original Arabic version, see ʿAbbādī, Aḥmad Mukhtār, *Dirāsāt fī l-Maghrib wa-l-Andalus*, Alexandria: Mu'assasat Shabāb al-Jāmiʿa, 1967, 471–484.

27 Viguera, "Las cartas" 340–374. See also Dandash, ʿIṣmat, "Dirāsa ḥawla rasā'il Ibn al-ʿArabī", in *al-Manāhil* 9 (1977), 149–191.

28 Lévi-Provençal, E., "Le titre souverain des Almoravides et sa légitimation par le califat ʿAbbāside", in *Arabica* 2 (1955), 265.

29 Viguera, "Las cartas" 345.

religious and cultural audience".[30] The number of studies on the Muslim West having greatly increased since the 1960s, not to mention the easier access to the works of Ibn al-ʿArabī with the editions of Ṭālbī and Slīmānī, the biography offered by Lagardère is unsurprisingly much more elaborate and documented than those of his predecessors. The ten pages of the article are divided into two unequal parts: the first, concerning his life and literary journey, represents two thirds of the study, while only four pages are devoted to the second part dealing with his other works. Once again, such an imbalance is explained by the fact that the work of Ibn al-ʿArabī remained very poorly known at the time. Two years earlier, in an article about a text attributed to al-Ghazālī, Lagardère had pointed out the influence that Ibn al-ʿArabī exercised and his major role in disseminating not only Ashʿarī doctrine but also the works of al-Ghazālī among Andalusī and Maghribī jurists.[31] The interest of the article lies in the fact that its author was able to rely on recently published editions of books by Ibn al-ʿArabī, which gave him access to precisely dated information: his date of birth, the stages of his journey, the scholars he was able to meet, and so forth. In his article, Lagardère refers mainly to the *Qānūn al-taʾwīl* as well as to the *ʿAwāṣim*, and aims to shed light on Ibn al-ʿArabī, however only through the prism of his stature as a qadi, as Lagardère's focus of interest was Andalusī qadis. As a result, he does not explore other interesting aspects of Ibn al-ʿArabī, such as his impact in Ashʿarī theology. Like Robson, he refers to the list established by al-Maqqarī in his *Nafḥ al-ṭīb*, but Lagardère's list also relies on the work of Brockelman,[32] which allowed him to point out several books attributed to Ibn al-ʿArabī that are absent from al-Maqqarī.[33] To sum up, the article offers the first relatively complete biography of Ibn al-ʿArabī, even if, for lack of sources, it could not highlight the major role of Ibn al-ʿArabī in the transmission of al-Ghazālī's work in al-Andalus and the Maghrib. Lagardère concluded his article with the suggestion that the Ibn al-ʿArabī "is considered to have imported Ghazālian mysticism into the West: he is even credited with having given Abū Yaʿzā (d. 572/1177) and ʿAlī b. Ḥirzihim (d. 559/1165) the Sufi

30 Lagardère, Vincent, "Abû Bakr b. al ʿArabi, grand cadi de Seville", in *Revue des mondes musulmans et de la Méditerranée* 40 (1985), 91–102.

31 Lagardère, Vincent, "À propos d'un chapitre du *Nafḫ wal-taswiya* attribué a Ġāzalī", in *Studia Islamica* 60 (1984), 119–136.

32 Brockelmann, Carl, *Geschichte der arabischen Litteratur*, Leiden: Brill, 1937–1949, i, 412–413, S i, 632–633, 732–733.

33 Lagardère, "Abû Bakr b. al ʿArabi" 99. An error should, however, be noted: Lagardère cites as his tenth and last book *al-Qawāṣim wa-l-ʿawāṣīm wa-l-maḥṣūl fī uṣūl al-fiqh*. These are, in fact, two separate works: *al-Qawāṣim wa-l-ʿawāṣim* (an alternative title for *al-ʿAwāṣim min al-qawāṣim*) and *al-Maḥṣūl fī ūṣūl al-fiqh*.

investiture as he had received it himself from al-Ghazālī." Unfortunately, Largardère does not provide further justifications or details regarding this.

A year later, Saʿīd Aʿrāb, a Moroccan researcher, discovered a late manuscript of an abridgment of *Tartīb al-riḥla* copied in 741/1338, kept at the Ḥājī Sālim Āghā Library.[34] In 1987, he decided to publish it for Dār al-Gharb al-Islāmī, in a book entitled *Maʿa al-Qāḍī Ibn al-ʿArabī*. The book is divided into two unequal parts: the first (188 pages) is a detailed biography of Ibn al-ʿArabī, and the second (44 pages) is the edited text itself. The interest of the book is twofold. Firstly, by publishing the book, Aʿrāb contributed to making the *riḥla* known to a wider scientific audience. Secondly, Aʿrāb offered a list of Ibn al-ʿArabī's works which was no longer based only on the *Nafḥ al-ṭīb* of al-Maqqarī, but also on personal investigations in the manuscripts attributed to Ibn al-ʿArabī himself. Aʿrāb's contribution to the research on Ibn al-ʿArabī was quite remarkable and all subsequent researchers, working on the *riḥla* or more generally on the scholar, systematically refer to this book.

In 1992, Dār al-Gharb al-Islāmī published Wuld Karīm's thesis, defended in 1986 at the Meccan University Umm al-Qurā, which consists of an edition of one of the commentaries of Mālik b. Anas's *al-Muwaṭṭaʾ* written by Ibn al-ʿArabī: *al-Qabas fī sharḥ Muwaṭṭaʾ Mālik ibn Anas*. Despite its partisan aspect (Karīm deplores the Ashʿarī positions of Ibn al-ʿArabī),[35] this edition provides researchers with a publication in three volumes contributing to the study of the legal thought of Ibn al-ʿArabī. In the same year, Maktabat al-Thaqāfa al-Dīniyya published the thesis of ʿAlawī Madgharī, which is an edition of another work of Ibn al-ʿArabī, *Kitāb al-Nāsikh wa-l-mansūkh*. With this publication, it became possible to understand Ibn al-ʿArabī's position on the subject of abrogation in the Qurʾān and to examine another facet of his career, namely his specialisation in *uṣūl al-fiqh*.

In 1996, in Madrid, Muḥammad Yaʿlā published his thesis, supervised by María Viguera, which consisted of the editing and study of two collated manuscripts kept at the National Library of Rabat, shelfmarks K1275 and D1020. Viguera had already noticed the importance of the first manuscript, and Lévi-Provençal had pointed out the importance of the second in 1934. The shelfmarks K1275 and D1020 refer to two copies of the same collection of texts, including one, *Shawāhid al-jilla wa-l-aʿyān fī mashāhid al-Islām wa-l-buldān*,

34 Aʿrāb, Saʿīd, *Maʿa al-Qāḍī Ibn al-ʿArabī*, Tunis: Dār al-Gharb al-Islāmī, 1987.

35 Ibn al-ʿArabī, Abū Bakr, *al-Qabas fī sharḥ Muwaṭṭaʾ Mālik ibn Anas*, ed. Wuld Karīm, Beirut: Dār al-Gharb al-Islāmī, 1992, i, 2.

attributed to Ibn al-ʿArabī.[36] This text provides the precise historical framework in which the Almoravids, a Berber dynasty, gained, via Ibn al-ʿArabī, definitive legitimisation from the Abbasids. In the introduction, Yaʿlā traces a brief history of research around the famous letters of al-Ghazālī and al-Ṭurṭūshī and indicates the studies, partial editions, and translations that have been made of them.[37] Yaʿlā surmises that Ibn al-ʿArabī's lost *riḥla* would have comprised two discrete parts. The first is a "cultural" work, dealing with his meetings with scholars and the books he studied, entitled *Tartīb al-riḥla*. The text entitled *Shawāhid al-jilla* would then represent the second part and be dedicated to the official documents and missives sent to the Abbasid caliph by his father. *Shawāhid al-jilla* is therefore the "political counterpart" of *Tartīb al-riḥla*, forming with it one single book.

Since the 2000s, researchers interested in Ibn al-ʿArabī have focused on his *riḥla*. Several works, based on those of Aʿrāb and Yaʿlā, give particular importance to the *Tartīb al-riḥla*, which would represent, according to them, an Andalusī model reflecting an important stage in the development of this genre of travel literature. In his study in historical anthropology on scholarly journeys in the first centuries of Islam, Houari Touati devoted the seventh and last chapter to journey and travel accounts, and offered an analysis of the style of Ibn al-ʿArabī's *riḥla* without dwelling too much on its contents.[38] According to Touati, Ibn al-ʿArabī revolutionised travel writing. With his *riḥla*, he inaugurated an original literary form[39] which combines an inventory of scholars, epistolary narrations, and adventure story, all in rhymed prose (*sajʿ*) such as one finds in overtly literary productions like al-Ḥarīrī's *Maqāmāt*. With his study, Touati highlighted that in addition to being a theologian, Ibn al-ʿArabī was well versed in *adab*, to the point of giving an identity to a new literary genre, of which the *Tartīb al-riḥla* would be the very first exemplar.

In 2006, the proceedings of a colloquium were published in which Maribel Fierro took an interest in Andalusī scholars who sought to "reform" Mālikī law. She cites among them Ibn al-ʿArabī, who, as soon as he returned to his native land, felt like a stranger amongst his own people (*gharīb bayna qawmihi*) due to his exertions in this project.[40]

36 Yaʿlà, Muḥammad, *Tres textos árabes sobre beréberes en el Occidente Islámico*, Madrid: CSIC, 1996, 275–376.

37 Yaʿlà, *Tres textos árabes* 33–37.

38 Touati, *Islam et voyage* 287–299.

39 Touati, *Islam et voyage* 287.

40 Fierro, Maribel, "Proto-Mālikīs, Mālikīs and reformed Mālikīs in al-Andalus", in Peri Bearman, Rudolph Peters, and Frank E. Vogel (eds.), *The Islamic school of law: Evolution, devolution and progress*, Cambridge, MA: Harvard Series in Islamic Law, 2006, 57–76, at p. 73.

In 2007, Slīmānī, cited above for the edition of *Qānūn al-ta'wīl*, edited, with the help of his sister ʿĀ'isha, another commentary by Ibn al-ʿArabī on the *Muwaṭṭa'*, much larger than *al-Qabas*, namely *al-Masālik fī sharḥ Muwaṭṭa' Mālik*, in eight volumes. This work stands out in that it offers, in the introduction, a biography that is more documented than previous studies. Over more than seventy pages, Slīmānī covered a record number of thirty-seven biographical sources for Ibn al-ʿArabī, from various periods and geographical areas, while incorporating certain findings published by his predecessors, on the letters of al-Ghazālī and al-Ṭurṭūshī in particular. In addition, he proposed a larger list of works than that of Aʿrāb, indicating some errors made in the reading of certain titles.[41] However, the list is still incomplete: the *Talkhīṣ ṭuruq al-mutaqaddimīn*, for example, is still absent, even though Slīmānī refers to it in the edited text.[42] The *Masālik* indicates that Ibn al-ʿArabī was an ardent defender of Mālikism. Indeed, in the preamble, he specifies that he wrote the commentary in response to contemporary Ẓāhirīs (that is, supporters of Ibn Ḥazm), who disregarded the *Muwaṭṭa*ʿ.[43]

In 2008, a study of one of Ibn al-ʿArabī's most famous books, *Aḥkām al-Qur'ān*, was finally published. The editor, Delfina Serrano, described in her commentary how the author approached the Qur'ān as a legal source.[44]

In 2009, Cano Ávila, García Sanjuán, and Tawfiq wrote in a specialised encyclopaedia on al-Andalus published by the Ibn Ṭufayl Foundation (Almeria) a new entry on "Ibn al-ʿArabī, Abū Bakr".[45] This is the most complete biography of him to date, in that it investigates various significant events in his life, from his childhood to the period after his return from the East. Many aspects overlooked in other studies are specifically addressed here, such as the privileged social rank of his maternal family in Seville, and the role of his uncle at the end of the Abbasid dynasty. This biography includes the various journeys that Ibn al-ʿArabī made, before, during, and after his famous *riḥla*, all precisely dated and transcribed on a very well-designed map. Finally, the entry closes with a

41 Ibn al-ʿArabī, Abū Bakr, *al-Masālik fī sharḥ Muwaṭṭa' Mālik*, ed. ʿĀ'isha and Muḥammad Slīmānī, Tunis: Dār al-Gharb al-Islāmī, 2007, i, 115.

42 Ibn al-ʿArabī, *al-Masālik* vi, 312, 352; vii, 123.

43 Ibn al-ʿArabī, *al-Masālik* i, 331.

44 Serrano Ruano, Delfina, "El Corán como fuente de legislación islámica: Abū Bakr Ibn al-ʿArabī y su obra *Aḥkām al-Qur'ān*", in Miguel Hernando de Larramendi Martínez and Salvador Peña Martín (eds.), *El Corán ayer y hoy: Perspectivas actuales sobre el islam. Estudios en honor al profesor Julio Cortés*, Madrid: La biblioteca del Polo Norte, 2008, 251–275.

45 Cano Ávila, García Sanjuán, and Tawfiq, "Ibn al-ʿArabī".

list of works attributed to Ibn al-ʿArabī reaching 115 titles: four times larger than that of Slīmānī.

The same year, in his book on the theological contribution of al-Ghazālī, Frank Griffel considered it necessary to write a chapter on his most influential disciples, as "the writings of his students are an important source for our understanding of al-Ghazālī's theology."[46] In this list of disciples, the first named after, of course, al-Ghazālī's brother, is Ibn al-ʿArabī. Indeed, Griffel continues, "Among his contemporaries, Abū Bakr Ibn al-ʿArabī is the most important source of information about al-Ghazālī's life and his teachings."[47] After a paragraph on Ibn al-ʿArabī's journey to the East, Griffel examines passages from his work that relate to his meeting with al-Ghazālī, in particular two extracts: the first from *Qānūn al-taʾwīl*, and a second from *al-ʿAwāṣim*. The latter is especially interesting as it enumerates various questions that Ibn al-ʿArabī had, as a young student at the time, about the reality of the human soul (*rūḥ*) and stages a dialogue of questions and answers on this subject with al-Ghazālī that Ibn al-ʿArabī transcribed once he became established as a scholar. This exchange would not only have influenced Ibn al-ʿArabī: al-Ghazālī would have drawn from these answers the material necessary for his book *al-Qānūn al-kullī fi-l-taʾwīl*.[48] Without disturbing the frame of the master–disciple relationship, Griffel remarks that Ibn al-ʿArabī shows himself in *al-ʿAwāṣim* to be critical of al-Ghazālī regarding certain positions, for which he accuses him of having been influenced by *falsafa*.[49] Griffel laments that several important works by Ibn al-ʿArabī that could tell us more about al-Ghazālī's teachings remain unpublished, including *al-Amad al-aqṣā* and *Sirāj al-murīdīn*.

After a doctoral dissertation focusing on the culture of travel among the scholars of al-Andalus from the 5th/11th century to the 7th/13th century, Yann Dejugnat wrote an article published in 2011 about the *Tartīb al-riḥla*. In this article, Dejugnat described Ibn al-ʿArabī's *riḥla* as covering all "the meaning that Andalusī scholars assigned to the traditional journey to the East in search

46 Griffel, *Al-Ghazālī's philosophical theology* 61.

47 Griffel, *Al-Ghazālī's philosophical theology* 62.

48 Griffel, *Al-Ghazālī's philosophical theology* 70. On al-Ghazālī's *Qānūn kullī* and its connection with Ibn al-ʿArabī's questions, see Griffel, Frank, "Al-Ghazālī at his most rationalist: *The universal rule for allegorically interpreting revelation* (*al-Qānūn al-kullī fī t-taʾwīl*)", in Georges Tamer (ed.), *Islam and rationality: The impact of al-Ghazālī. Papers collected on his 900th anniversary*, Leiden: Brill, 2015, 89–120, at p. 91. In 2012, ʿAbduh edited on behalf of Dār al-Kutub al-ʿIlmiyya the manuscript Rabat, National Library, Q555, 1b–14b, entitled *Ajwibat al-Ghazālī ʿan asʾilat Ibn al-ʿArabī*, taking up the famous questions and answers, which allowed Griffel to realize a confusion he had made between this text and another, entitled *Nafkh al-rūḥ wa-l-taswiya*. See Griffel, "Al-Ghazālī at his most rationalist" 92n11.

49 Griffel, *Al-Ghazālī's philosophical theology* 66.

of science in the thirteenth century."[50] According to Dejugnat, this journey encompassed not only knowledge and politics, but also action and mysticism. Indeed, Dejugnat uses the title of the *riḥla* (he translates *targhīb fī l-milla* as "awakening of the desire of religion") to establish a bold comparison with al-Ghazālī's famous *Iḥyāʾ ʿulūm al-dīn* (*The revival of the religious sciences*) and sees in the book an allegorical narrative staging Ibn al-ʿArabī's journey as a response to a double *fitna*: that of the Almoravids and that of his desire for science. A certain tendency to extrapolation runs through the article, especially in the process of "symbolisation" of the journey.

This claimed parallel to the *Iḥyāʾ* did not convince Kenneth Garden,[51] who, for his part, tried in an article to prove that the *riḥla* had not only a simple political or purely scientific goal, but that it also allowed a "self-reinvention" of Ibn al-ʿArabī.[52] Garden shows how this journey forged the personality of Ibn al-ʿArabī, and how his training in the East allowed him, on his return to Seville, to regain the wealth and elite status that his father had lost with the advent of the Almoravids.

In 2015, the Tangier publishing house Dār al-Ḥadīth al-Kattāniyya inaugurated a new book series entitled Aʿlāq Andalusiyya (Andalusian Jewels), which consists of editions of rare manuscripts of scholars from the al-Andalus area. Among the first publications in this series was a study conducted by the Saudi professor Alzamil Alsulaim on *Muljiʾat al-mutafaqqihīn ilā maʿrifat ghawāmiḍ al-naḥwiyyīn wa-l-lughawiyyīn*, a book in which Ibn al-ʿArabī announces that he will solve the linguistic mysteries and difficulties of the Qurʾān and the Sunna, which have caused many problems for jurists. With no access to any manuscript of *Muljiʾat al-mutafaqqihīn*, Sulaym undertook the tedious work of compiling those extracts from the book which Ibn al-ʿArabī himself quoted in his other works.

Other than *Muljiʾat al-mutafaqqihīn*, the series published editions of three important titles by Ibn al-ʿArabī, namely *al-Amad al-aqṣā fī sharḥ asmāʾ Allāh al-ḥusnā*, *al-Mutawaṣṣiṭ fī l-iʿtiqād wa-l-radd ʿalā man khālaf al-sunna min ahl al-bidaʿ wa-l-ilḥād*, and finally, *Sirāj al-murīḍīn fī sabīl al-dīn*, all three edited by ʿAbd Allāh al-Tawrātī.

The first, published in 2015, is a detailed commentary on the divine names according to the theological positions of the Ashʿarī school. The second, published the same year, is an abridgment of a book of *kalām* entitled *al-Awsaṭ*

50 Dejugnat, Yann, "À l'ombre de la *fitna*, l'émergence d'un discours du voyage : À propos du *Tartîb al-rihla* d'Abû Bakr ibn al-ʿArabî (m. 1148)", in *Mediévales* 60 (2011), 85–101, at p. 85.

51 Garden, "The *riḥla* and self-reinvention" 2n4.

52 Garden, "The *riḥla* and self-reinvention" 16–17.

written by the Ashʿarī Abū l-Muẓaffar al-Asfarāyīnī, to which Ibn al-ʿArabī grafts additions of his own, in particular his opinons concerning certain definitions (*ʿaql, ʿilm, taklīf*) and a virulent refutation of the Ẓāhirīs. The interest of this edition, in addition to exposing how Abū l-Muẓaffar was a favoured reference for Ibn al-ʿArabī, is also its introduction, in which Tawrātī exploits three unpublished biographical sources, absent in previous studies.[53] The last publication is a four-volume work published in 2017, which focuses on the themes of Sufism in Ibn al-ʿArabi's work, and in particular his nostalgia for the East, his feelings about his Andalusī contemporaries, his exchanges with al-Ghazālī, and his opinion on the didactics of traditional education. This edition presents very interesting details about the books that Ibn al-ʿArabī introduced from the East into al-Andalus.

Finally, the Aʿlāq Andalusiyya series has also published two very brief epistles edited by Yūsuf Baqayān, namely *al-Risāla al-ḥākima fī masʾalat al-aymān al-lāzima* (2015) which is a fatwa concerning a judgment on oaths (*aymān*) and *al-Risāla fī l-kalām ʿalā mushkil ḥadīth wa-l-subuḥāt wa-l-ḥijāb* (2019) which is a succinct text in which Ibn al-ʿArabī interprets the term *ḥijāb* ("veil") mentioned in several eschatological *ḥadīth*s. In 2019, Niẓām al-Yaʿqūbī published a collection of six epistles, each of them bringing together *ḥadīth*s with common themes, and one of which is attributed to Ibn al-ʿArabī. The latest edition of this collection is an epistle published in 2020 by Tawrātī titled *Tanbīh al-ghabī ilā miqdār al-nabī*, which elucidates delicate questions related to the notion of the impeccability (*ʿiṣma*) of the Prophet Muḥammad.

For its part, the Rābiṭa Muḥammadiyya published in 2017 Nabila Zakri's edition of the *Kitāb al-Afʿāl*, in which Ibn al-ʿArabī examines Qurʾanic verses dealing with divine acts. An introduction dealing with the *kalām* texts of the author precedes the work.

These last years have seen an increase in the number of university theses on Ibn al-ʿArabī in the Arab world, some of them giving rise to publications which bear upon on the legal thought of Ibn al-Arabī, for instance *al-Ikhtiyārāt al-uṣūliyya li-l-Qāḍī Abī Bakr ibn al-ʿArabī fī kitābihi al-Maḥṣūl fī ūṣūl al-fiqh* by Sawsān al-Kūsh (2017), in which the opinions expressed by Ibn al-ʿArabī in his *al-Maḥṣūl fī ūṣūl al-fiqh* are examined, and *Kitāb Nukat al-Maḥṣūl fī uṣūl al-fiqh*, which is the first truly critical edition of the text, compiled by Qāsim Bāy (2017).

In 2018, García Sanjuán published a new biographical note for the *Encyclopaedia of Islam*, considerably more extensive and updated than the now obsolete entry of Robson. The discussion of the scientific relationship between our

53 Ibn al-ʿArabī, Abū Bakr, *al-Mutawaṣṣiṭ fī l-iʿtiqād wa-l-radd ʿalā man khālaf al-sunna min ahl al-bidaʿ wa-l-ilḥād*, ed. A. Tawrātī, Tangier: Dār al-Ḥadīth al-Kattāniyya, 2015, 19–24.

author and al-Ghazālī is not deep enough, but García refers to recent studies and newly available sources to conclude that "Ibn al-ʿArabī was a reformist Mālikī supporting the normalisation of Mālikism in Sunni Islam, that is, of its adaptation to the rules of the classical methodology of law as established by al-Shāfiʿī."[54]

Returning to *Nukat al-Maḥṣūl*, Muḥammad Ḥajj ʿĪsā, in a 2019 article, was interested in Ibn al-ʿArabī's legal methodology[55] and its positioning in the "path of theologians/path of jurists" dichotomy inherited from a reading of the history of *uṣūl al-fiqh* under the influence of Ibn Khaldūn, a reading whose limits I demonstrated in my thesis published by Gorgias Press in early 2023, which offers a critical edition of the *Nukat al-Maḥṣūl*, along with a French translation and commentary upon it.[56]

A few months before the writing of this introduction, a treatise on theology by Ibn al-ʿArabī entitled *al-Wuṣūl ilā maʿrifat al-uṣūl* was edited by Ḥamza Nuhayrī for Dār al-Fatḥ (Jordan) and a set of collections of short *ḥadīth*s compiled by Ibn al-ʿArabī, was edited by Nourredine Elhmidy, which he prefaced with historicising comments in which he asserted that Ibn al-ʿArabī was "the *muḥaddith* of the jurists, and the jurist of the *muḥaddithīn*".[57] Tawrātī has also finished his critical edition of al-Ghazālī's *al-Istishfāʾ min Kitāb al-Shifāʾ*.

At the time of the rise of digital humanities, two of the most important works of Ibn al-ʿArabī are currently being republished: *Aḥkām al-Qurʾān* and *ʿĀriḍat al-aḥwadhī bi-sharḥ Ṣaḥīḥ al-Tirmidhī*.[58]

Organised in Rabat on 28 and 29 January 2022, on the occasion of the 900th anniversary of his death with the support of the Center Jacques Berque, the international symposium "The Qāḍī Abū Bakr Ibn al-ʿArabī: Journey, heritage, and transmission" aimed to draw up an assessment of the research and academic study carried out on the person, the work, and the posterity of one of the most influential scholars of the Muslim West, research which continues after nearly a century and a half. The issues discussed during the meeting were diverse: What information on scientific activity in the Muslim West from the 5th/11th century does the work of Ibn al-ʿArabī and its transmission bring us?

54 García Sanjuán, Alejandro, "Ibn al-ʿArabī, Abū Bakr" in *EI3*.

55 Ḥājj ʿĪsā, "Manhaj al-imām Abī Bakr ibn al-ʿArabī fī kitabihi Nukat al-maḥṣūl", in *al-Mishkāt* 19 (2019), 197–232.

56 Amharar, *Langage et théologie*.

57 Al-Ḥumaydī, Nūr al-dīn, *al-Qāḍī Abū Bakr ibn al-ʿArabī, faqīh al-muḥaddithīn wa-shaykh bi-l-Andalus*, Tunis: Dār al-Gharb al-Islāmī, 2020.

58 The *ʿĀriḍa* edition has just been published in Libya. It is a collective work directed by Ḥamza Abū Fāris published by the Shaykh Ṭāhir al-Zāwī Centre. As for the *Aḥkām*, Tawrātī is working on a new critical edition.

What are the modalities of his influence on the spread of Ashʿarism in particular? How much was his work influenced by his relationship with al-Ghazālī? Does the exploitation of private manuscript collections allow the discovery of new texts by Ibn al-ʿArabī?

In her contribution entitled "The writings of Ibn al-ʿArabī: A quantitative approach", Maribel Fierro offers a study of the scientific output of Ibn al-ʿArabī compared to other great intellectual figures, including Ibn Ḥazm, whom Ibn al-ʿArabī virulently criticised, even though he was one of his father's teachers.

In her "Law, theology, and language in Ibn al-ʿArabī's thought: A personal approach (2002–2022)", Delfina Serrano-Ruano gives us the results of several studies carried out on different works of our subject: the *Qabas*, his *Aḥkām al-Qurʾān* and his *Qānūn al-taʾwīl*, which touch on different cases such as stoning, the Mālikī legal position concerning rape, the voluntary recourse of the *dhimmīs* (non-Muslim residents of Islamic states) to Islamic judicial authorities, the thorny subject of the *takfīr* (theological anathematisation) of the Almoravids under the Almohads on grounds of *tajsīm* (anthropomorphising God), as well as more strictly theological matters, such as the diffusion of Muʿtazilī or Qadarī doctrines in al-Andalus in the twelfth century.

Jaafar Ben El Haj Soulami focuses in his "Mafhūm al-ilḥād fī fikr Abī Bakr Ibn al-ʿArabī" ("The concept of *ilḥād* in the thought of Ibn al-ʿArabī") on the lexicon used by Ibn al-ʿArabī to disqualify his theological opponents, in particular the term *ilḥād* and its derivatives such as *mulḥid*,[59] lexicon which he considers as a key element allowing us to access more precisely the theological thought of the Qadi.

Kenneth Garden's contribution, entitled "Ibn al-ʿArabī and al-Ṭurṭūshī: Are their al-Ghazālī critiques related?" deals with a double relationship: that between Ibn al-ʿArabī and Abū Bakr al-Ṭurṭūshī, who taught him during his years in Jerusalem; and that of both Ibn al-ʿArabī and Abū Bakr al-Ṭurṭūshī to Abū Ḥamīd al-Ghazālī, whom each met and about whom each wrote.

As previously noted, recent studies of Ibn al-ʿArabī focus, among other things, on his impact on the development of *uṣūl al-fiqh* in the Muslim West, an impact long neglected in Western orientalist research, which is perhaps a little too influenced by Ibn Khaldūn, and the negative picture he paints of scientific activity in *uṣūl al-fiqh* in the Muslim West. Abdelghani Daikal's ambition in his contribution is to analyse all aspects of Ibn al-ʿArabī's "*uṣūlī* personality".

---

59 According to Lane, the *mulḥid* is "one who deviates, or swerves, from the truth [...] an impugner of religion [...]. Some apply the appellation especially to the Bāṭiniyya." See Lane, Edward, *An Arabic–English lexicon*, London: Williams & Norgate, 1863 s.v. *l-ḥ-d*.

In his contribution on *al-Istishfāʾ*, Abdallah Taourati aims to show that despite al-Ghazālī's definite influence on Ibn al-ʿArabī's legal thought, the two scholars had two quite distinct scientific projects, which explains why at certain points of *uṣūl* our subject differed from the positions of his teacher.

In my contribution I aim to outline the transmission of Ashʿarism from its eponymous founder to Ibn al-ʿArabī, which allows me to highlight some blind spots in the theological orientation of the Qadi, especially before his *riḥla*.

As we have already pointed out, the influence of Ibn al-ʿArabī in the transmission of knowledge in the Muslim West was not limited to religious disciplines alone. Fareed Abdulaziz Alzamil Alsulaim's contribution thus shows that Ibn al-ʿArabī was also a grammarian and lexicographer who could develop his own theories in these disciplines. Moreover, as evidenced by Bruna Soravia's article, he was an important transmitter of *adab* and poetry in the region, in particular that of Abū al-ʿAlāʾ al-Maʿarrī. After a careful consultation of four Almoravid sources, Soravia analyses the extent and variety of this transmission. These four texts make it possible to discover a little-studied aspect of the figure of Ibn al-ʿArabī that can perhaps be found more generally in the Andalusī scholarly elites: Ibn al-ʿArabī was a personality respected and listened to his local literary world, having the power to launch the career of a poet or to render him forgotten.

If these contributions confirm to us that Ibn al-ʿArabī was certainly the polymath of his time in al-Andalus, they should not make us forget that his notoriety outside the peninsula is due to the reputation he has gained in a particular discipline: the Ḥadīth. The contribution of Nourredine Elhmidy reminds us of this. It is known that Ibn al-ʿArabī was qualified by all sources as being *ḥāfiẓ*, that is to say, the highest degree in this science, granting him *taṣḥīḥ*, the right to declare on the authenticity of Ḥadīth reports. In addition, he was the author of the most famous commentary on *al-Jāmiʿ al-ṣaḥīḥ* of al-Tirmidhī, a collection of prophetic traditions, which was first introduced into al-Andalus by Ibn al-ʿArabī's grandfather, Abū Ḥafṣ al-Hawzanī (d. 460/1068). What has not been realised, however—and this is the focus of Elhmidy's work—is that Ibn al-ʿArabī was the first to have inaugurated a sub-genre of Ḥadīth literature in the Muslim West: the *musalsalāt*.[60]

60 A *ḥadīth* is said to be *musalsal* when it presents on several links of the same *isnād* a common factor which can be textual (such as each reporter being preceded or succeeded by the same formula) or gestural (such each reporter holding his beard after transmitting the *ḥadīth*).

Finally, Mohammed Aalouane's article offers a survey of some private collections in the Souss and what they contain in the way of precious and unexploited manuscripts of Ibn al-ʿArabī's texts.

Bibliography

ʿAbbādī, Aḥmad Mukhtār, *Dirāsāt fī l-Maghrib wa-l-Andalus*, Alexandria: Mu'assasat Shabāb al-Jāmiʿa, 1967.

ʿAbbās, Iḥsān, "al-Jānib al-siyāsī min riḥlat Ibn al-ʿArabī ilā l-Mashriq", in *al-Abḥāth* 16 (1963), 217–236.

ʿAbbās, Iḥsān, "Riḥlat Ibn al-ʿArabī ilā l-Mashriq kamā ṣawwarahā Qānūn al-ta'wīl", in *al-Abḥāth* 21 (1968), 59–91.

Alzamil Alsulaim, Fareed, *Masā'il min Mulji'at al-mutafaqqihīn*, Tanger: Dār al-Ḥadīth al-Kattāniyya, 2017.

Amharar, Ilyass, "Abū ʿImrān al-Ǧawrā'ī et sa profession de foi (ʿaqīda): À la découverte d'un théologien ašʿarite maghrébin encore inconn. Édition critique et traduction", in *Mélanges de l'Institut dominicain d'études orientales* 35 (2020), 265–300.

Amharar, Ilyass, *Langage et théologie chez Ibn al-ʿArabī: Les informations subtiles de la somme de théorie légale* (*Nukat al-Maḥṣūl fī ʿilm al-uṣūl*), Piscataway NJ: Gorgias Press, 2023.

Aʿrāb, Saʿīd, *Maʿa al-Qāḍī Ibn al-ʿArabī*, Tunis: Dār al-Gharb al-Islāmī, 1987.

Ben ʿAbd al-Salām, *ʿAfīf, La vie littéraire dans l'Espagne musulmane sous les Mulūk al-Ṭawā'if : v^e^/x^ie^ siècle*, Damas: Presses de l'Ifpo, 2001.

Brockelmann, Carl, *Geschichte der arabischen Litteratur*, Leiden: Brill, 1937–1949.

Cabanelas, Darío, "Notas para la historia de Algazel en España", in *al-Andalus* 17 (1952), 223–232.

Cano Ávila, P., A. García Sanjuán, and A. Tawfiq, "Ibn al-ʿArabī", in *Biblioteca de al-Andalus*, Almería: Fundación Ibn Ṭufayl de Estudios Árabes, 2009, 129–158.

Dandash, ʿIṣmat, "Dirāsa ḥawla rasā'il Ibn al-ʿArabī ", in *al-Manāhil* 9 (1977), 149–191.

Dejugnat, Yann, "À l'ombre de la fitna, l'émergence d'un discours du voyage: À propos du Tartîb al-rihla d'Abû Bakr ibn al-ʿArabî (m. 1148)", in *Mediévales* 60 (2011), 85–101.

EI2 = P. Bearman, Th. Bianquis, C.E. Bosworth, E. van Donzel, and W.P. Heinrichs (eds.), *The Encyclopaedia of Islam, second edition*, Leiden: Brill, 1954–2004.

EI3 = Kate Fleet, Gudrun Krämer, Denis Matringe, John Nawas, and Devin J. Stewart (éds.), *The Encyclopaedia of Islam, Three*, Leiden: Brill, 2007–.

Fierro, Maribel, "Proto-Mālikīs, Mālikīs and reformed Mālikīs in al-Andalus", in Peri Bearman, Rudolph Peters, and Frank E. Vogel (eds.), *The Islamic school of law: Evolution, devolution and progress*, Cambridge, MA: Harvard Series in Islamic Law, 2006, 57–76.

García Sanjuán, Alejandro, "Ibn al-ʿArabī, Abū Bakr" in Kate Fleet et al(eds.), *Encyclopaedia of Islam, THREE*.

Garden, Kenneth, "The riḥla and self-reinvention of Abū Bakr Ibn al-ʿArabī", in *Journal of the American Oriental Society* 135 (2015), 1–17.

Griffel, Frank, *Al-Ghazālī's philosophical theology*, New York: Oxford University Press, 2009.

Griffel, Frank, "Al-Ghazālī at his most rationalist: The universal rule for allegorically interpreting revelation (al-Qānūn al-kullī fī t-taʾwīl)", in Georges Tamer (ed.), *Islam and rationality: The impact of al-Ghazālī. Papers collected on his 900th anniversary*, Leiden: Brill, 2015, 89–120.

Ḥājj ʿĪsā, "Manhaj al-imām Abī Bakr ibn al-ʿArabī fī kitabihi Nukat al-maḥṣūl", in *al-Mishkāt* 19 (2019), 197–232.

al-Ḥumaydī, Nūr al-dīn, *al-Qāḍī Abū Bakr ibn al-ʿArabī, faqīh al-muḥaddithīn wa-shaykh bi-l-Andalus*, Tunis: Dār al-Gharb al-Islāmī, 2020.

Ibn al-ʿArabī, Abū Bakr, *Qānūn al-taʾwīl*, ed. Muḥammad Slīmānī, Tunis: Dār al-Gharb al-Islāmī, 1990.

Ibn al-ʿArabī, Abū Bakr, *al-Qabas fī sharḥ Muwaṭṭaʾ Mālik ibn Anas*, ed. Wuld Karīm, Beirut: Dār al-Gharb al-Islāmī, 1992.

Ibn al-ʿArabī, Abū Bakr, *al-Masālik fī sharḥ Muwaṭṭaʾ Mālik*, ed. ʿĀʾisha and Muḥammad Slīmānī, Tunis: Dār al-Gharb al-Islāmī, 2007.

Ibn al-ʿArabī, Abū Bakr, *al-Mutawaṣṣiṭ fī l-iʿtiqād wa-l-radd ʿalā man khālaf al-sunna min ahl al-bidaʿ wa-l-ilḥād*, ed. A. Tawrātī, Tangier: Dār al-Ḥadīth al-Kattāniyya, 2015,

Lagardère, Vincent, "À propos d'un chapitre du Nafḫ wal-taswiya attribué a Ġazalī", in *Studia Islamica* 60 (1984), 119–136.

Lagardère, Vincent, "Abû Bakr b. al ʿArabi, grand cadi de Seville", in *Revue des mondes musulmans et de la Méditerranée* 40 (1985), 91–102.

Lagardère, Vincent, *Histoire et société en occident musulman au moyen âge: Analyse du Miʿyār d'al- Wanšarīšī*, Madrid: Casa de Velazquez, 1995.

Lane, Edward, *An Arabic–English lexicon*, London: Williams & Norgate, 1863.

Lévi-Provençal, E., *Nubadh taʾrīkhiyya fī akhbār al-barbar fī l-qurūn al-wusṭā, muntakhaba min al-majmūʿ al-musammā bi-Kitāb Mafākhir al-barbar*, Rabat: al-Maṭbaʿa al-Jadīda, 1934.

Lévi-Provençal, E., "Le titre souverain des Almoravides et sa légitimation par le califat ʿAbbāside", in *Arabica* 2 (1955), 265.

al-Maqqarī, Aḥmad, *Analectes sur l'histoire et la littérature des Arabes d'Espagne*, trans. Reinhart Dozy, Amsterdam: Oreintal Press, 1967.

Pons Boigues, Fransisco, *Ensayo bio-bibliográfico sobre los historiadores y arábigo-españoles geógrafos*, Madrid: Establecimiento Tipográfico de San Fransisco de Sales, 1898.

al-Samlālī, ʿAbbās b. Ibrāhīm, *al-Iʿlām bi-man ḥalla Marrākush wa-Aghmāt min al-aʿlām*, Rabat: Imprimerie Royale, 1993.

Serrano Ruano, Delfina, "El Corán como fuente de legislación islámica: Abū Bakr Ibn al-ʿArabī y su obra Aḥkām al-Qurʾān", in Miguel Hernando de Larramendi Martínez and Salvador Peña Martín (eds.), *El Corán ayer y hoy: Perspectivas actuales sobre el islam. Estudios en honor al profesor Julio Cortés*, Madrid: La biblioteca del Polo Norte, 2008, 251–275.

Serrano Ruano, Delfina, "La diffusion de l'ashʿarisme et la réforme du crédo malikite à l'époque almoravide: Ibn Rushd al-Jadd, Abū Bakr Ibn al-ʿArabī et le Qāḍī ʿIyāḍ", in Aillet Cyrille, Tuil Leonetti Bulle (eds.), *Dynamiques religieuses et territoires du sacré au Maghreb médiéval*, Madrid: CSIC, 2015, 79–102.

Touati, Houari, *Islam et voyage au Moyen Âge. Histoire et anthropologie d'une pratique lettrée*, Paris: Le Seuil, 2000.

Urvoy, Dominique, *Le monde des ulémas andalous du* V*ème*/XI*ème au* VII*ème*/XIII*ème siècle*, Geneva: Droz, 1978.

Viguera, María, "Las cartas de al-Gazālī y al-Ṭurṭūšī al soberano almorávid Yūsuf b. Tāšufīn", in *al-Andalus* 42 (1977), 340–374.

Yaʿlà, Muḥammad, *Tres textos árabes sobre beréberes en el Occidente Islámico*, Madrid: CSIC, 1996.

1

The Writings of Ibn al-ʿArabī

A Quantitative Approach

Maribel Fierro

Abū Bakr Ibn al-ʿArabī (468/1076–543/1148) was one of the most prolific scholars in the history of al-Andalus, with some 115 works to his name, according to information collected in the database *Historia de los Autores y Transmisores de al-Andalus* (HATA) and the *Biblioteca de al-Andalus* (BA).*[1] This number can only be approximate, as some titles may be false attributions or refer to the same work, issues that can only be resolved by specific studies. However, for works that have not been preserved or on which no study has been carried out, it is sometimes difficult to reach a clear conclusion. This caveat—that the figures are approximate in the case of Ibn al-ʿArabī's output and in other cases here considered—should be kept in mind.

The number of works written by Ibn al-ʿArabī, even if approximate, remains impressive. His output was, however, much lower than that of the most prolific Andalusī author, the mystic Muḥyī l-dīn Ibn ʿArabī (560/1165–638/1240), to whom around 481 works are attributed. Still, Ibn al-ʿArabī is not far behind the second most prolific, Ibn Ḥazm al-Ẓāhirī (384/994–456/1064), with 156

* This paper was read at the conference *Le Qāḍī Abū Bakr Ibn al-ʿArabī (m. 543/1148) : Parcours, héritage et transmission*, Colloque international à l'occasion du 900ème anniversaire de son décès, organized by Ilyass Amharar, Centre Jacques Berque, Rabat, 28–29 January 2022. It is part of the project *Al-Andalus y el Magrib en el Oriente islamico: Movilidad, migracion y memoria*, directed by Mayte Penelas and Maribel Fierro (2021–2025) PID2020-116680GB-I00, financed by MCIN/AEI/10.13039/501100011033, and of the project *El proceso pedagógico en el Occidente islámico medieval* (PIE 202010E135). This work received additional funding from the ILC's strategy programme for Supporting, Promoting and Disseminating Research (2023), under the aegis of the grant TR2023/XX. I wish to thank Luis Molina for his help.

1 See *HATA*, an online and free-access bio-bibliographical resource, at https://www.eea.csic.es/red/hata/. For the *BA*, see Jorge Lirola Delgado and José Miguel Puerta Vílchez (eds.), *Biblioteca de al-Andalus*, 7 vols., Almería: Fundación Ibn Tufayl de Estudios Árabes, 2004–2012, together with an *Apéndice*, ed. Jorge Lirola Delgado, Almería: Fundación Ibn Tufayl de Estudios Árabes, 2012, and *La producción intelectual andalusí: balance de resultados e índices*, Almería: Fundación Ibn Tufayl de Estudios Árabes, 2013. Abū Bakr Ibn al-ʿArabī's entry in the *BA* by P. Cano Ávila, A. García Sanjuán and A. Tawfiq is found in ii, 129–158, no. 298.

 | DOI:10.1163/9789004706859_003

TABLE 1.1 The most prolific Andalusī scholars

Authors	Number of works
Muḥyī l-dīn Ibn ʿArabī (560/1165–638/1240)	481
Ibn Ḥazm (384/994–456/1064)	156
Ibn al-ʿArabī (468/1076–543/1148)	115
Ibn Rushd al-Ḥafīd (520/1126–595/1198)	102

works, and he has more titles to his name than Ibn Rushd al-Ḥafīd (Averroes; 520/1126–595/1198), who penned 102 (Table 1).

To illustrate the fact that Ibn al-ʿArabī's case, together with those of the other three scholars listed in Table 1, is to be considered extraordinary—as part of the very small number of Andalusī scholars[2] who authored more than one hundred works—Table 2 below offers a survey of other highly reputed Andalusī scholars whose output, even if abundant or influential, was far less extensive (the scholars mentioned are ordered by date of death).[3]

Among the scholars listed in Table 2, Abū l-Walīd al-Bājī was a forerunner of Ibn al-ʿArabī with regards to his involvement in the study of legal methodology or "the fundamentals of the law" (*uṣūl al-fiqh*), but his intellectual interests were not as wide-ranging as those of the younger scholar. In this respect, Ibn al-ʿArabī was closer to Ibn Ḥazm, whom he also resembled in terms of his combination of both belles-lettres (*adab*) and religious sciences. The breadth and depth of Ibn al-ʿArabī's scholarship likewise comes to the fore when compared with some of his contemporaries (Table 3) and with the scholars of the following generation (Table 4). The comparison clearly shows how the sheer quantity of Ibn al-ʿArabī's output was unusual for his context, being comparable only to that of Ibn Rushd al-Ḥafīd.

As mentioned, the information regarding the number of books these scholars authored is taken from both the *Biblioteca de al-Andalus* and the database *HATA*. The aim of *HATA* is to help researchers find information on the works written and transmitted in al-Andalus as a first step toward carrying out further

2 Authors qualify as "Andalusī" if they are included in a biographical dictionary dealing with the scholars of al-Andalus. Such works often have a section devoted to the *ghurabāʾ*, that is, to scholars from outside al-Andalus who settled in or visited al-Andalus. Moreover, for a significant period of time al-Andalus was integrated into the North African empires, meaning that the category "Andalusiī" is not always easy to distinguish from that of "Maghribī".

3 The caveat above about the difficulty of establishing a "fixed" number of works should be kept in mind.

TABLE 1.2 Number of works by renowned Andalusī scholars

Authors	Number of works
ʿAbd al-Malik b. Ḥabīb (174/790–238/853)[a]	45
Abū ʿAmr al-Dānī (371/981–444/1053)[b]	73
Abū ʿUmar b. ʿAbd al-Barr (368/978–463/1071)[c]	32
Abū l-Walīd al-Bājī (403/1013–474/1081)[d]	41
al-Ḥumaydī (420/1029–488/1095)[e]	22
Ibn Rushd al-Jadd (450/1058–520/1126)[f]	19
Ibn al-Sīd al-Baṭalyawsī (444/1052–521/1127)[g]	29
ʿIyāḍ b. Mūsā al-Yaḥṣubī (476/1083–544/1149)[h]	28
Ibn Bashkuwāl (494/1101–578/1183)[i]	30
Ibn Saʿīd al-Maghribī (605/1208 or 610/1213–673/1274 or 685/1286)[j]	38
Abū Ḥayyān al-Gharnāṭī (654/1256–745/1344)[k]	68
Ibn al-Khaṭīb (713/1313–776/1374)[l]	71

a Arcas Campoy, M. and D. Serrano Niza, "Ibn Ḥabīb al-Ilbīrī, ʿAbd al-Malik", in *BA* iii, 219–227, no. 509.
b Vizcaíno Plaza, J.M. and M. Fakhri al-Wasif, "al-Dānī, Abū ʿAmr", *BA* i, 308–322, no. 96.
c Fierro, M. with J. Haremska, A. Hernández López, and E. Samba Campos, "Ibn ʿAbd al-Barr, Abū ʿUmar", in *BA* i, 575–585, no. 179.
d Fierro, M. with J. Haremska and A. Hernández López, "al-Bāŷī, Abū l-Walīd", in *BA* i, 233–243, no. 75.
e Rosselló Bordoy, G. with J. Haremska, "al-Ḥumaydī, Abū ʿAbd Allāh", in *BA* i, 460–464, no. 162.
f Serrano Ruano, D., "Ibn Rushd al-Ŷadd, Abū l-Walīd", in *BA* iv, 617–626, no. 1007.
g Peña, S., "Ibn al-Sīd al-Baṭalyawsī, Abū Muḥammad", in *BA* v, 304–337, no. 1151.
h Serrano Ruano, D., "ʿIyāḍ, Abū l-Faḍl", in *BA* vi, 404–434, no. 1479.
i De la Puente, C., "Ibn Bashkuwāl, Abū l-Qāsim", in *BA* ii, 550–565, no. 393.
j Cano Ávila, P. and A. Tawfik, with M. Comes, "Ibn Saʿīd al-ʿAnsī, Abū l-Ḥasan", in *BA* v, 137–566, no. 1067.
k Puerta Vílchez, J.M., "al-Garnāṭī, Abū Ḥayyān", in *BA* i, 361–396, no. 120.
l Lirola Delgado, J., with R. Ariè, et al, "Ibn Jaṭīb al-Salmānī, Lisān al-dīn", in *BA* iii, 643–698, no. 705.

studies. All researchers who have been involved in the study of the life and activity of a specific scholar have asked questions such as: What did the scholar learn from his Andalusī teachers? If he travelled to the East, what did he study there? What were the works circulating in al-Andalus at the time? What were the intellectual concerns prevalent among the scholar's contemporaries? To what extent did the scholar share those concerns, and to what extent could

TABLE 1.3 Number of works by Ibn al-ʿArabī's contemporaries

Authors	Number of works
Abū ʿAlī al-Ṣadafī (454/1062–514/1120)[a]	9
Ibn Rushd al-Jadd (450/1058–520/1126)[b]	19
Ibn al-Sīd al-Baṭalyawsī (444/1052–521/1127)[c]	28
Ibn Barrajān al-Ishbīlī al-Ṣūfī (b. 450/1058–470/1077, d. 536/1141)[d]	5
Ibn al-ʿArīf al-Ṣanhājī al-Mariyyī (481/1088–536/1141)[e]	5
ʿIyāḍ b. Mūsā al-Yaḥṣubī (476/1083–544/1149)[f]	28
Ibn Qasī (d. 546/1151)[g]	1

a De la Puente, C., "al-Ṣadafī, Abū ʿAlī", in *BA* vii, 231–738, no. 1680.
b Serrano Ruano, "Ibn Rushd al-Ŷadd".
c Peña, "Ibn al-Sīd al-Baṭalyawsī".
d González, A., "Ibn Barraŷān, Abū l-Ḥakam (Abuelo)", in *BA* ii, 524–538, no. 387.
e Lirola Delgado, J., "Ibn al-ʿArīf, Abū l-ʿAbbās", in *BA* ii, 335–343, no. 303.
f Serrano Ruano, "ʿIyāḍ, Abū l-Faḍl".
g El Hour, R., "Ibn Qasī, Abū l-Qāsim", in *BA* iv, 365–371, no. 946.

TABLE 1.4 Number of works by various scholars from the generation of Ibn al-ʿArabī's pupils

Authors	Number of works
Ibn Khayr al-Ishbīlī (502/1109–575/1179)[a]	1
Ibn Bashkuwāl (494/1101–578/1183)[b]	30
Ibn Ṭufayl (494/1100 ó 506/1110–581/1185)[c]	4
Ibn Rushd al-Ḥafīd (520/1126–595/1198)[d]	102

a Vizcaíno Plaza, J.M., "Ibn Jayr, Abū Bakr", *BA* iii, 719–723, no. 712.
b De la Puente, "Ibn Bashkuwāl".
c Puerta Vílchez, J.M. and J. Lirola Delgado, "Ibn Ṭufayl, Abū Bakr", in *BA* v, 498–503, no. 1260.
d Lirola Delgado, J. with I. Ferrando et al, "Ibn Rushd al-Ḥafīd, Abū l-Walīd", *BA* iv, 517–617, no. 1006.

his own work be considered innovative? What books made up his library and those of his contemporaries?

HATA's materials have been drawn from sources that include historical chronicles and literary compilations, but the most important sources are biographical and bibliographical dictionaries, particularly those dealing with al-Andalus.[4] Regarding the bibliographical dictionaries (*fahāris, barāmij*), the analysis of the *riwāyas* written in certain cases by modern editors has proved of great help in identifying the Andalusī authors and transmitters; it has also aided in completing existing references and judging the content of the works more accurately, thus making it possible to assign them better to particular genres and disciplines. *HATA* is divided into sections, each one devoted to a field of knowledge, and items are ordered chronologically within each section. The section-based arrangement was inspired by Fuat Sezgin's *Geschichte des Arabischen Schrifttums* (GAS),[5] with some changes.

The sections of *HATA* dealing with the different fields of knowledge (with the details on the section's contents provided at the beginning of each one) are as follows:

I. Qur'ān. Qur'anic sciences.
II. Ḥadīth.
III. *Fiqh.*
IV. Dogmatics. Religious polemics.
V. Asceticism. Mysticism. Works of religious content.
VI. Geography. History.
VII. Poetry.
VIII. *Adab.*
IX. Grammar. Lexicography.
X. Pharmacy. Gastronomy. Medicine. Veterinary science. Zoology.
XI. Astrology. Astronomy. Mathematics. Meteorology.
XII. Agriculture. Alchemy. Botany. Chemistry.
XIII. Philosophy. Music. Politics.
XIV. *Fahāris.*
XV. Others. (Bookbinding. Games. Interpretation of dreams. *Kutub al-ʿilm.* Occult sciences and magic. War. Unspecified works.)

In Ibn al-ʿArabī's case, the works he authored and the works he transmitted belong to the following fields:

4 These are listed in detail in the *Sources* section of *HATA*.

5 Sezgin, Fuat, *Geschichte des Arabischen Schrifttums*, Leiden: Brill, 1967–2010, available at: https://brill.com/view/db/sezo.

TABLE 1.5 Number of Ibn al-ʿArabī's works and transmissions according to the fields of knowledge in *HATA*

Fields of knowledge	Works	Transmissions
I. Qurʾān. Qurʾanic sciences.	10	17
II. Ḥadīth.	26	45
III. *Fiqh*.	27	15
IV. Dogmatics. Religious polemics.	11	10
V. Asceticism. Mysticism. Works of religious content.	15	39
VI. Geography. History.	8	12
VII. Poetry.	2	16
VIII. *Adab*.	7	23
IX. Grammar. Lexicography.	2	24
XI. Astrology. Astronomy. Mathematics. Meteorology.		1
XIII. Philosophy. Music. Politics.	1	1
XIV. *Fahāris*.	2	1
XV. Others.	4	12

This means that his intellectual activity was related to thirteen out of fifteen fields of knowledge.[6] Along these lines, the modern scholar Saʿīd Aʿrāb has classified Ibn al-ʿArabī's own works (omitting works transmitted) into nine fields of knowledge: 1. *al-tafsīr wa-ʿulūm al-Qurʾān* (exegesis and the Qurʾanic sciences); 2. *ʿulūm al-ḥadīth* (Ḥadīth sciences); 3. *al-fiqh wa-uṣūluhu* (law and its foundations); 4. *ʿilm al-kalām* (dogmatic theology); 5. *al-zuhd wa-ādāb al-ṣāliḥīn* (asceticism and the behaviour of the pious); 6. *al-tarbiya wa-l-taʿlīm* (education and teaching); 7. *al-lugha wa-l-adab* (grammar and belles-lettres); 8. *al-riḥlāt wa-l-siyar* (travelogues and biographies); 9. *ārāʾ* (his own opinions).[7]

Returning to the number of works written by the scholars mentioned in the tables above, although they may be approximate figures, they nevertheless offer us a broad idea of how productive those scholars were, and help us to rank them in terms of their productivity. The results as listed in Table 1 raise a number of questions. Why were these four scholars—Muḥyī l-dīn Ibn ʿArabī, Ibn Ḥazm, Ibn al-ʿArabī, and Ibn Rushd al-Ḥafīd—so productive? Was it just a reflection of their personalities, or the result of how they conceived their

6 The only two absent are X (Pharmacy. Gastronomy. Medicine. Veterinary Science. Zoology.) and XII (Agriculture. Alchemy. Botany. Chemistry.).

7 Aʿrāb, S., *Maʿa al-Qāḍī Abī Bakr ibn al-ʿArabī*, Beirut: Dār al-Gharb al-Islāmī, 1987.

intellectual role? Did it have to do with the times in which they lived? Did it have to do with their education? There does not seem to be just one answer to these questions for all four cases.

Ibn Ḥazm lived in the context of the Taifa kingdoms, that is, a context of political fragmentation, and his life was characterised by high mobility inside al-Andalus, intense political involvement in the first part of his life, and no long-term professional stability. He never performed the *riḥla* and never left al-Andalus: he was able to write his 156 works using the resources that were available to him in the Iberian Peninsula. What is peculiar to him is that he abandoned the prevailing Mālikī legal school and adopted Ẓāhirī legal methodology, scarcely represented until then in al-Andalus.[8] His extensive output might, therefore, have been related to his desire to spread the new school of thought he was promoting.

Like Ibn Ḥazm, Ibn Rushd al-Ḥafīd did not perform the *riḥla* to the East, his intellectual training taking place within the confines of al-Andalus and the Maghrib. In his case, the rulers' patronage influenced his written output. He was commanded by the Mu'minid prince Abū Yaʿqūb Yūsuf, later the second Almohad caliph, to produce his commentaries on Aristotle's works. His legal treatise, *Bidāyat al-mujtahid wa-nihāyat al-muqtaṣid*, can also be linked to the caliphs' need to forge a new "Almohad" society.[9] Most of Ibn Rushd al-Ḥafīd's written output is related to the Almohad intellectual and educational project, which he helped shape in numerous ways.[10]

Muḥyī l-dīn Ibn ʿArabī, on the other hand, did travel to the East, as did Ibn al-ʿArabī, but, unlike the latter, Muḥyī l-dīn Ibn ʿArabī never returned to al-Andalus. He was older than his predecessor when he set out (he was 36, whereas Ibn al-ʿArabī was only 16), and did so of his own accord (in the case of Ibn al-ʿArabī, it was his father's decision). Moreover, most of Muḥyī l-dīn Ibn ʿArabī's writings were produced while in the East, whereas most of Ibn al-ʿArabī's writings were produced in al-Andalus. Also, unlike Ibn al-ʿArabī, Muḥyī l-dīn Ibn ʿArabī acted as a transmitter of the knowledge of the Maghrib

8 Adang, C., M. Fierro, and S. Schmidtke, *Ibn Ḥazm of Cordoba: The life and works of a controversial thinker*, Leiden: Brill, 2012.

9 Fierro, M., "The legal policies of the Almohad caliphs and Ibn Rushd's *Bidāyat al-mujtahid*," in *Journal of Islamic Studies* 10 (1999), 226–248. Repr. in Carool Kersten (ed.), *The caliphate and Islamic statehood: Formation, fragmentation and modern interpretations*, 3 vols., Berlin: Gerlach Press, 2015, iii, 252–271.

10 Fierro, M., "The facilitation of knowledge in Almohad times", in R. Veit and C. Schliwski (eds.), *Averroes as a physician: Intellectual and social contexts of medicine in al-Andalus*, forthcoming; Fierro, M., "Ibn Rushd's (Averroes) 'disgrace' and his relation with the Almohads", in Abdelkader Al Ghouz (ed.), *Islamic philosophy from the 12th to the 14th century*, Göttingen: Bonn University Press and V&R unipress, 2018, 73–118.

to the Mashriq.[11] Moreover, his output falls almost entirely within the field of mysticism, whereas the other three had a more varied output covering different fields of knowledge.

Like Ibn Ḥazm, Ibn al-ʿArabī lived in difficult times, witnessing in his youth the end of the Abbadid Taifa kingdom of Seville and, in his old age, the end of the Almoravid empire and the beginnings of the Almohad caliphate. The Almoravids' intervention in the affairs of al-Andalus decisively shaped Ibn al-ʿArabī's intellectual trajectory through a decision taken by his father Abū Muḥammad ʿAbd Allāh Ibn al-ʿArabī (d. 493/1099). Having been stripped of his properties because of his collaboration with the Taifa ruler of Seville, his father decided to leave al-Andalus in 485/1092, taking the young Abū Bakr with him on a journey to the East (his father would die in Alexandria on the way back to al-Andalus). Given his youth, Ibn al-ʿArabī's literary production would be indebted to the formative teachings he received in the East. After his return to al-Andalus in 495/1102–3, he became very vocal in informing his contemporaries there that true knowledge was to be found in the East, and that Andalusīs who failed to make the journey would forever lack proper training. As Kenneth Garden has highlighted, Ibn al-ʿArabī often wrote about his admiration for the East, and his conviction that the West simply could not compare. There is a telling anecdote in this regard in his *Qanūn al-taʾwīl*, a work written in 533/1139 (that is, almost forty years after his return to al-Andalus). Abū Bakr was with his father in a caravan taking them from Jerusalem to Baghdad and, as translated by Garden:

> When the crescent moon that signaled the beginning of Ramadan was sighted, the members of the caravan began calling out "*Allāhu akbar!*" ʿAbd Allāh joined them and turned to face his son. In one of the most dramatic statements of his disdain for the Maghrib and accompanying proclamations of the superiority of the East, Abū Bakr writes that he did not return his father's gaze because his father was standing to the west of him, a direction he despised. He preferred to gaze in the opposite direction, toward his aspirations.[12]

For Garden, "The fact that the *riḥla* to the East was undertaken by many Andalusī religious scholars is a testimony to the East's primacy in the production of

---

11 Addas, Cl., *Ibn ʾArabī ou La quête du Soufre Rouge*, Paris: Editions Gallimard, 1989.

12 Garden, K., "The *riḥla* and self-reinvention of Abū Bakr Ibn al-ʿArabī," in *Journal of the American Oriental Society* 135 (2015), 1–17, at p. 8, quoting Ibn al-ʿArabī, Abū Bakr, *Qānūn al-taʾwīl*, ed. Muḥammad al-Slīmānī, Beirut: Dār al-Gharb al-Islāmī, 1990, 107–8.

religious knowledge; but not every traveling Andalusī treated the Maghrib with such disdain. His [i.e. Ibn al-ʿArabī's] exile from al-Andalus seems to have resulted in genuine bitterness toward his homeland, though another motive for including this detail in his memoir is to remind his Andalusī readers of the superiority of his Eastern intellectual pedigree."[13] However, Ibn al-ʿArabī travelled to the East in a period in which travel was not necessarily considered an indispensable requisite to build a scholarly reputation: previous luminaries such as Ibn Ḥazm and Ibn ʿAbd al-Barr never left al-Andalus. Disparaging the West helped Ibn al-ʿArabī highlight the value of his own training in the East, and bolster his own reputation as a scholar. This came in a period—that of the Almoravids—in which political considerations once again gave centrality to the Mashriq, as the Almoravids legitimised their rule by paying allegiance to the Abbasid caliph in the East. The Banū l-ʿArabī played an important role in this process, which ultimately helped the young Abū Bakr secure the restitution of the family's stolen properties, as during their stay in the East they had obtained letters documenting the "Eastern" recognition of the Almoravids' legitimacy.[14] In extolling the merits of his trip to the East and his studies there, Ibn al-ʿArabī wanted to impress on his Andalusī readers the intellectual superiority not so much of the East but of himself: his Eastern scholarly pedigree put him above the Andalusīs who had never left their homeland. This also allowed him to stress his proximity to the Mālikī Ashʿarī scholar Abū l-Walīd al-Bājī, a figure who seems to have acted as the model on which Ibn al-ʿArabī modelled himself, although with some differences. This comes through quite clearly in another anecdote narrated by Ibn al-ʿArabī, also in his *Qanūn al-taʾwīl.* One day, while he was still in al-Andalus, a bookseller (*simsār*) stopped by to show his father some books. They included works by the Iraqi Abū Jaʿfar al-Simnānī (d. 444/1052) on theology and legal controversy (*ʿilm al-khilāf*), disciplines that had not yet been cultivated in al-Andalus at that time. Al-Simnānī had been the teacher of Abū l-Walīd al-Bājī, who was responsible for introducing his works and doctrines in al-Andalus.

> Those present—father, bookseller, and teacher—agreed that these were "great books and lofty sciences that al-Bājī brought from the East." Abū Bakr writes, "This word (i.e., *al-mashriq*) pierced my liver and struck my heart. They began to recount [al-Bājī's] memory, and to say that the *fuqahāʾ* of our land do not comprehend him. How remarkable that this small quantity [of knowledge] is brought to a community (*umma*) and there

13 Garden, "The *riḥla* and self-reinvention" 9.

14 Viguera, "Las cartas".

is not a single one among them who can add to it except weakly and incompetently."[15]

Contrary to what had happened to Abū l-Walīd al-Bājī's knowledge as portrayed in the anecdote, Ibn al-ʿArabī had many students,[16] and his teachings were not lost in the following generations. One such student, Ibn Bashkuwāl, reported that no one had ever brought to Seville such a vast quantity of knowledge from the East.[17] Ibn al-ʿArabī constantly insisted on the importance of his stay in the East, especially in Jerusalem, Damascus, and Baghdad, even writing a poem addressed to the Baghdadis in which he expressed his nostalgia for the city.[18]

But do these anecdotes related by Ibn al-ʿArabī in his *Qanūn al-taʾwīl* reflect what he had really felt during his *riḥla*, or should they be understood as reflecting what he needed to convey at the time he wrote the book many years later (533/1139)?

Interestingly, his father had studied not under the Mālikī Abū l-Walīd al-Bājī, but his main opponent, the Ẓāhirī Ibn Ḥazm, and owned copies of several of the latter's books.[19] As noted, Ibn Ḥazm did not travel to the East. However, for Ibn al-ʿArabī, who lived in Almoravid al-Andalus where Mālikism reigned supreme and where the legitimacy of the rulers rested upon their acknowledgment of and by a caliph residing in Baghdad, the East necessarily had to play a major role in a way that was absent in Ibn Ḥazm's day. Not only did Ibn Ḥazm not acquire any of his knowledge in the East, he was also loyal to the local

---

15 Garden, "The *riḥla* and self-reinvention" 4, quoting Ibn al-ʿArabī, *Qānūn al-taʾwīl* 76–77. On this episode see also Schmidtke, S., "Ibn Ḥazm's sources on Ashʿarism and Muʿtazilism", in C. Adang, M. Fierro, and S. Schmidtke (eds.), *Ibn Ḥazm of Cordoba: The life and works of a controversial thinker*, Leiden: Brill, 2013, 373–401.

16 Lucini Baquerizo, M.M., "Discípulos de Abū Bakr Ibn al-ʿArabī en *al-Dhayl wa-l-Takmila* de al-Marrākushī," in *Estudios onomástico-biográficos de al-Andalus VII*, ed. Manuela Marín and Helena de Felipe, Madrid: CSIC, 1995, 191–202.

17 Garden, "The *riḥla* and self-reinvention" 16.

18 On the relationship that Andalusī scholars had with the East, and especially with Baghdad, see Fierro, M., "Entre Bagdad y Córdoba: Centro y periferia en el mundo del saber islámico (siglos III/IX–VI/XII)", in Salvador Peña (ed.), *Iraq y al-Andalus: Oriente en el Occiente islámico*, Almería: Fundación Ibn Tufayl, 2009, 63–90; ʿĀmirī, Muḥammad Bashīr and Ḥasan Rāḍī, *Baghdād fī ʿuyūn al-ʿulamāʾ wa-qulūb al-ʿāmma fī l-Andalus*, Amman: Dār al-Manāhij li-l-Nashr wa-l-Tawzīʿ, 2015.

19 While in the East, father and son are said to have had contact with a student of Ibn Ḥazm, al-Ḥumaydī (420/1029–488/1095), who had left al-Andalus never to return. The Banū l-ʿArabī left al-Andalus in 485/1092 and arrived in Baghdad in 489/1096, when al-Ḥumaydī was already dead, so it seems that they could not have entered into contact with him. This impossibility was perhaps one of the reasons why Ibn al-ʿArabī's narrative of his stay in the East came under suspicion.

Umayyad caliphate that had developed in al-Andalus against the Abbasids, and in his writings often expressed his conviction that the West could easily hold its weight against the East in a comparison.[20] By praising the East as the place where knowledge resided and where he had developed his scholarship, Ibn al-ʿArabī was distancing himself from Ibn Ḥazm, who was not popular with the Almoravids. At the same time, he was also reinforcing his own value as one who, like Abū l-Walīd al-Bājī before him, had brought from the East books and knowledge that had been lacking in the West. He himself, in his *Sirāj al-murīdīn*, listed the books (64 in total) that he had brought with him from the East.[21] Thanks to the data found in *HATA*, it is possible to independently identify some twenty works that Ibn al-ʿArabī introduced for the first time in al-Andalus.[22] Thus, while there is no doubt that he was in the East and studied there,[23] doubts were expressed about the specific details of his stay in the East. The fact that Ibn al-ʿArabī mentioned that he had lost his travelogue (the *Tartīb al-riḥla*) before returning to al-Andalus, having in his possession only a few remaining pages, and that he had to use his memory to reconstruct what he had lost, leaves room for speculation: Was this a cover-up story for changes introduced in the narrative of his travels that were justified by gaps in his

20 Pellat, Ch., "Ibn Hazm, bibliographe et apologiste de l'Espagne musulmane", in *al-Andalus* 19 (1954), 53–101.

21 Ibn al-ʿArabī, Abū Bakr, *Sirāj al-murīdīn*, ed. ʿAbd Allāh al-Tawrātī, 6 vols., Tangier: Dār al-Ḥadīth al-Kattāniyya, 2017, iv, 396.

22 Some examples in the fields of Qur'anic sciences, Ḥadīth, and *fiqh*: Abū Bakr al-Ṭurṭūshī's *Kitāb Ikhtiṣār al-kashf wa-l-bayān ʿan tafsīr al-Qurʾān*; al-Rummānī's *Kitāb*; ʿAbd al-Jabbār al-Hamadhānī's *Kitāb al-Muḥīṭ*; Abū l-Ḥasan al-Ashʿarī's *Kitāb al-Mukhtazin*; ʿAbd al-Ghanī al-Azdī's *al-Aḥādīth allatī ijtamaʿa fīhā arbaʿa min al-ṣaḥāba*; Abū l-Ḥusayn al-Ṭuyūrī's *Fawāʾid, Ḥadīth al-luqma, al-Ḥadīth al-musalsal bi-l-suʾāl ʿan al-ism wa-tawābiʿihi*; Ibn al-Madīnī's *Maʿrifat man yadūr ʿalayhi al-isnād*; Abū l-Fawāris Ṭirād b. Muḥammad al-Zaynabī's *Majlisān, Majlisān min ḥadīth al-Ṣābunī*; al-Khaṭīb al-Baghdādī's (*juzʾ*) *al-Naṣīḥa li-ahl al-ḥadīth*; al-Bukhārī's *Rubāʿiyyāt*; Sulaym al-Rāzī's *Kitāb Taqrīb al-Gharībayn li-Ibn ʿUbayd wa-Ibn Qutayba*; Ibn al-Ṭuyūrī's (*juzʾ min*) *Ḥadīth*; al-Dāraquṭnī's *al-Aḥādīth allatī khūlifa fīhā … Mālik ibn Anas*; Ibn Fāris's *Kitāb Ḥilyat al-fuqahāʾ*; al-Ghazālī's *Kitāb al-Mustaṣfā*; ʿAbd al-Wahhāb al-Baghdādī's *Sharḥ al-Mudawwana* and *Sharḥ al-Risāla*; Abū Bakr al-Ṭurṭūshī's *Kitāb al-Taʿlīqa fī masāʾil al-khilāf*.

23 Ibn al-ʿArabī's case is thus different from that of Ibn Tūmart: the Eastern sojourn of the Almohad founder is discussed in Almohad sources, but cannot be confirmed by non-Almohad sources. When Ibn al-ʿArabī travelled to Marrakesh with the other delegates from Seville to pledge obedience to the Almohad caliph, he was asked whether he had met with Ibn Tūmart in the East, and answered that he had not, but that he had heard of him. This was chronologically impossible, as Ibn Tūmart's journey to the East, which is difficult to date, would in any case have taken place after Ibn al-ʿArabī's return to al-Andalus.

memory? Does this reflect that he had to reshape the narrative of his stay in the East according to changing needs?

While it is true that Ibn al-ʿArabī's travels to the East allowed him to bring back many books, this alone does not explain his high productivity. His teacher Abū ʿAlī al-Ṣadafī also travelled to the East, but instead wrote very little. In the case of the mystic Muḥyī l-dīn Ibn ʿArabī, the striking number of works he wrote may be related to his need, as an émigré, to find patrons in the East and secure a reputation outside his homeland. In the cases of Ibn Ḥazm and Ibn Rushd al-Ḥafīd, their productivity is not connected with the *riḥla* to the East, as they never travelled there. As mentioned above, Ibn Ḥazm's many works reflect his desire to spread Ẓāhirism in a Mālikī territory, while Ibn Rushd al-Ḥafīd's productivity can be linked to his proximity to the Almohad caliphs, who had a sweeping agenda of educational and intellectual reform.

Ibn al-ʿArabī was also close to the state apparatus, at least for a time. One of his students, described as a pious ascetic (*zāhid ʿābid*), stopped attending his lectures after three months because Ibn al-ʿArabī was always ready to leave class on short notice if summoned by the ruler, so much so that he always kept his mule hitched by the door of the mosque where he taught.[24] Ibn al-ʿArabī directly discussed the relationship between knowledge and political power, writing, for example: "What an aid for knowledge is leadership!" (*wa-niʿmat al-ʿawn ʿalā l-ʿilm al-riʾāsa*).[25] Ibn Ḥazm before him had said:

> Two *madhhab*s spread at the beginning of their existence through political leadership (*al-riʾāsa*) and political power (*al-sulṭān*). This was the case of the *madhhab* of Abū Ḥanīfa. When Abū Yūsuf [a pupil of Abū Ḥanifa] was named *qāḍī l-quḍāt,* the other judges were named by him and he only assigned to the judicature of any place, from the easternmost Mashriq to the farthest district of Ifrīqiya, his close companions and those who were affiliated to his *madhhab*. The other case is the *madhhab* of Mālik b. Anas in al-Andalus. Yaḥyā b. Yaḥyā [a pupil of Mālik] was highly influential with the sultan, and his opinion was accepted by the judges. No judge was named in our lands except by consulting him and by his choice. He only suggested his close companions and those who followed his *madhhab*. People are easily attracted to this world and to leadership, and turn toward that which they anticipate will help them attain their goals. However, Yaḥyā b. Yaḥyā was never named judge and never agreed

24 Al-Maqqarī, *Nafḥ al-ṭīb min ghuṣn al-Andalus al-raṭīb*, ed. Iḥsān ʿAbbās, 8 vols., Beirut, 1968.

25 Ibn al-ʿArabī, *Qānūn al-taʾwīl* 116.

> to be one [when offered]. This added to his veneration by the people of al-Andalus and was an incentive for the judges to take Yaḥyā's opinions into account. The same happened in Ifrīqiya with Saḥnūn b. Saʿīd. The people then followed what had already become widespread.[26]

From this text it is clear that *riʾāsa* (leadership) is what certain scholars are able to achieve when they manage to get the rulers to put them in positions of power and influence. Ibn al-ʿArabī did manage to exercise *riʾāsa* when he was named judge of Seville in 528/1134, but he only lasted in the post for a little over a year, as some of his decisions met with backlash. He then sought another type of *riʾāsa* by moving to Cordoba and devoting himself to teaching.[27]

Because Ibn Ḥazm did not manage to exert same the type of *riʾāsa* as Abū Yūsuf and Yaḥyā b. Yaḥyā, he tried to make Ẓāhirism known and available via a different path: an extensive literary output aimed at spreading his teachings in different formats and to different audiences. Ibn al-ʿArabī tried to combine the precedents of both Yaḥyā b. Yaḥyā and Ibn Ḥazm to spread what I have called elsewhere his "reformed" Mālikism.[28] Like Yaḥyā b. Yaḥyā, he influenced the judges in his activity as a jurist advisor (*mushāwar*); like Abū Yūsuf, he managed to have himself named judge, although it proved only a short stint; like Ibn Ḥazm, he was very active in producing works that fostered what he thought was proper belief and practice. It is thus not surprising that the disciplines in which the two men wrote are strikingly similar, as shown in Table 6.

Although Ibn al-ʿArabī was closer in his doctrinal outlook to the Mālikī Ashʿarī scholar Abū l-Walīd al-Bājī than to the Ẓāhirī Ibn Ḥazm, he engaged in an intense intellectual activity that involved writing extensively in different fields of knowledge, as his aim—like that of Ibn Ḥazm—was to change the status quo through education. Ibn al-ʿArabī was one of those who complained of feeling himself a "stranger" (*gharīb*) in his own times, given the gap between what he was convinced was the true religion, and how people behaved in practice; his extensive written output was a way to remedy the situation.[29] He was thus indebted to the legacy of both of his predecessors.

---

26 Al-Ḥumaydī, *Jadhwat al-muqtabis*, ed. Muḥammad b. Tāwīt al-Ṭanjī, Cairo: Maktab Nashr al-Thiqāfa al-Islāmiiyya, 1952. See also Ibn Ḥazm, *al-Iḥkām fī uṣūl al-aḥkām*, ed. A.M. Shākir, 8 vols. in 1, Cairo: Maṭbaʿa al-Saʿāda, 1345–1348 H [1926–1929], iv, 230.

27 We do not yet have a clear overview of the dating of Ibn al-ʿArabī's scholarly production, which was probably spread out over the course his life, with some individual books even being written over a long period of time.

28 Fierro, M., "Proto-Mālikīs, Mālikīs and reformed Mālikīs", in P. Bearman, R. Peters, F.E. Vogel (eds.), *The Islamic school of law: Evolution, devolution, and progress*, Cambridge, MA: Harvard University Press, 2005, 57–76.

29 Fierro, M. "Spiritual alienation and political activism: the *Ġurabāʾ* in al-Andalus during the sixth/twelfth century", in *Arabica* 47 (2000), 230–260. Repr. in M. Fierro, *The Almohad*

TABLE 1.6 The fields of knowledge of the works written by Ibn al-ʿArabī and Ibn Ḥazm

Fields of knowledge	Abū Bakr b. al-ʿArabī	Ibn Ḥazm
Total number of works	115	156
I. Qur'ān. Qur'anic sciences.	10	5
II. Ḥadīth.	26	13
III. *Fiqh.*	27	37
IV. Dogmatics. Religious polemics.	11	39
V. Asceticism. Mysticism. Works of religious content.	15	6
VI. Geography. History.	8	21
VII. Poetry.	2	4
VIII. *Adab.*	7	2
IX. Grammar. Lexicography.	2	4
X. Pharmacy. Gastronomy. Medicine. Veterinary science. Zoology.		10
XIII. Philosophy. Music. Politics.	1	6
XIV. *Fahāris.*	2	1
XV. Others.	4	8

Bibliography

Adang, C., M. Fierro, and S. Schmidtke, Ibn Ḥazm of Cordoba: The life and works of a controversial thinker, Leiden: Brill, 2012.

Addas, Cl., Ibn 'Arabī ou La quête du Soufre Rouge, Paris: Editions Gallimard, 1989.

ʿĀmirī, Muḥammad Bashīr and Ḥasan Rāḍī, Baghdād fī ʿuyūn al-ʿulamāʾ wa-qulūb al-ʿāmma fī l-Andalus, Amman: Dār al-Manāhij li-l-Nashr wa-l-Tawzīʿ, 2015.

Aʿrāb, S., Maʿa al-Qāḍī Abī Bakr ibn al-ʿArabī, Beirut: Dār al-Gharb al-Islāmī, 1987.

Arcas Campoy, M. and D. Serrano Niza, "Ibn Ḥabīb al-Ilbīrī, 'Abd al-Malik", in BA iii, 219–227, no. 509.

BA = Biblioteca de al-Andalus, ed. Jorge Lirola Delgado and José Miguel Puerta Vílchez, 7 vols., Almería: Fundación Ibn Tufayl de Estudios Árabes, 2004–2012, together with an Apéndice, ed. Jorge Lirola Delgado, Almería: Fundación Ibn Tufayl de Estudios Árabes, 2012, and La producción intelectual andalusí: balance de resultados e índices, Almería: Fundación Ibn Tufayl de Estudios Árabes, 2013.

revolution: Politics and religion in the Islamic West during the twelfth–thirteenth centuries, Ashgate: Variorum, 2012, ch. 1.

Cano Ávila, P., A. García Sanjuán, and A. Tawfiq, "Abū Bakr b. al-ʿArabī", in BA ii, 129–158, no. 298.

Cano Ávila, P. and A. Tawfik, with M. Comes, "Ibn Saʿīd al-ʿAnsī, Abū l-Ḥasan", in BA v, 137–166, no. 1067.

El Hour, R., "Ibn Qasī, Abū l-Qāsim", in BA iv, 365–371, no. 946.

Fierro, M. with J. Haremska, A. Hernández López, and E. Samba Campos, "Ibn ʿAbd al-Barr, Abū ʿUmar", in BA i, 575–585, no. 179.

Fierro, M., with J. Haremska and A. Hernández López, "al-Bāŷī, Abū l-Walīd", in BA i, 233–243, no. 75.

Fierro, M. "Spiritual alienation and political activism: the Ġurabāʾ in al-Andalus during the sixth/twelfth century", in Arabica 47 (2000), 230–260. Repr. in M. Fierro, The Almohad revolution: Politics and religion in the Islamic West during the twelfth–thirteenth centuries, Ashgate: Variorum, 2012, ch. 1.

Fierro, M., "Proto-Mālikīs, Mālikīs and reformed Mālikīs", in P. Bearman, R. Peters, F.E. Vogel (eds.), The Islamic school of law: Evolution, devolution, and progress, Cambridge, MA: Harvard University Press, 2005, 57–76.

Fierro, M., "Entre Bagdad y Córdoba: Centro y periferia en el mundo del saber islámico (siglos III/IX–VI/XII)", in Salvador Peña (ed.), Iraq y al-Andalus: Oriente en el Occiente islámico, Almería: Fundación Ibn Tufayl, 2009, 63–90.

Fierro, M., "The legal policies of the Almohad caliphs and Ibn Rushd's Bidāyat al-mujtahid", in Journal of Islamic Studies 10 (1999), 226–248. Repr. in Carool Kersten (ed.), The caliphate and Islamic statehood: Formation, fragmentation and modern interpretations, 3 vols., Berlin: Gerlach Press, 2015, iii, 252–271.

Fierro, M., "Ibn Rushd's (Averroes) 'disgrace' and his relation with the Almohads", in Abdelkader Al Ghouz (ed.), Islamic philosophy from the 12th to the 14th century, Göttingen: Bonn University Press and V&R unipress, 2018, 73–118.

Fierro, M., "The facilitation of knowledge in Almohad times", in R. Veit and C. Schliwski (eds.), Averroes as a physician: Intellectual and social contexts of medicine in al-Andalus, forthcoming.

Garden, K., "The riḥla and self-reinvention of Abū Bakr Ibn al-ʿArabī", in Journal of the American Oriental Society 135 (2015), 1–17.

González, A., "Ibn Barraŷān, Abū l-Ḥakam (Abuelo)", in BA ii, 524–538, no. 387.

HATA = Historia de los autores y transmisores de al-Andalus, ed. M. Fierro, available at: https://www.eea.csic.es/red/hata/.

al-Ḥumaydī, Jadhwat al-muqtabis, ed. Muḥammad b. Tāwīt al-Ṭanjī, Cairo: Maktab Nashr al-Thiqāfa al-Islāmiiyya, 1952.

Ibn al-ʿArabī, Abū Bakr, Qānūn al-taʾwīl, ed. Muḥammad al-Slīmānī, Beirut: Dār al-Gharb al-Islāmī, 1990.

Ibn al-ʿArabī, Abū Bakr, Sirāj al-murīdīn, ed. ʿAbd Allāh al-Tawrātī, 6 vols., Tangier: Dār al-Ḥadīth al-Kattāniyya, 2017.

Ibn Ḥazm, al-Iḥkām fī uṣūl al-aḥkām, ed. A.M. Shākir, 8 vols. in 1, Cairo: Maṭbaʿa al-Saʿāda, 1345–1348 H [1926–1929].
Lirola Delgado, J. with R. Ariè, I. Garijo Galán, E. Molina López, J.M. Puerta Vílchez, and M.C. Vázquez de Benito, "Ibn Jaṭīb al-Salmānī, Lisān al-dīn", in BA iii, 643–698, no. 705.
Lirola Delgado, J., "Ibn al-ʿArīf, Abū l-ʿAbbās", in BA ii, 335–343, no. 303.
Lirola Delgado, J. with I. Ferrando, M. Forcada, J.M. Puerta Vílchez, J. Puig Montada, D. Urvoy and M.C. Vázquez de Benito, "Ibn Rushd al-Ḥafīd, Abū l-Walīd", BA iv, 517–617, no. 1006.
Lucini Baquerizo, M.M., "Discípulos de Abū Bakr Ibn al-ʿArabī en al-Dhayl wa-l-Takmila de al-Marrākushī", in Estudios onomástico-biográficos de al-Andalus VII, ed. Manuela Marín and Helena de Felipe, Madrid: CSIC, 1995, 191–202.
al-Maqqarī, Nafḥ al-ṭīb min ghuṣn al-Andalus al-raṭīb, ed. Iḥsān ʿAbbās, 8 vols., Beirut, 1968.
Pellat, Ch., "Ibn Hazm, bibliographe et apologiste de l'Espagne musulmane", in al-Andalus 19 (1954), 53–101.
Peña, S., "Ibn al-Sīd al-Baṭalyawsī, Abū Muḥammad", in BA v, 304–337, no. 1151.
de la Puente, C., "Ibn Bashkuwāl, Abū l-Qāsim", in BA ii, 550–565, no. 393.
de la Puente, C., "al-Ṣadafī, Abū ʿAlī", in BA vii, 231–238, no. 1680.
Puerta Vílchez, J.M., "al-Garnāṭī, Abū Ḥayyān", in BA i, 361–396, no. 120.
Puerta Vílchez, J.M. and J. Lirola Delgado, "Ibn Ṭufayl, Abū Bakr", in BA v, 498–503, no. 1260.
Rosselló Bordoy, G. with J. Haremska, "al-Ḥumaydī, Abū ʿAbd Allāh", in BA i, 460–464, no. 162.
Schmidtke, S., "Ibn Ḥazm's sources on Ashʿarism and Muʿtazilism", in C. Adang, M. Fierro, and S. Schmidtke (eds.), Ibn Ḥazm of Cordoba: The life and works of a controversial thinker, Leiden: Brill, 2013, 373–401.
Serrano Ruano, D., "Ibn Rushd al-Ŷadd, Abū l-Walīd", in BA iv, 617–626, no. 1007.
Serrano Ruano, D., "ʿIyāḍ, Abū l-Faḍl", in BA vi, 404–434, no. 1479.
Sezgin, Fuat, Geschichte des Arabischen Schrifttums, Leiden: Brill, 1967–2010, available at: https://brill.com/view/db/sezo.
Viguera, M.J., "Las cartas de al-Gazālī y al-Ṭurṭūšī al soberano almorávid Yūsuf b. Tāšufīn", in al-Andalus. 42 (1977), 341–374.
Vizcaíno Plaza, J.M. and M. Fakhri al-Wasif, "al-Dānī, Abū ʿAmr", BA i, 308–322, no. 96.
Vizcaíno Plaza, J.M., "Ibn Jayr, Abū Bakr", BA iii, 719–723, no. 712.

2

Law, Theology, and Language in Ibn al-ʿArabī's Thought

Steps in a Personal Approach (2002–2022)

Delfina Serrano-Ruano

This chapter reviews Abū Bakr Ibn al-ʿArabī's thought on different questions of Islamic legal and theological doctrine, such as unlawful sex (*zinā*), sexual violence, conciliation (*ṣulḥ*), arbitration (*taḥkīm*), the *dhimma* covenant and *dhimmīs*' voluntary recourse to Islamic justice (a practice known in English as "forum shopping"), the theological anathematisation (*takfīr*) of political and religious opponents, theological anthropomorphism (*tajsīm*), Sufism, debates about the number and entity of God's best and most beautiful names (*asmāʾ Allāh al-ḥusnā*), and the dissemination of doctrines associated with the Qadarīs and the Muʿtazilīs in 5th–6th/12th century al-Andalus.

With one exception, the results I shall present in what follows have already been published, starting in 2002 until present. For reasons I will explain later, I intend to disseminate these results among the interested researchers. Using the opportunity to contribute to a collective volume on the legacy of Ibn al-ʿArabī to spread these results has an additional advantage. As I will show in the conclusions, however scattered the different surveys of Ibn al-ʿArabī's thought I have undertaken so far may seem, they provide some interesting clues to approach the man and his mindset from a comprehensive perspective. Exploring Ibn al-ʿArabī's works within the historical context of Almoravid al-Andalus sheds light on the insertion of Arabic language and literature into the development of his intellectual project aimed at disseminating the knowledge on Islamic legal methodology (*uṣūl al-fiqh*) and *kalām* he acquired during his stay in the East. By doing so, he aspired to raise the educational level of his contemporaries and hence their capacity to identify religious ideas carrying a potential ideological or political risk. At the same time, he also sought to assert his scholarly pedigree over that of his peers and to enhance his public profile.

My interest in a scholar like Ibn al-ʿArabī was sparked by the nature of the sources I had to deal with at the beginning of my research career, and which have been the focus of my work ever since. By the time I came to work on my doctoral thesis, historians of al-Andalus were becoming aware of the advantages of looking at legal sources to compensate for the limitations of chronicles

 | DOI:10.1163/9789004706859_004

and the almost complete lack of archive materials to reconstruct social and economic history. One of the first Spanish scholars who drew attention to the benefit of legal sources in researching the history of al-Andalus was María Jesús Viguera. In a piece published in 1993, she reviewed a number of relevant works, including collections of fatwas and actual legal cases written during the Almoravid period, which have been preserved and which by then were being or had already been the subject of critical editions.[1] She recommended me to focus on one such collection and agreed to supervise my dissertation, eventually devoted to *Madhāhib al-ḥukkām fī nawāzil al-aḥkām*, a collection of legal cases by Muḥammad b. ʿIyāḍ (d. 575/1179–80) that features the activity as both qadi and mufti of his father, ʿIyāḍ b. Mūsā (d. 544/1149).[2]

A substantial portion of the materials transmitted by Muḥammad b. ʿIyāḍ consist of fatwas requested by his father from one of his most important masters, Ibn Rushd al-Jadd (d. 520/1126), which made me turn my attention to this great intellectual as well.[3] An examination of the lives and works of Ibn Rushd and his disciple ʿIyāḍ b. Mūsā offers a unique glimpse into the history of the Almoravid period, ranging from political events and sophisticated intellectual debates to the daily life of the average people, including male and female, free and slave, Muslim and non-Muslim. Yet the resulting image would not be comprehensive enough without turning the gaze towards the protagonist of this volume, whose life and activities constitute another fundamental landmark of the Almoravid period, given his involvement in the most relevant intellectual debates and political events of the period.

1 Viguera, María Jesús, "En torno a las fuentes jurídicas de al-Andalus", in *La civilisation d'al-Andalus dans le temps et dans l'espace: Actes du colloque international, [Université Hassan II, Maroc], 16–18 avril 1992*, Muhammadiyya: Imprimerie de Fedala, 1993, 71–78.

2 Ibn ʿIyāḍ, *Madhāhib al-ḥukkām fī nawāzil al-aḥkām li-l-qāḍī ʿIyāḍ wa-waladihi Muḥammad*, ed. Mohamed Bencherifa, Beirut: Dār al-Gharb al-Islāmī, 1990. My dissertation was published in book form, including a complete Spanish translation of *Madhāhib al-ḥukkām* preceded by an introductory study. See Serrano Ruano, Delfina, *Madhāhib al-ḥukkām fī nawāzil al-aḥkām* (*La actuación de los jueces en los procesos judiciales*) *de Muḥammad b. ʿIyāḍ*, Madrid: CSIC-AECI, 1998. In addition, I am the author of the entries on ʿIyāḍ b. Mūsā and his son, Muḥammad b. ʿIyāḍ, in the bibliographical dictionary *Biblioteca de al-Andalus*. See, respectively, Serrano Ruano, Delfina, "ʿIyāḍ b. Musà", in Jorge Lirola Delgado (ed.), *Biblioteca de al-Andalus*, Almería: Fundación Ibn Tufayl de Estudios Árabes, 2009, 404–434, no. 1479 and Serrano Ruano, Delfina, "Ibn ʿIyāḍ al-Sabtī, Abū ʿAbd Allāh", in Jorge Lirola Delgado and José Miguel Puerta Vílchez (eds.), *Biblioteca de al-Andalus: de Ibn al-Dabbāg a Ibn Kurz*, Almería: Fundación Ibn Tufayl de Estudios Árabes, 2004, 542–544, no. 661.

3 See Serrano Ruano, Delfina, "Ibn Rushd al-Jadd", in David Powers, Susan Spectorsky, and Oussama Arabi (eds.), *Islamic legal thought: A compendium of Muslim jurists*, Leiden: Brill, 2013, 295–322.

The relevance of Ibn al-ʿArabī to approaching the intellectual and political history of the Almoravid period and of the Islamic West as a whole does not need to be emphasised here. Allowing myself to adopt a more personal tone, I would rather stress Ibn al-ʿArabī's relevance for my own evolution in the study of Mālikism and Ashʿarism in the pre-modern Islamic West. Exploring the views of such multifaceted, prolific, and singular scholars like Ibn Rushd al-Jadd, Ibn al-ʿArabī, and ʿIyāḍ b. Mūsā is always worth the effort. Their intellectual output, with its marked educational tone, opens a window into cutting edge Islamic scholarship. Islamic legal methodology was used by them to translate the universal message of the revealed texts to specific circumstances. In this regard, they engaged different religious sciences against sheer legalism, anti-intellectualism, sectarianism, extremism, and discord. They promoted intellectual effort and education, and strived to convince the masses that compliance with the Sharīʿa and salvation in the afterlife were available to the average rational mind. In the specific case of Ibn al-ʿArabī, reading *al-Qabas fī sharḥ Muwaṭṭaʾ Mālik ibn Anas*, *Aḥkām al-Qurʾān*, *al-ʿAwāṣim min al-qawāṣim*, and *Qanūn al-taʾwīl* has encouraged me to develop new objectives beyond my initial interest in legal sources as repositories of social and economic data and extended my perspective beyond the historical and spatial outlines of the Almoravid period.

I am very grateful for the opportunity to participate first in a colloquium devoted to Ibn al-ʿArabī's intellectual legacy with experts from different Arab and Western countries, and then in the resulting publication. It is a great pleasure to see this project realised thanks to the enthusiasm, the perseverance, and the overwhelming energy of Ilyass Amharar. Academic activities of this kind help to bridge the economic and academic gaps that hinder communication among all the interested experts. Be that as it may, the exchange of information about dissertations, new publications, and research projects among us all continues to be less fluid than would be desired. For this reason, rather than contributing results of new unpublished research, I saw fit to present a general overview of the data from Ibn al-ʿArabī's works that I have collected in the last years. I thus intended to draw the attention of other researchers on my work about Ibn al-ʿArabī and disseminate its results among them. During the debates that followed each individual contribution to the colloquium, the existence of the kind of communication problems I've described before were exposed even within a single academic circle. This fact, together with the email exchange with other participants in the aftermath of the conference, convinced me even more that my initial proposal was worth pursuing.

The following summary is divided into nine sections according to a thematic order rather than the chronological order of publication of the reviewed articles and book chapters.

1 The Controversy between Ibn al-ʿArabī and Ibn al-Sīd al-Baṭalyawsī (d. 521/1127) on the Meaning of the Poetry of Abū l-ʿAlāʾ al-Maʿarrī (d. 449/1057)

While I was writing an article on the life and works of the renowned Andalusī grammarian Ibn al-Sīd al-Baṭalyawsī, I learned of his controversy with Ibn al-ʿArabī. The conflict was motivated by the latter's criticisms of the mistakes which he claimed that Ibn al-Sīd had committed in his *Sharḥ Dīwān Abī l-ʿAlāʾ al-Maʿarrī* and *al-Iqtiḍāb fī sharḥ Adab al-kuttāb*, a commentary on Ibn Qutayba's *Adab al-kuttāb*.

This moved Ibn al-Sīd to write a tract entitled *al-Intiṣār mimman ʿadala ʿan al-Istibṣār* with the aim of refuting, question by question, the shortcomings of Ibn al-ʿArabī's criticisms.[4] Ibn al-ʿArabī's writing against Ibn al-Sīd does not appear to have been preserved. Yet Ibn al-Sīd's reaction is extant and has even been edited twice by Ḥamīd ʿAbd al-Majīd.[5] Salvador Peña points to a manuscript copy preserved at El Escorial Library in Madrid with a title indicating that *al-Iqtiḍāb* was not part of the contents: *Radd Ibn al-Sīd ʿala iʿtirāḍāt Ibn al-ʿArabī ʿalayhi fī sharḥ al-Maʿarrī.*"[6] This impression is confirmed by ʿAbd al-Majīd's edition; it is thus pertinent to ask whether Ibn al-Sīd's commentary on Ibn Qutayba's *Adab al-kuttāb* was the subject of an independent polemic of which no other trace has appeared so far. Be that as it may, since Ibn al-Sīd claims to have replied to Ibn al-ʿArabī's criticisms one by one, *al-Intiṣār* allows us to recover the bulk of them, which in many cases Ibn al-Sīd appears to quote verbatim.

Ibn al-ʿArabī became acquainted with Abū l-ʿAlāʾ al-Maʿarrī's poetry through Abū Zakariyā al-Tabrīzī, a disciple of the poet whom Ibn al-ʿArabī met during his stay in the East; whereas the source for Ibn al-Sīd—who never travelled to the East—was another disciple of al-Maʿarrī named Abū l-Faḍl al-Baghdādī.[7]

In a recent survey of *al-Intiṣār*, Kevin Blankinship echoes Ibn al-Sīd's complaints about Ibn al-ʿArabī's criticisms of him, which he finds excessive and

4 See Serrano Ruano, Delfina, "Ibn al-Sīd al-Baṭalyawsī (444/1052–521/1127): De los reinos de taifas a la época almorávide a través de la biografía de un ulema polifacético", in *al-Qanṭara* 23 (2002), 53–92.

5 Cairo, 1955 and 1996.

6 See Peña, Salvador, "Ibn al-Sīd al-Baṭalyawsī, Abū Muḥammad", in Jorge Lirola Delgado (ed.), *Biblioteca de al-Andalus: de Ibn Saʿāda a Ibn Wuhayb*, Almeria Fundación Ibn Tufayl de Estudios Árabes, 2007, v, 304–337. For more information on the extant copies of Ibn al-Sīd's *Intiṣār* see ʿAbd al-Majīd, Ḥamīd, "Taʿrīf bi-l-muʾallif wa-l-kitāb", introd. to Ibn al-Sīd, *al-Intiṣār mimman ʿadala ʿan al-istibṣār*, ed. Ḥamīd ʿAbd al-Mājid, Cairo: al-Maṭbaʿa al-Amīriyya, 1955, pp. ض, ظ, and غ.

7 See ʿAbd al-Majīd, "Taʿrīf bi-l-muʾallif wa-l-kitāb", p. خ.

unfair.[8] According to Ibn al-Sīd, his detractor treated him as if he were a coarse and ignorant beginner.[9] Salvador Peña has observed that it is easy to appreciate the seriousness of the accusations levelled at Ibn al-Sīd by Ibn al-ʿArabī, as most were related to the transmission of the text of al-Maʿarrī's poems. No wonder, then, that Ibn al-Sīd's response was angry, as the attack affected something for which he always showed particular concern.[10]

The polemic was not restricted to textual criticism from a literary and linguistic point of view.[11] Ibn al-ʿArabī reproached Ibn al-Sīd for having used philosophical doctrines to explain al-Maʿarrī's poetry. Ibn al-Sīd reacted by showing that certain verses of the great poet cannot be explained but by means of philosophical doctrines.[12] Not that Ibn al-Sīd had overlooked and failed to object against the—in his view—undesirable implications of some of al-Maʿarrī's verses such as representing God as a body, echoing nonsense and false beliefs entertained by the Shīʿīs, questioning the immortality of the souls of evil humans and fostering divinatory practices.[13] However, this was to no avail against Ibn al-ʿArabī's animosity towards him.

In one case, Ibn al-Sīd examines a line that Ibn al-ʿArabī thought questionable, not for its language but for its content (metre: *khafīf*.:

هذه الشهب خلتها شبك الدهر لها فوق أهلها إلماء

> These stars—I imagined them as the nets of eternity,
> Taking in all of its [the Earth's] folk from above.

"Here, the poet meant that the heavens surround creation [*al-falak muḥīṭ bi-l-khalq*]," says Ibn al-ʿArabī in Ibn al-Sīd's second-hand report, "and that creation is in its grasp and cannot escape it. But such a statement [*lafẓ*] applies only to God." To counter this high-stakes objection, Ibn al-Sīd ups the ante by quoting Q 55:33: "O tribe of jinn and of men, if you are able to pass through the

8 See Kevin Blankinship, "Al-Maʿarrī's esteem in the Islamic West: A preliminary overview", in *al-Masāq: Journal of the Medieval Mediterranean* 31 (2019), 253–271, at pp. 260–261.

9 See Ibn al-Sīd, *al-Intiṣār mimman ʿadala ʿan al-istibṣār*, ed. Ḥamīd ʿAbd al-Mājid, Cairo, 1955, 1–2.

10 See Peña, Salvador, *Maʿarrī según Baṭalyawsī: Crítica y poética en al-Andalus, siglo XI*, Granada: Universidad de Granada, 1990, 150–152.

11 See ʿAbd al-Majīd, "Taʿrīf bi-l-mu'allif wa-l-kitāb", p. خ. See also several examples in Ibn al-Sīd, *al-Intiṣār* 41–52.

12 See ʿAbd al-Majīd, "Taʿrīf bi-l-mu'allif wa-l-kitāb", p. ذ. Also see Elamrani-Jamal, Abdelali, "Ibn al-Sīd al-Baṭalyūsī et l'enseignement d'al-Fārābī", *Bulletin d'études orientales* 48 (1996), 155–161, at p. 157. For several examples see also pp. 158–159, drawing on Ibn al-Sīd, *al-Intiṣār* 48–49. I thank the author for pointing me to this reference.

13 See Peña, *Maʿarrī según Baṭalyawsī* 11.

confines of heaven and earth, pass through them! You shall not pass through except with an authority." For Ibn al-Sīd, the verse posits "creation within the heavens' grasp", which justifies the metaphysical soundness of al-Maʿarrī's line.[14]

Another accusation Ibn al-ʿArabī made against Ibn al-Sīd was for saying that the renowned early mystic Dhū l-Nūn [al-Miṣrī] al-Ikhmīmī—whom Ibn al-Sīd refers to as "Dhū l-Nūn al-Zāhid ['the Renunciant'] al-Ikhmīmī"—was an esotericist (*bāṭinī*). Ibn al-ʿArabī found the label did not correspond to a virtuous man like this since, in his view, the *bāṭinī*s were heretics (*zanādiqa*). He was ready to acknowledge that in its literal sense, *bāṭiniyya* referred to all those who diverged from the literalists (*al-ẓāhiriyya*), but the term had become a label applied specifically to the Qarmatians and the Ismaʿīlīs.[15]

Given that some of Ibn al-ʿArabī's criticisms seem to be rather forced, adding credit to Ibn al-Sīd's complaint that they were exaggerated, we can speculate that Ibn al-ʿArabī's attack was a means to emphasise his intellectual superiority to Ibn al-Sīd (who was a very successful teacher, to judge by the high number of his students), asserting his direct contact with the sources of the knowledge that they were debating, whereas Ibn al-Sīd never left al-Andalus, and received al-Maʿarrī's poetry from one of his disciples only by license of transmission (*ijāza*) or through written correspondence.[16] It was thus a question of discrediting a rival whom Ibn al-ʿArabī saw as a potential threat to his prestige and influence. In fact, although Ibn al-Sīd was roughly two decades his senior, Ibn al-ʿArabī never studied with him upon his return to al-Andalus, unlike ʿIyāḍ b. Mūsā, for example, who did study with him. This impression is reinforced by al-Qurṭubī's characterisation of Ibn al-ʿArabi's polemical style as being in bad taste with bad professional ethics.[17]

The polemic must also be seen in the light of a general hostility towards philosophy which also moved Ibn al-ʿArabī to assert that, in his attempt to refute the philosophers, al-Ghazālī had ended up contaminated by their arguments.[18] This is a paradox, because Ibn al-ʿArabī had studied directly under al-Ghazālī

---

14 Quotation from Blankinship, Kevin, "Al-Maʿarrī's esteem", 260–261.

15 See Ibn al-Sīd, *al-Intiṣār* 46.

16 This, as Ibn al-Sīd himself takes care to remark, did not prevent him from acquiring several copies of al-Maʿarrī's poetry and from comparing them to write his commentaries. See ʿAbd al-Majīd, "Taʿrīf bi-l-muʾallif wa-l-kitāb", pp. خ to ذ.

17 See below, section 4.

18 See Asín Palacios, Miguel, *La espiritualidad de Algazel y su sentido cristiano*, Madrid: Escuelas de Estudios Árabes de Madrid y Granada, 1934, i, 43–45; Fierro, Maribel, *El libro de las novedades y las innovaciones*, Madrid: Consejo Superior de Investigaciones Científicas-Instituto de Cooperación con el Mundo Árabe, 1993, 64, 86.

and brought his books to the West, thereby contributing to the assimilation of his master's thought by his many disciples. In fact, Ibn al-ʿArabī's teaching was, according to Dominique Urvoy, instrumental for the consolidation of *kalām*, and the remarkable progress of philosophy and mysticism observed over the same period due to the adoption of Aristotelian logic and much of the Neoplatonic metaphysics which al-Ghazālī had borrowed from Ibn Sīnā and other thinkers to fight them on their own turf.[19]

2 Ibn al-ʿArabī's Refutation of Theological Literalism[20]

A chapter of Ibn al-ʿArabī's *al-ʿAwāṣim min al-qawāṣim* is devoted to showing the evil consequences of theological literalism, a hotly debated topic in his time, which the then emergent Almohad movement had adopted as a central motive of the ideological war against the Almoravids which was evolving before Ibn al-ʿArabī's eyes.[21]

According to Ibn al-ʿArabī, the most serious consequence of reading the Qurʾān and Prophetic Ḥadīth too literally was that one would end up assigning human qualities to God and comparing Him with His creation. For this reason, he qualified the literalists not just as whimsical or ignorant, but as abominable and vile people who must be rejected like "the offspring of extramarital relationships." The latter remark anticipated his concern to avoid the proliferation of unlawful intercourse that is evident in his doctrine on *zinā* and rape (on which see below). Eventually, he likened the literalists to the Jews and to the infidels (*kuffār*). At the opposite end of this spectrum was esotericism (*bāṭiniyya*) which, for Ibn al-ʿArabī, represented the exact counterpoint of the evil caused by literalism since "the literalists consider evident what is hidden while the esotericists consider hidden what is evident."

19 See Urvoy, Dominique, *El mundo de los ulemas andaluces del siglo V/XI al VII/XIII: Estudio sociológico*, Madrid: Ediciones Pegaso, 1983, 144, 196, 198–201; Urvoy, Dominique, *Pensers d'al-Andalus: La vie intellectuelle à Cordoue et à Séville au temps des empires berbères (fin XIe siècle-début XIIIe siècle)*, Toulouse: Presses Universitaires du Miraill, 1990, 168.

20 This section draws on Serrano Ruano, Delfina, "¿Por qué llamaron los almohades antropomorfistas a los almorávides?" in Maribel Fierro, Patrice Cressier, and Luis Molina (eds.), *Los almohades: Problemas y perspectivas*, Madrid: CSIC/Casa de Velázquez, 2005, ii, 815–852.

21 See Ibn al-ʿArabī, Abū Bakr, *al-Naṣṣ al-kāmil li-kitāb al-ʿAwāṣim min al-qawāṣim li-l-qāḍī Abī Bakr ibn al-ʿArabī*, ed. ʿAmmār Ṭālibī, Doha: Dār al-Thaqāfa, 1413/1992, 208–248. The chapter includes one *qāṣima* and two *ʿāṣimas*.

Like other Ashʿarīs of the Almoravid period, Ibn al-ʿArabī had overcome the *bi-lā kayf wa-lā takyīf* position attributed to the eponymous founder of the Mālikī school and expressed confidence that both learning and reasoning may help in discerning the sense of problematic or ambiguous passages in revelation. He drew a distinction between juridical and theological literalism, and took issue with the latter, thus excluding the followers of Ibn Ḥazm (d. 456/1064), whom he considered legal but not theological literalists.[22]

Thanks to his previously mentioned direct style, he referred to the specific groups of scholars who, in his view, were guilty of anthropomorphism, even going as far as to name individual scholars. These included the Ḥanbalīs, as might be expected, but others as well. According to Ibn al-ʿArabī, the threat posed by theological literalism—and the one about which he was most concerned—lay also within the very ranks of the Mālikīs. As an example of this, he pointed to the opening chapters of Ibn Abī Zayd al-Qayrawānī's (d. 386/996) compilation of Mālikī legal doctrine known as *al-Risāla*, which, he claimed, bordered on anthropomorphism and thus could no longer be considered the Mālikī profession of faith par excellence.

After explaining the main negative consequences of literalism (i.e., the *qāṣima* in question), Ibn al-ʿArabī came up with his own method to avoid venturing into anthropomorphism (i.e., *ʿāṣima*). This method relies on the basic premise that homonymy and polysemy are found in the sacred texts, as they are present in the Arabic language itself. He proposed the following steps. First, one should ascertain whether the corporeal descriptions of God in question come from the Qurʾān itself, or from Prophetic tradition, as any description of God in the Qurʾān is considered solid from a textual point of view whereas only recurrent (*mutawātir*) Prophetic reports provide true knowledge. However, reports of this kind, he observed, are very scarce.

The Qurʾān is solid from a textual point of view, he asserted, but before a given description of God can be considered His attribute it must be put to a grammatical and lexicological test to check whether or not it can be considered a complement of His, or something that can be annexed to Him (*iḍāfa*). The same applies to recurrent Prophetic reports. Solitary (*āḥād*) reports, on the other hand, cannot be used to establish the reality of divine attributes, as the esotericists and the literalists do. He also rejected consensus as a valid source to establish the reality of divine attributes.

Further, Ibn al-ʿArabī applied the three intellectual categories of obligatory, possible, and impossible (*wājib wa-jāʾiz wa-mustaḥīl*) to the analysis of

22 See Ibn al-ʿArabī, *al-ʿAwāṣim min al-qawāṣim* 248ff.

revelation. If the grammatical and lexicological test is passed, but the literal meaning of a given passage comes into conflict with reason, he argued, in the lack of contextual indications to the contrary, it must be assigned a metaphorical sense. More specifically, according to Ibn al-ʿArabī, in the case of conflict between reason and a textual piece of Qurʾanic or Prophetic evidence, the former prevails. It may happen that both the literal and the allegorical meaning of a given text are absurd; in this case, he added, that text cannot be taken as support of a principle of belief.

Though Muslims are all obliged to believe in God and obey His commands, they are not in the same position to understand the meaning of the texts from which the principles of Islamic faith are drawn, since different Muslims have different intellectual capacities. Consequently, Ibn al-ʿArabī stated that "he who is qualified to understand and explain the meaning of Qurʾanic or Prophetic texts the literal senses of which come into conflict with reason, is entitled to carry out this task, and, indeed, will be praised for it. However, he who is short of the capacity to reflect on theological matters must cling to his faith, deny that God can be likened to His creatures, believe that God must be praised, and dismiss the idea that He may be subject to misfortunes or that anything similar to Him may exist."

In his *Qānūn al-taʾwīl* Ibn al-ʿArabī argued further against anthropomorphism, and explained why this approach to conceiving of God undermines the fundamental principle of His oneness. If there were anything similar to God, he claimed, neither would qualify to be the Supreme Creator more than the other. The fact that he was writing on the eve of the Almohad conquest of al-Andalus renders his remark that "for this reason, all those who abandon the prescription of the unitarians (*al-muwaḥḥidūn*) partake in comparison of God with His creatures (*mushabbihūn*)" all the more noteworthy.[23]

Chronology—the *Qānūn* was written in 533/1139[24]—and his proven loyalty towards the Almoravids show that Ibn al-ʿArabī's overall initiative was primarily aimed at thwarting the Almohads' self-promotion as both the exclusive champions of the fight against anthropomorphism and the supreme preachers of God's absolute oneness.

---

23 See Ibn al-ʿArabī, Abū Bakr, *Qānūn al-taʾwīl*, ed. M. Slīmānī, Tunis: Dār al-Gharb al-Islāmī, 1990, 127–128.

24 See Ibn al-ʿArabī, *Qānūn al-taʾwīl* 27.

3 Ibn al-ʿArabī on Extremist Sufis[25]

By the first half of the twelfth century CE, most Western Mālikī—Ashʿarīs had come to accept the existence of a superior class of believers (*ṣāliḥīn, muttaqīn, awliyāʾ Allāh*) who had reached knowledge of God (*maʿrifa bi-Llāh taʿālā*) thanks to their extraordinary piety, and their capacity to perform miracles (*karāmāt*). However, in a long fatwa, Ibn Rushd al-Jadd argued that saints' capacity to perform miracles did not put saints and prophets on an equal footing, since prophets, contrary to saints, have knowledge of the unseen (*ʿilm al-ghayb*). Meanwhile, a widely accepted and authoritative text, the *Risāla fī ʿilm al-taṣawwuf* of al-Qushayrī (d. 465/1072–73), identified the Sufis with the saints.

Like his contemporaries Abū Bakr al-Ṭurṭūshī (d. 520/1126), Ibn Ḥamdīn (d. 508/1114), and the above-mentioned ʿIyāḍ b. Mūsā, Ibn al-ʿArabī held that certain forms of Sufism represented a potentially disrupting force. They attributed to radical Sufis the claim that the kind of knowledge acquired through intellectual effort did not play a decisive role in the path to spiritual purification but, rather, it was the regular performance of religious rituals and good deeds that did so. According to the same contention, ritual practice and virtuous conduct could by themselves bring knowledge of the Almighty God (*maʿrifa bi-Llāh taʿālā*) and open the doors to sainthood, described in terms of "nearness to God" but also—and this was the above-mentioned scholars' most serious concern—to prophecy. They blamed al-Ghazālī for the risk of syncretism involved in some of his most controversial doctrines, as well as for fostering antinomianism and these forms of extremist Sufism. As has already been pointed out, Ibn al-ʿArabī's involvement in these accusations is suprising, as he had studied directly under al-Ghazālī and had brought the latter's books to the West, contributing to the assimilation of his master's thought by his many disciples. Yet his attitude may be explained, at least in part, by the contemporary political circumstances in al-Andalus and the Maghrib. These led some of the aforementioned scholars to suspect that mystics were mobilising a significant part of the discontent with the Almoravids, with the intention of removing them from power either directly or indirectly, the latter by lending

25 This section draws on Casassas Canals, Xavier, and Delfina Serrano-Ruano, "Putting criticisms against al-Ghazālī in perspective: New materials on the interface between law, rational theology and mysticism in Almoravid and Almohad al-Andalus (Ibn Rushd al-Jadd and al-Qurṭubī)", in Ayman Shihadeh and Jan Thiele (eds.), *Philosophical theology in Islam: Later Ashʿarism East and West*, Leiden: Brill, 2020, 254–297.

their support to alternative candidates—as was eventually the case with Ibn Qasī and the Almohads.

4 Ibn al-ʿArabī's Position on God's Best and Most Beautiful Names (*asmāʾ Allāh al-ḥusnā*) according to al-Qurṭubī (d. 671/1273)[26]

The reality of Allāh's best and most beautiful names is established by Q 7:180, where God enjoins the faithful to name and evoke Him by means of them. Their exact number, namely ninety-nine, is specified in a Prophetic *ḥadīth* transmitted by Abū Hurayra (d. 58/678), according to which he who manages to grasp their profound meaning (*aḥṣāhā*) will enter Paradise. There are two extended versions of this *ḥadīth*, both of which include a list of the ninety-nine best and most beautiful names of God, although the names in the lists do not coincide. Additionally, these two lists do not include names of God mentioned by the Prophet according to other sound (*ṣaḥīḥ*) traditions coming from him.

At the beginning of chapter 33 of his *al-Asnā fī sharḥ asmāʾ Allāh al-ḥusnā wa-ṣifātihi al-ʿulā* al-Qurṭubī reviews Ibn al-ʿArabī's criticisms of al-Ghazālī and Ibn Ḥazm concerning the identification of God's best and most beautiful names.[27] Al-Qurṭubī recounts that in his *al-Maqṣad al-asnā fī sharḥ asmāʾ Allāh al-ḥusnā*, al-Ghazālī had found it surprising that Ibn Ḥazm—to whom he respectfully referred as "a certain *ḥāfiẓ* from the Maghrib called ʿAlī b. Ḥazm"—had managed to find only "some 80 names" after searching for them in the Qurʾān and in the sound Prophetic *ḥadīth*s, whereas the rest of the names, in al-Ghazāli's view, could be derived by means of intellectual interpretive effort (*ijtihād*) from the traditions (*al-akhbār*). Further, according to the same account, al-Ghazālī hinted that either Ibn Ḥazm didn't know the *ḥadīth* in which the exact number of God's best and most beautiful names is specified, or else he found that its chain of transmitters was defective. Otherwise, he could not explain why he set this *ḥadīth* aside in favor of other sound *ḥadīth*s, on the basis of which he likely elaborated his position on the best and most beautiful divine names.

Subsequently al-Qurṭubī mentions that, according to al-Uqlīshī (d. 549/1154), the source from which al-Ghazālī became acquainted with Ibn Ḥazm's doctrine on the best and most beautiful divine names was the latter's *Kitāb al-Muḥallā fī sharḥ al-mujallā*. The contents of this book make clear that Ibn Ḥazm was

26 This section draws on Casassas Canals and Serrano-Ruano, "Putting criticisms" 271–276.

27 See al-Qurṭubī, *al-Asnā fī sharḥ asmāʾ Allāh al-ḥusnā wa-ṣifātihi al-ʿulā*, ed. ʿA. b. S. al-ʿAshā Ḥassūna al-Dimashqī, Cairo: al-Maktaba al-Ḥaḍariyya, 1427/2006.

aware of the *ḥadīth* of Abū Hurayra as transmitted by al-Bukhārī that established the figure of ninety-nine names. Although al-Qurṭubī does not explicitly say so, it seems clear that Ibn Ḥazm did not fully trust Abū Hurayra's *ḥadīth*. Al-Qurṭubī proceeds that al-Uqlīshī, for his part, acknowledged that only those names that God gave to Himself can be applied to Him and that their number cannot exceed the figure mentioned in the Abū Hurayra *ḥadīth*. Yet he also stated that the names mentioned in the Qur'ān inform us about God's actions, so therefore the reports transmitting that the names amount to ninety-nine cannot be relied upon. This led him to the conclusion that, according to the Qur'ān and sound *ḥadīth*s, God's best and most beautiful names number eighty-four, as many as those he claimed had also been collected by Ibn Ḥazm. This implies that al-Uqlīshī agreed with Ibn Ḥazm and that al-Ghazālī's guess was accurate. Al-Uqlīshī makes his agreement with Ibn Ḥazm applicable to a "group of scholars" for whom ninety-nine was a maximum number, rather than the exact figure, thus refusing to give full credit to the tradition that established said figure.

After presenting these facts, al-Qurṭubī joins in al-Ghazālī's surprise that Ibn Ḥazm refused to add to the list—and so obtain a total of ninety-nine—names mentioned in the Qur'ān; all the more so when God specified that He had not "missed anything out of the record" (Q 6:38). At the same time, al-Qurṭubī remarks that Ibn al-ʿArabī went too far in criticising Abū Ḥāmid al-Ghazālī and Ibn Ḥazm—in this order—in both his *Aḥkām al-Qur'ān* and his *al-Amad al-aqṣā*. In the first work, itself an important source for al-Qurṭubī's *al-Jāmiʿ li-aḥkām al-Qur'ān*, Ibn al-ʿArabī referred to Ibn Ḥazm as a "fool from the Maghrib, who claimed to have counted God's names and to have obtained only eighty as a result." Ibn al-ʿArabī added, according to al-Qurṭubī, "No wonder. What is really surprising is that the man from Ṭūs [i.e., al-Ghazālī] said that a certain *ḥāfiẓ* from the Maghrib counted God's names and obtained only eighty, giving credit to what a man of the likes of al-Ḥumaydī [d. 488/1095] transmitted to him. And if Abū Ḥāmid committed the error of giving credit to an Andalusī outcast [he had been exiled from Mallorca], it was due to his [own] ignorance, since if on the one hand he [i.e., al-Ḥumaydī] was eloquent, scathing, and too wordy, he was [also] short of the basic rule (*al-qanūn*)." And, al-Qurṭubī proceeds further, "[Ibn al-ʿArabī] added that the names Ramaḍān, Sulṭān, and Dayyān mentioned by al-Ghazālī, draw on a *ḥadīth* that they [Ibn Ḥazm and his peers] used to transmit but which nobody else takes into consideration."

In al-Qurṭubī's opinion, there was no ground for Ibn al-ʿArabī's excessive precaution. Al-Qurṭubī reiterates that those names explicitly mentioned in the Qur'ān and which were not obtained through deduction or addition do not reach the number ninety-nine, as already noted by al-Uqlīshī and al-Ḥaṣṣār

(d. 611/1214–15). Furthermore, al-Qurṭubī observes, "Ibn al-ʿArabī mentions "Naẓīf" and other divine names on the authority of Ibn Barrajān, whereas other scholars like Ibn al-Ḥaṣṣār did not accept this name, arguing that it has no basis in the Qurʾān or the Sunna." After rejecting the latter argument, al-Qurṭubī adds that reciting the names mentioned in a weak *ḥadīth* to which Ibn al-ʿArabī, contrary to al-Ghazālī, had refused to lend full credibility is good and beautiful. Al-Qurṭubī's irritation with Ibn al-ʿArabī is particularly evident concerning Ibn al-ʿArabī's remark that Sufyān b. ʿUyayna (d. 196/811) had omitted names mentioned in the Qurʾān because, when he was asked about the number of names and after counting them, he did not consider the divine names Shākir, Ilah, Wāḥid, and Qarīb, which appear in the *Sūrat al-Baqara*. Yet, an indignant al-Qurṭubī replies, "this is sheer invention, foolishness, and speculation" and suggests that Ibn al-ʿArabī contradicted himself again for, according to his *al-Amad al-aqṣā*, the only name omitted by Sufyān b. ʿUyayna in that very same connection was Qarīb. Moreover, he adds, Ibn al-ʿArabī ignored that, according to Abū l-Qāsim al-Zujjājī (d. 340/952) and others, the list at stake—with a number of identified omissions—did not originate with Sufyān b, ʿUyayna but, rather, with Abū Zayd (Qays b. al-Sakan al-Anṣārī; d. after 2/624), and provides the list of omitted names, which amounts to 19.

5 Ibn al-ʿArabī on Free Will and Predestination[28]

This section deals with the chapters on both "The prohibition of the doctrine of free will (*qadar*)" and "Miscellaneous reports that have come down regarding people who uphold the doctrine of free will" in Ibn al-ʿArabī's *al-Qabas* which, as is well known, consists of a commentary to the *Muwaṭṭaʾ* by Mālik b. Anas.[29] Ibn al-ʿArabī's commentary on *qadar* has three features.

First, he shares Mālik's outright rejection of the Qadarīs. Yet he is critical of Mālik's treatment of *qadar*. Ibn al-ʿArabī declares that Mālik's use of the heading "The prohibition of the doctrine of free will" is problematic because it appears to conflict with another well-established Prophetic dictum enjoining

28 This section draws on Serrano Ruano, Delfina, "Debates on free will and predestination in the 12th century Islamic West: Abū Bakr Ibn al-ʿArabī's (468/1076–542/1147 or 543/1148) *Kitāb al-Qabas fī Sharḥ Muwaṭṭaʾ Mālik Ibn Anas*", in *Islamic Law Blog*, special feature on "*Muwaṭṭaʾ* Roundtable", available at: https://islamiclaw.blog/2019/12/06/muwaṭṭaʾ-roundtable-debates-on-free-will-and-predestination-in-the-12th-century-islamic-west-abu-bakr-ibn-al-ʿarabis-468-1076-542-1147-or-543-1148-kitab-al-qabas/.

29 See Ibn al-ʿArabī, Abū Bakr, *al-Qabas fī sharḥ Muwaṭṭaʾ Mālik ibn Anas*, ed. Muḥammad ʿAbd Allāh Wuld Karīm, Beirut: Dār al-Gharb al-Islāmī, 1992, iii, 1091–1095.

Muslims "to believe in *qadar* with its good, evil, sweet, and bitter aspects." How then is it possible to subject *qadar* to an absolute proscription? To explain this divergence Ibn al-ʿArabī suggests that the term "prohibition" might reflect the Companions' reticence towards free will because the Prophet had previously prohibited them from debating it.

Second, he shows his scepticism that there is a rational way to reconcile the arguments for free will and predetermination, each of which has a reasonable basis in both reason and revelation. This scepticism concerns Muslim rational theologians and the Prophet himself. Explaining why the Prophet forbade speculative debate regarding free will, Ibn al-ʿArabī asserts that because rational investigation (*naẓar*) cannot yield a clear and unequivocal understanding of the issue, discussion of this problem only yields more confusion, confusion that not even the Prophet could conclusively resolve (*al-khawḍ lā yaʾūl fīhi ilā bayān li-anna al-nabiyya idhā taʿarraḍa li-bayānihi fasada wa-kharaja ʿan ḥaddihi*).

Third, he omits any reference to the distinctively Ashʿarī doctrine of *kasb* (humans' "acquisition of liability" for acts originated by God), which the Ashʿarīs offered as a medial solution to the problem of free will and predestination. His silence regarding *kasb* is surprising for an Ashʿarī theologian of his stature who, based on our knowledge of his other works, upheld that doctrine. In fact, he says that he exhaustively reviewed the relevant Prophetic evidence in another—unfortunately non-extant—work of his, namely the *Kitāb al-Mushkilayn*, (that is, *Mushkil al-Qurʾān wa-mushkil al-sunna*). Perhaps he thought that a commentary on a legal text like the *Muwaṭṭaʾ* was not an appropriate context in which to raise one of the most difficult and contentious issues in Islamic theology.

Instead, in the *Qabas*, Ibn al-ʿArabī provides his disciples with a few simple instructions: Qadarism is evil and the Qadarīs are wrong. He does not address how Qadarīs should be treated. He is more interested in emphasising that God is eternal, has no progenitor, and determines His creatures' actions in a way that preserves their responsibility for those actions, even if humans cannot understand how. Ibn al-ʿArabī willingly concedes some agency to created beings but not at the cost of questioning God's omnipotence, or limiting it.

Despite the omission of *kasb*, he tries to mediate between extreme predestination and free will by commenting on a series of *ḥadīth*s which, in Ibn al-ʿArabī's view, affirm predestination somewhat unqualifiedly and differ as to their level of soundness. Apart from explaining the import of the *ḥadīth*s included by Mālik in the corresponding chapters of his *Muwaṭṭaʾ*, Ibn al-ʿArabī provides sounder relevant ones in the hope that they may either qualify or solve the alleged shortcomings of Mālik's *ḥadīth*s.

6 Ibn al-ʿArabī's Doctrine on *zinā*[30]

As could be expected from a Sunni jurist, Ibn al-ʿArabī assigned the Qurʾān and sound Prophetic *ḥadīth*s an equivalent legal weight. The punishment of *zinā* (fornication) provides a clear example in this regard since stoning (*rajm*) is based on *ḥadīth*s, whereas the relevant Qurʾanic text (Q 24:2) only refers to flogging and banishment.

In the *Qabas*, Ibn al-ʿArabī insists that stoning relies upon sound Prophetic reports, which he lists. In this way, he aims at stressing the legality and legitimacy of that form of punishment.[31] Commenting on Q 24:2 in his *Aḥkam al-Qurʾān*, Ibn al-ʿArabī selects one of the *ḥadīth*s mentioned in the *Qabas* which, in his view, helps best to clarify the role played by the Qurʾān and Prophetic *ḥadīth*s in determining the punishment of *zinā*.[32] This is known as the ʿUbāda *ḥadīth*, by the name of its main transmitter. The problem addressed by Ibn al-ʿArabī is that, read literally, the *ḥadīth* implies that the punishment corresponding to a *muḥṣan* who commits *zinā* with another *muḥṣan* is one hundred lashes plus stoning to death.[33] Yet in this case, most Sunni jurists do not add the hundred lashes to the stoning, and they establish one hundred lashes plus one year's banishment for unmarried culprits of *zinā*. Thus, they do not establish an analogy between the double punishment of *zinā* when committed by an unmarried person and the punishment of the *muḥṣan*s. They hold that various statutory sanctions may concur only when they apply to different crimes, and even in this case, a penalty involving death cancels the rest. In another section of his *Aḥkam al-Qurʾān* corresponding to the commentary on Q 4:15, Ibn al-ʿArabī asserts, contrary to Aḥmad b. Ḥanbal and others clinging to the literal sense of the ʿUbāda *ḥadīth*, that the correct doctrine was not to combine flogging with stoning, as "the Prophet did not order to flog those he sentenced to stoning." Contradicting the Shāfiʿīs, he adds that banishment

30 This section draws on Serrano Ruano, Delfina, "El Corán como fuente de legislación islámica: Abū Bakr Ibn al-ʿArabī y su obra *Aḥkām al-Qurʾān*", in Miguel Hernando de Larramendi and Salvador Peña Martín (eds.), *El Corán ayer y hoy: Perspectivas actuales sobre el islam*, Córdoba: Berenice, 2008, 251–275; Serrano Ruano, Delfina, "La lapidación como castigo de las relaciones sexuales no legales (*zinà*) en el seno de la escuela malikí: Doctrina, práctica legal y actitudes individuales frente al delito (ss. XI y XII)", in *al-Qanṭara* 26 (2005), 449–473, at pp. 466–467; and Serrano Ruano, Delfina, "Rape in Maliki legal doctrine and practice (8th–15th centuries CE)", in *Hawwa* 5 (2007), 166–207.

31 See Ibn al-ʿArabī, *al-Qabas* iii, 1002.

32 See Ibn al-ʿArabī, Abū Bakr, *Aḥkām al-Qurʾān*, 2 vols., Cairo, 1331/1914, ii, 83–119.

33 A *muḥṣan* (fem. *muḥṣana*) is an adult and free Muslim who is married, divorced, or widowed, or a free Muslim man who has experienced sexual intercourse with a concubine slave. See Burton, John, "Muḥṣan", in *EI2*.

does not apply to the case of the unmarried woman guilty of *zinā*, as this would predispose her to commit the crime again.[34]

Muslim jurists treated rape as "forced *zinā*" and considered the perpetrator subject to the same punishments that would apply in case of "consensual *zinā*". That rape is treated as a form of *zinā* means, as a counterpart, that victims risked being accused of calumny of *zinā* (*qadhf*), if they did not manage to prove their claims. An unproven accusation of rape could also be treated as implicit recognition of *zinā*. Moreover, the Mālikīs consider pregnancy outside of marriage as evidence of *zinā*. Therefore, a victim who could not manage to prove her accusation risked being punished for *zinā* if she became pregnant as a consequence of the act.

Ibn al-ʿArabī conceived of the crime of rape somewhat differently. He was the first Mālikī who classified rape as a combination of *zinā* and banditry (*ḥirāba*). In his *Aḥkām al-Qurʾān* he recounted that during his tenure as Qadi of Seville, a group of men was denounced before him for having performed acts of banditry against another group of men, and for having abducted the wife of one of them. Given the seriousness of the accusation, they were captured and brought to his presence. "I consulted the muftis with whom God afflicted me," he said, "and they held that the accused could not be assimilated with bandits since the [legal classification of] banditry is applied to [acts involving robbery or destruction of] properties rather than women's private parts (*li-anna al-ḥirāba innamā takūn fī l-amwāl lā fī l-furūj*)." This statement moved an infuriated Ibn al-ʿArabī to reply: "Don't you know that banditry perpetrated over women's private parts is more abominable than that perpetrated over patrimonial property and that people are ready to give up their properties, if need be, but a man would never surrender his wife or his daughter? Certainly, the harshest of those punishments prescribed by God should correspond to plundering women's private parts."[35]

He was also concerned that rape committed in the territory of the enemy was considered a legitimate act of war and remained unpunished. It is true that according to Ibn al-Mājishūn (d. 214/829), if someone commits *zinā* in the land of the enemy (*dār al-ḥarb*) with one of their women, he is not subject to the *ḥadd* (statutory) punishment. Yet Ibn al-ʿArabī warns that this opinion cannot be understood as legitimising the crime, for this would amount to ignoring

34 See Ibn al-ʿArabī, *Aḥkām al-Qurʾān* (Cairo, 1331/1914), ii, 150.

35 See Abou El Fadl, Khaled, *Rebellion and violence in Islamic law*, Cambridge: Cambridge University Press, 2001, 253, where the episode is ascribed to Ibn Rushd al-Jadd. Abou El Fadl draws on Ibn al-ʿArabī, Abū Bakr, *Aḥkām al-Qurʾān*, ed. ʿAlī Muḥammad al-Bajawī, 4 vols., Beirut: Dār al-Maʿrifa, 1972, ii, 597.

the fundamental principles of the Sharīʿa. There is difference of opinion as to whether the *ḥadd* for *zinā* is in force in the territory of the enemy; however, all agree that committing *zinā* in the land of the enemy with one of their women is forbidden.[36]

7 Ibn al-ʿArabī on Arbitration (*taḥkīm*) and Conciliation (*ṣulh*)[37]

There is no specific chapter, section, or subsection dedicated to *ṣulḥ* in Ibn al-ʿArabī's *al-Qabas*. The same goes for arbitration, if the Qur'anic practice of engaging two arbiters (*al-ḥakamān*) to settle marital disputes is excluded. Ibn al-ʿArabī simply defines imparting justice to the people as "the pillar of the Sharīʿa, the axis of legal norms and God's succession in humankind (*al-qaḍāʾ bayna al-nās aṣl al-sharīʿa wa-madār al-aḥkām wa-khilāfat Allāh fī l-khalq*)."[38] Arbitration is discussed in the commentary to Q 5:41–44 of his *Aḥkām al-Qurʾān* mainly to emphasise that, once the two parties agree on engaging an arbitrator, neither of them can withdraw and they are obliged to abide by the arbitrator's judgement.[39]

Ibn al-ʿArabī recounted that when the Jews alluded to in Q 5:42 asked the Prophet to impart justice to them, the Prophet granted their request but did not give them the opportunity to withdraw (*al-rujūʿ*). Hence, the verse is the legal referent for all those who voluntarily ask someone to impart divine justice (*al-dīn*) to them, that is, as an arbitrator. Mālik b. Anas asserted that when someone asks someone else to impart justice to him or her, the latter's decision is binding (*fa-ḥukmuhu māḍin*). If the person who requested arbitration is not satisfied with the result and elevates the case to an officially invested religious magistrate (that is, a qadi), the latter must endorse the arbitrator's decision unless it becomes evident that he is not a just person (*jawr*). However, according to Saḥnūn [b. Saʿīd al-Tanūkhī; d. 240/854–55], the qadi is not

36 See Ibn al-ʿArabī, *Aḥkām al-Qurʾān*, ed. al-Bajawī, i, 516.

37 This section draws on Serrano Ruano, Delfina "Bringing arbitration (*taḥkīm*) and conciliation (*ṣulḥ*) under the *qāḍī*'s purview in Mālikī al-Andalus (10th to 12th centuries CE)", in *Revue des mondes musulmans et de la Méditerranée* 140 (2016), 73–100.

38 See Ibn al-ʿArabī, al-*Qabas* ii, 758–759, 869.

39 What follows draws from Ibn al-ʿArabī, Abū Bakr, *Aḥkām al-Qurʾān*, ed. Muḥammad ʿAbd al-Qādir ʿAṭā, Beirut: Dār al-Kutub al-ʿIlmiyya, 2003, ii, 121–127, question (*masʾala*) number 6. See also Serrano Ruano, Delfina, "Forum shopping in al-Andalus (II): Discussing Coran V, 42 and 49 (Ibn Ḥazm, Ibn Rushd al-Jadd, Abū Bakr Ibn al-ʿArabī and al-Qurṭubī)" in Ana Echevarría, Juan Pedro Monferrer, and John Tolan (eds.), *Law and religious minorities in medieval societies: Between theory and praxis*, Turnhout: Brepols Publishers, 2016, 173–196, at pp. 186–188.

obliged to endorse it unless he sees it fit. These opinions, Ibn al-ʿArabī points out, apply solely to legal issues concerning private rights (lit. "those rights that are exclusive to the claimant", *al-ḥuqūq allatī takhtaṣṣ bi-l-ṭālib*). As to God's rights—that is, public rights—like statutory sanctions (*ḥudūd*), they can only be decided by an executive authority (*al-ṣulṭān*). What is firmly established (*al-ḍābiṭ*), he concludes, is that arbitration is permitted for claims on private rights only and that the decision taken by the arbitrator is binding (*wa-nafdh taḥkīm al-muḥakkam bihi*).

Next Ibn al-ʿArabī refutes al-Shāfiʿī's opinion that arbitration is legitimate but not binding (*al-taḥkīm jāʾiz wa-huwa ghayr lāzim*) and that the arbitrator's judgment is like a legal assessment (*fatwā*). Al-Shāfiʿī reasoned that the individuals do not have the capacity per se to appoint governors and judges, nor can they become rulers by themselves. According to Ibn al-ʿArabī, al-Shāfiʿī put the following example to illustrate the issue of arbitration: "He who is asked to arbitrate a conflict is the creator of a decision and if he says 'I have taken a decision', then there are two options: either this happens with no predicate [*laghwan*] or it yields a predicate [*mufīdan*]. However it cannot but yield a predicate [*mufīdan*] in which case there are two options: either a complex sentence made up of two predicates separated by a coordinating conjunction, or an addition, as in when you say [*fa-idhā afāda fa-lā yakhlū an-yufīda al-takthīr ka-qawlaka*]: 'I pronounced it and I shortened it', or a complex sentence in which one predicate creates a new one, as in when you say [*aw yakūna bi-maʿnā jaʿaltu lahu, ka-qawlaka*]: 'I rode it [in a way that] I improved it'—that is, you caused its being ridden and good with the result that you made the ruling (*ay jaʿaltu lahu markūban ḥasanan*). [...] And this is verified by the fact that imparting justice to the people is their right, not the judge's right, whereas entrusting someone with the task of adjudicating a litigation tears a hole in the principle underlying political authority (*al-wilāya*) and leads people to revolt in the same way as the consumption of wine. There is no choice but to add a separative particle (*fāṣil*); the sacred law must be enjoined by decree of the ruler to cut the basis of sedition. The option to ask someone to settle a dispute is permitted only to ease the functioning of ordinary justice by alleviating officially invested judges from the hardships of people filing suits against each other (*mashaqqat al-tarāfuʿ*) to both parties' satisfaction."

Ibn al-ʿArabī agreed with al-Shāfiʿī that imparting justice to the people is the latter's right, not the judge's right. Yet, as we have seen, he rejected the idea that the arbitrator's decision was not binding. Interestingly, he remarked that the Shāfiʿīs do not approach the Sharīʿa as Mālik did, for "they do not take into consideration those criteria that are of public interest (*al-maṣāliḥ*) let alone the objectives [of the sacred law] (*al-maqāṣid*). They just consider the literal expressions (*al-ẓawāhir*) contained in the sacred texts and the rulings they derive from them." Ibn al-ʿArabī claimed that he explained this issue in more detail in his treatises on legal methodology and in *al-Qabas*. In a remarkable exercise of self-assertion, he thereby claimed for the Mālikīs the credit of having pioneered the incorporation of the notions of *maṣlaḥa* and *maqāṣid al-sharīʿa* to Islamic legal methodology.

Subsequently Ibn al-ʿArabī stated that, to the best of his knowledge, there was no *ḥadīth* relevant to assess the issue of *taḥkīm*, with the exception of the following account transmitted to him by an Iraqi qadi: A Jew named Hāniʾ b. Shurayḥ b. Yazīd b. al-Miqdām b. Shurayḥ reported that whenever the Messenger of God entrusted to him a mission with his people, he listened to what they had to say, so that they started to call him the judge par excellence (Abū l-Ḥukm). One day, the Messenger of God summoned him and asked him, "Verily, God is the Ruling and the Ruling belongs to Him. Why do they call you Abū l-Ḥukm?" He replied, "Certainly my people, when they disagree about something, approach me and I decide their case to both parties' satisfaction." The Prophet said: "What is better than this? Who are your children?" He replied, "I have Shurayḥ, ʿAbd Allāh, and Muslim." Then the Prophet asked, "Who is the eldest?" He replied, "Shurayḥ," to which the Prophet remarked, "Then you are Abū Shurayḥ," and he called him so, and invited him and his sons [to adopt this *kunya* instead of Abū l-Ḥukm, given the theological problem it posed].

8 Ibn al-ʿArabī on the Practice of "Forum Shopping"[40]

Apart from the right to keep their own religious beliefs and rituals in exchange for the payment of the special taxes known as *jizya* and *kharāj*, *dhimmī* communities had autonomy to manage their affairs internally, according to their own laws and under the supervision of their specific religious and civil authorities. Q 5:42 and 5:49 provide textual arguments for the practice known as "forum shopping", whereby *dhimmī*s were granted the freedom to bring their litigation before Muslim judicial authorities as well. In most cases, this freedom was exercised against the will of *dhimmī* judges and to the detriment of their authority.

40 This section draws on Serrano Ruano, "Forum shopping in al-Andalus" 174–178 and 182–192.

Ibn al-ʿArabī discusses Q 5:42 in a joint section devoted to Q 5:41–44 in *Aḥkām al-Qurʾān.*[41] He asserts that "the People of the Book came to an agreement (*ṣulḥ*) [with the Muslims], relying on the fundamental principle that they cannot be obliged to submit [to Islamic justice] in anything. If they voluntarily wish to submit to our [justice] and report their disputes to us, it is possible that what they submit to us falls into the category of an unjust act (*ẓulm*) that is not permitted in [any] religious law, or a case the legal qualification of which in the Sharīʿa differs [from its legal qualification in their respective sacred laws]. In the first instance—that is, an act which is forbidden in all the [involved] legal systems such as usurpation, homicide, and similar issues—it is not possible to give power to some of them over the rest in [order to fix] these matters [internally] (*lam yumakkan baʿḍuhum min baʿḍ fīhi*). However, if they voluntarily submit to us the decision to adjudicate a matter on which the Sharīʿa rules differently to what [is established by their respective confessional laws], the *imām* has the option to choose between imparting justice among them or rejecting the assignment."

Ibn al-ʿArabī seemed to be inclined towards the option of accepting the assignment. Unlike Ibn Rushd al-Jadd, for example, Ibn al-ʿArabī was not particularly worried by the potential risk of social and religious syncretism carried by *dhimmīs* mingling and interacting with the qadi, his assistants, and the other Muslim frequenters of the court; nor did he attempt to restrict *dhimmīs*' right to litigate before Muslim authorities.[42] He was more interested in rebuking an unspecified group of Mālikīs who, availing themselves of an opinion by Ibn al-Qāsim (d. 191/806) transmitted by ʿĪsā b. Dīnār (d. 212/828), appeared to reject the argument that Q 5:42 supported the practice of "forum shopping". Ibn al-ʿArabī also warned against adducing Q 5:41 as an argument to treat all non-Muslims as spies; were any of them suspect of espionage, he says, commission of the crime should first be established and then treated according to the relevant Islamic rules introduced after the revelation of the verse in question.

In his commentary on Q 5:49, after discussing the circumstances of the revelation of the verse, Ibn al-ʿArabī addressed the claim of "some interpreters" that the verse abrogates the capacity of the Muslim judge to opt between judging or

41 See Ibn al-ʿArabī, *Aḥkām al-Qurʾān*, ed. ʿAbd al-Qādir ʿAṭā, ii, 121–127 (Q 5:41–44) and 136–137 (Q 5:49).

42 For Ibn Rushd al-Jadd's position on "forum shopping" see Serrano Ruano, Delfina, "*Lā yajūz li-ḥukm al-muslimīn an yaḥkum bayna-humā*: Ibn Rushd al-Jadd (Cordoba, d. 1126 CE) and the restriction on dhimmis shopping for Islamic judicial forums in al-Andalus" in Nora Berend, Youna Hameau-Masset, Capucine Nemo-Pekelman, and John Tolan (eds.), *Religious minorities in Christian, Jewish and Muslim law (5th–15th centuries)*, Turnhout: Brepols, 2017, 395–411.

not doing so (*al-takhyīr*) established by Q 5:42. In his view, that claim was too broad, since the conditions for abrogation are four, one of them being knowledge of the chronology of the revelation of each verse. However, as pointed out by Ibn al-ʿArabī, it is not possible to establish which verse of the two in question was revealed first, so the suggestion that one of them abrogates the other is moot.

Next, he deals with the claim that the segment "And take good care that they do not tempt you away from *any* (*baʿḍ*) of what God has sent down to you" in Q 5:49 means that the intention of the Jews who approached the Prophet asking him to impart justice to them was to tempt Muḥammad to abandon the law revealed to him by God altogether. Yet the correct interpretation, according to Ibn al-ʿArabī, is that "any" does not mean "everything". Rather, it must be understood literally, that is, as something specific referring to the kind of punishment—be it stoning or any other penalty—the Jews wanted the Prophet to pronounce.

9 Conclusions

Abū Bakr Ibn al-ʿArabī was a great Muslim intellectual whose writings characterise him as a passionate and outspoken person concerned with the ideological and political circumstances of his time. He was proud of the scholarly competences he acquired during his long stay in the East, including the opportunity to attend the lessons of such an outstanding master as al-Ghazālī, whose influence, however, he did not hesitate to renounce when the political circumstances so warranted. Moreover, he was the son of a high-ranking official in service of the Abbadid Taifa rulers of his native Seville, who managed to obtain letters of recognition for Almoravid rule in al-Andalus from Abū Bakr al-Ṭurṭūshī, al-Ghazālī himself, and, last but not least, the Abbasid caliph himself.

No doubt these facts helped fuel his ambition and convince him that he was the best political advisor and religious scholar of the time. The soundness of his scholarly training and his social pedigree appear to have inclined him to develop a certain superiority complex and to overestimate his capacity to defeat those whom he thought antagonised or rivalled him. His self-perceived position of privilege compounded doctrinal confrontation with both the Ẓāhirīs—he says that they made life insufferable for him in Seville[43]—and with other Mālikīs, and, together with his undeniable relationship with al-Ghazālī,

43 See Serrano Ruano, "¿Por qué llamaron?" 820*n*17, drawing on Ibn al-ʿArabī, *al-ʿAwāṣim min al-qawāṣim* 248–249.

submitted him to strong suspicion. His high social background and his excellent training in Islamic religious sciences do not seem to have helped him face the ensuing stress with the serenity and self-confidence that would have spared him the informed criticisms of Ibn al-Sīd al-Baṭalyawsī and al-Qurṭubī. However, his frequent references to his immediate mileau—be they political events, social and legal practices, or specific persons (even if inappropriate on occasion)—provide a "gold mine" for contemporary scholars, who must often cope with the gaps and omissions imposed by the etiquette and the stylistic and social conventions common among pre-modern ulemas. His rare frankness led him to challenge stylistic and social conventions of the day—in the same way in which he challenged received wisdom—and to eschew a hinting or indirect style. Thanks to this approach, specific details about persons, places, and facts involved in the ideas which Ibn al-ʿArabī strived to refute or to promote have been preserved for us.

By applying the principles of Islamic legal methodology, which required reading a fundamental text like the *Muwaṭṭaʾ* from a critical point of view, Ibn al-ʿArabī hoped to solve its inconsistencies and in doing so, to strengthen the the doctrine of the Mālikī school.

As I have pointed out elsewhere, the advantage offered by focusing on sources compiled by an author like Ibn al-ʿArabī—who not only witnessed but also suffered from the consequences of the power shift from the Almoravids to the Almohads—is that in his efforts to avoid its consequences—which he was able to anticipate—he counter-argued and in this way provided us with important clues with which to approach elements of the nascent Almohad ideology—with all its apparent contradictions and fluctuations—that are not explicit in the literature produced in the wake of the movement and within its own ranks.

Ibn al-ʿArabī embodies and advocates a self-confident, self-sufficient Mālikism capable of purging itself of the shortcomings and contradictions that made the school an easy target for its opponents.

Ibn al-ʿArabī's personal battle against the crime of *zinā* is an important topos in his doctrine on "commanding right and forbidding evil". He upheld the legitimacy of the Sharīʿa by disclosing the inconsistencies between sound legal doctrine and practice, rather than projecting an idealised portrayal of them as if they were in total accord. He was both outspoken and manipulative, but just enough. He was driven by a sincere desire for reform, which involved a critical examination and recognition of reality.

Ibn al-ʿArabi's nuanced, at times inclusive and even apologetic perspective on the validity and unexpired continuity of the *dhimma* covenant should not be a surprise in a Sunni Mālikī jurist like him. Yet this attitude is in sharp contrast with his reputation as a staunch anti-Christian polemicist who adopted harsh measures against the consumption of wine during his tenure as qadi of Seville.[44] In addition, he vehemently rejected other exegetes' tendency to use traditions coming from Judaism (*isrāʾīliyyat*) to illustrate the meaning of the Qurʾān. Moreover, and as I've mentioned before, Ibn al-ʿArabī identified Muslim anthropomorphists with the Jews. Last, but not least, he did not limit himself to simply encouraging the rulers and his fellow Andalusis to practice *jihād* against north Iberian Christians, but took an active part as a combatant in several of the ensuing military campaigns. Such an emphasis on the continued validity of the *dhimma* covenant is absent from the approach of other authors I have dealt with—such as Ibn Ḥazm, Ibn Rushd al-Jadd, and al-Qurṭubī—to the issue of forum shopping. Given the chronology, Ibn al-ʿArabī appears to have anticipated Almohads' plans to abolish the *dhimma* pact, and tried his best to undermine them.

Bibliography

ʿAbd al-Majīd, Ḥamīd, "Taʿrīf bi-l-muʾallif wa-l-kitāb", introd. to Ibn al-Sīd, *al-Intiṣār mimman ʿadala ʿan al-istibṣār*, ed. Ḥamīd ʿAbd al-Mājid, Cairo: al-Maṭbaʿa al-Amīriyya, 1955.

Abou El Fadl, Khaled, *Rebellion and violence in Islamic law*, Cambridge: Cambridge University Press, 2001.

Asín Palacios, Miguel, *La espiritualidad de Algazel y su sentido cristiano*, Madrid: Escuelas de Estudios Árabes de Madrid y Granada, 1934.

Blankinship, Kevin, "Al-Maʿarrī's esteem in the Islamic West: A preliminary overview", in *al-Masāq: Journal of the Medieval Mediterranean* 31 (2019), 253–271.

Casassas Canals, Xavier, and Delfina Serrano-Ruano, "Putting criticisms against al-Ghazālī in perspective: New materials on the interface between law, rational theology and mysticism in Almoravid and Almohad al-Andalus (Ibn Rushd al-Jadd and al-Qurṭubī)", in Ayman Shihadeh and Jan Thiele (eds.), *Philosophical theology in Islam: Later Ashʿarism East and West*, Leiden: Brill, 2020, 254–297.

EI2 = Peri Bearman, Thierry Bianquis, Clifford Edmund Bosworth, Emery van Donzel, and Wolfhart P. Heinrichs (eds.), *The Encyclopaedia of Islam, second edition*, Leiden: Brill, 1954–2004.

44 Dominique Urvoy characterises him as one of the main moral and political authorities of his time and asserts that he and ʿIyāḍ b. Mūsā were the most relevant anti-Christian polemicists of the Almoravid period. See Urvoy, *Pensers d'al-Andalus* 83, 166. See also Fierro, Maribel, "La religión", in María Jesús Viguera et al (eds.), *El retroceso territorial de al-Andalus: Almorávides y almohades, siglos XI al XIII*, Madrid: Espasa Calpe, 1997, 435–546, at pp. 531, 533.

Elamrani-Jamal, Abdelali, "Ibn al-Sīd al-Baṭalyūsī et l'enseignement d'al-Fārābī", *Bulletin d'études orientales* 48 (1996), 155–161.

Fierro, Maribel, "La religión", in María Jesús Viguera et al. (eds.), *El retroceso territorial de al-Andalus: Almorávides y almohades, siglos XI al XIII*, Madrid: Espasa Calpe, 1997, 435–546.

Fierro, Maribel, *El libro de las novedades y las innovaciones*, Madrid: Consejo Superior de Investigaciones Científicas-Instituto de Cooperación con el Mundo Árabe, 1993.

Ibn al-ʿArabī, Abū Bakr, *Aḥkām al-Qurʾān*, 2 vols., Cairo, 1331/1914.

Ibn al-ʿArabī, Abū Bakr, *Aḥkām al-Qurʾān*, ed. ʿAlī Muḥammad al-Bajawī, 4 vols., Beirut: Dār al-Maʿrifa, 1972.

Ibn al-ʿArabī, Abū Bakr, *Qānūn al-taʾwīl*, ed. M. Slīmānī, Tunis: Dār al-Gharb al-Islāmī, 1990.

Ibn al-ʿArabī, Abū Bakr, *al-Naṣṣ al-kāmil li-kitāb al-ʿAwāṣim min al-qawāṣim li-l-qāḍī Abī Bakr ibn al-ʿArabī*, ed. ʿAmmār Ṭālibī, Doha: Dār al-Thaqāfa, 1413/1992.

Ibn al-ʿArabī, Abū Bakr, *al-Qabas fī sharḥ Muwaṭṭaʾ Mālik ibn Anas*, ed. Muḥammad ʿAbd Allāh Wuld Karīm, Beirut: Dār al-Gharb al-Islāmī, 1992.

Ibn al-ʿArabī, Abū Bakr, *Aḥkām al-Qurʾān*, ed. Muḥammad ʿAbd al-Qādir ʿAṭā, Beirut: Dār al-Kutub al-ʿIlmiyya, 2003.

Ibn ʿIyāḍ, *Madhāhib al-ḥukkām fī nawāzil al-aḥkām li-l-qāḍī ʿIyāḍ wa-waladihi Muḥammad*, ed. Mohamed Bencherifa, Beirut: Dār al-Gharb al-Islāmī, 1990.

Ibn al-Sīd, *al-Intiṣār mimman ʿadala ʿan al-istibṣār*, ed. Ḥamīd ʿAbd al-Mājid, Cairo: al-Maṭbaʿa al-Amīriyya, 1955.

Peña, Salvador, *Maʿarrī según Baṭalyawsī: Crítica y poética en al-Andalus, siglo XI*, Granada: Universidad de Granada, 1990.

Peña, Salvador, "Ibn al-Sīd al-Baṭalyawsī, Abū Muḥammad", in Jorge Lirola Delgado (ed.), *Biblioteca de al-Andalus: de Ibn Saʿāda a Ibn Wuhayb*, Fundación Ibn Tufayl de Estudios Árabes, 2007, v, 304–337.

al-Qurṭubī, *al-Asnā fī sharḥ asmāʾ Allāh al-ḥusnā wa-ṣifātihi al-ʿulā*, ed. ʿA. b. S. al-ʿAshā Ḥassūna al-Dimashqī, Cairo: al-Maktaba al-Ḥaḍariyya, 1427/2006.

Serrano Ruano, Delfina, *Madhāhib al-ḥukkām fī nawāzil al-aḥkām* (*La actuación de los jueces en los procesos judiciales*) *de Muḥammad b. ʿIyāḍ*, Madrid: CSIC-AECI, 1998.

Serrano Ruano, Delfina, "Ibn al-Sīd al-Baṭalyawsī (444/1052–521/1127): De los reinos de taifas a la época almorávide a través de la biografía de un ulema polifacético", in *al-Qanṭara* 23 (2002), 53–92.

Serrano Ruano, Delfina, "Ibn ʿIyāḍ al-Sabtī, Abū ʿAbd Allāh", in Jorge Lirola Delgado and José Miguel Puerta Vílchez (eds.), *Biblioteca de al-Andalus: de Ibn al-Dabbāg a Ibn Kurz*, Almería: Fundación Ibn Tufayl de Estudios Árabes, 2004, 542–544, no. 661.

Serrano Ruano, Delfina, "¿Por qué llamaron los almohades antropomorfistas a los almorávides?" in Maribel Fierro, Patrice Cressier, and Luis Molina (eds.), *Los almohades: Problemas y perspectivas*, Madrid: CSIC/Casa de Velázquez, 2005, ii, 815–852.

Serrano Ruano, Delfina, "La lapidación como castigo de las relaciones sexuales no legales (*zinà*) en el seno de la escuela malikí: Doctrina, práctica legal y actitudes individuales frente al delito (ss. XI y XII)", in *al-Qanṭara* 26 (2005), 449–473.

Serrano Ruano, Delfina, "Rape in Maliki legal doctrine and practice (8th–15th centuries CE)", in *Hawwa* 5 (2007), 166–207.

Serrano Ruano, Delfina, "El Corán como fuente de legislación islámica: Abū Bakr Ibn al-ʿArabī y su obra *Ahkām al-Qurʾān*", in Miguel Hernando de Larramendi and Salvador Peña Martín (eds.), *El Corán ayer y hoy: Perspectivas actuales sobre el islam*, Córdoba: Berenice, 2008, 251–275.

Serrano Ruano, Delfina, "ʿIyāḍ b. Musà", in Jorge Lirola Delgado (ed.), *Biblioteca de al-Andalus*, Almería: Fundación Ibn Tufayl de Estudios Árabes, 2009, 404–434, no. 1479.

Serrano Ruano, Delfina, "Ibn Rushd al-Jadd", in David Powers, Susan Spectorsky, and Oussama Arabi (eds.), *Islamic legal thought: A compendium of Muslim jurists*, Leiden: Brill, 2013, 295–322.

Serrano Ruano, Delfina"Bringing arbitration (*taḥkīm*) and conciliation (*ṣulḥ*) under the qāḍī's purview in Mālikī al-Andalus (10th to 12th centuries C.E.)", in *Revue des mondes musulmans et de la Méditerranée* 140 (2016), 73–100.

Serrano Ruano, Delfina, "Forum shopping in al-Andalus (II): Discussing Coran V, 42 and 49 (Ibn Ḥazm, Ibn Rushd al-Jadd, Abū Bakr Ibn al-ʿArabī and al-Qurṭubī)" in A. Echevarría, J.P. Monferrer, and J. Tolan (eds.), *Law and religious minorities in medieval societies: Between theory and praxis*, Turnhout: Brepols Publishers, 2016, 173–196.

Serrano Ruano, Delfina, "Lā yajūz li-ḥukm al-muslimīn an yaḥkum bayna-humā: Ibn Rushd al-Jadd (Cordoba, d. 1126 CE) and the restriction on dhimmis shopping for Islamic judicial forums in al-Andalus" in Nora Berend, Youna Hameau-Masset, Capucine Nemo-Pekelman, and John Tolan (eds.), *Religious minorities in Christian, Jewish and Muslim law (5th–15th centuries)*, Turnhout: Brepols, 2017, 395–411.

Serrano Ruano, Delfina, "Debates on free will and predestination in the 12th century Islamic West: Abū Bakr Ibn al-ʿArabī's (468/1076–542/1147 or 543/1148) Kitāb al-Qabas fī Sharḥ Muwaṭṭaʾ Mālik Ibn Anas", in *Islamic Law Blog*, special feature on "Muwaṭṭaʾ Roundtable", available at: https://islamiclaw.blog/2019/12/06/muwaṭṭaʾ-roundtable-debates-on-free-will-and-predestination-in-the-12th-century-islamic-west-abu-bakr-ibn-al-ʿarabis-468-1076-542-1147-or-543-1148-kitab-al-qabas/.

Urvoy, Dominique, *El mundo de los ulemas andaluces del siglo V/XI al VII/XIII: Estudio sociológico*, Madrid: Ediciones Pegaso, 1983.

Urvoy, Dominique, *Pensers d'al-Andalus: La vie intellectuelle à Cordoue et à Séville au temps des empires berbères (fin XIe siècle–début XIIIe siècle)*, Toulouse: Presses Universitaires du Miraill, 1990.

Viguera, María Jesús, "En torno a las fuentes jurídicas de al-Andalus", in *La civilisation d'al-Andalus dans le temps et dans l'espace: Actes du colloque international, [Université Hassan II, Maroc], 16–18 avril 1992*, Muhammadiyya: Imprimerie de Fedala, 1993, 71–78.

3

مَفهومُ الإلحادِ في فِكرِ أَبي بَكرٍ ابنِ العَرَبِيّ

جَعفَر ابن الحاج السُّلَميّ

[The Concept of *ilḥād* in the Thought of Ibn al-ʿArabī]

[*Jaafar Ben El Haj Soulami*]

1 اَلسُّنَّةُ وَالبِدعَةُ وَالإلحادُ وَالزَّندَقَة: مَفاتيحُ الفِكرِ الكَلامِيّ لِابنِ العَرَبِيّ

مِنَ اللاّفِتِ لِلنَّظَرِ في فِكرِ ابنِ العَرَبِيِّ الكَلامِيّ، اِستِعمالُهُ لِمَفاهيمَ كَلامِيَّةٍ سُنِّيَّةٍ أَشعَرِيَّةٍ تُنَظِّمُ فِكرَه، وَيَصدُرُ عَنها في تَقييمِ كُلِّ الأَفكارِ الكَلامِيَّةِ قَبولاً وَرَفضا. إنَّ بَعضَ هاذِهِ المَفاهيمَ مُخَصَّصٌ لِوَصفِ المُخالِفِ في تَفكيرِهِ الكَلامِيّ، وَإدانَتِهِ أَيضا، لِأَنَّ المُخالَفَةَ في التَّفكيرِ الدّينِيّ، كَثيرًا ما كانَت تُؤَدّي في العَصرِ الوَسيطِ إلى الإدانَة.

لَيسَت هاذِهِ المَفاهيمُ مِن إنشاءِ القاضي أَبي بَكر، وَلا هِيَ كانَت جَديدَةً في الأَندَلُسِ أَوِ المَغرِبِ وَأَدخَلَها هُو إلَيها. فَبَعضٌ مِنها استَعمَلَهُ مِن قَبلُ كُتّابُ كُتُبِ البِدَع، كَمُحَمَّدِ بنِ وَضّاحٍ القُرطُبِيّ، (287هـ\900م) في كِتابِه: "اَلبِدَعُ وَالنَّهيُ عَنها"، وَابنِ أَبي زَمَنين، (399هـ\1008م) في كِتابِه: "أُصولِ السُّنَّة"، وَالطُّرطوشِيّ، (520هـ\1126م) في كِتابِه: "الحَوادِثِ وَالبِدَع"، وَمُؤَرِّخو المِلَلِ وَالنِّحَل، كَابنِ حَزم، (458هـ\1066م) في كِتابِه: "اَلنَّصائِحُ المُنجِيَة، مِنَ الفَضائِحِ المُخزِيَة، وَالقَبائِحِ المُردِيَة، مِن أَقوالِ أَهلِ البِدَع، مِنَ الفِرَقِ الأَربَع، اَلمُعتَزِلَةِ وَالمُرجِئَةِ وَالخَوارِجِ وَالشِّيَع"، وَكِتابِه: "الفَصل، في المِلَلِ وَالأَهواءِ وَالنِّحَل".

أَمّا هُو، فَقَد سَمّى بَعضَ كُتُبِه: "اَلكِتابَ المُتَوَسِّطُ في الاِعتِقاد، وَالرَّدَّ عَلى مَن خالَفَ السُّنَّةَ مِن ذَوي البِدَعِ وَالإلحاد". وَقَد سَبَقَ لِلإمامِ أَبي الحَسَنِ الأَشعَرِيّ، إمامِ مَذهَبِه، (324هـ\935م) أَن أَلَّفَ كِتابَ "اَللُّمَع، في الرَّدِّ عَلى أَهلِ الزَّيغِ وَالبِدَع"، كَما كانَ لِمُعاصِرِهِ القاضي عِياض، (544هـ\1149م) تَوظيفٌ لَها في كِتابِهِ "تَرتيبِ المَدارِك".[1]

1 تَرتيبُ المَدارِك: 1/5.

 | DOI:10.1163/9789004706859_005

وَلاكِنَّ القاضِيَ أَبا بَكرٍ استَكثَرَ مِنها، وَجَعَلَها مَفاتيحَ لِفِكرِهِ الكَلامِيّ، وَمَوقِفِهِ مِنَ العالَمِ الَّذي كانَ يُحيطُ بِه. فَهُوَ يَستَعمِلُ لِوَصفِ المُسلِمينَ السُّنِّيّين، أَي ذاتِهِ الفِكريَّة، مُصطَلَحاتٍ جامِعَة. وَهِيَ ما يَلي:

1. "أَهلُ المِلَّة"،[2] "أَهلُ الحَقّ"،[3] "الأُمَّة".[4] بَل قَد يَعنيهِم بِـ"المُصَلّين"[5]، عَلى افتِراضٍ ضِمنِيٍّ بِأَنَّ مِن أَهَمِّ ما يُمَيِّزُ السُّنِّيَّ عَن غَيرِهِ مِنَ المُسلِمين، وَلا سِيَما الشّيعَةَ الإسماعيلِيَّة، إقامَةَ الصَّلاة. وَقَد يُسَمّي "أَهلَ السُّنَّةِ" اَلمُوَحِّدين، وَيَقصِدُ بِهِم عامَّةَ المُسلِمين.[6] أَمّا مُصطَلَحُ "المُوَحِّدين"، بِما هُم فِرقَةٌ خاصَّةٌ ظَهَرَت في المَغرِبِ وَالأَندَلُسِ عَلى يَدِ ابنِ تومَرت، وَكانَ هُوَ يَكتُبُ في بِدايَةِ زَمانِهِم، فَهُوَ غائِبٌ عِندَهُ غِيابًا لافِتًا لِلنَّظَر. وَمِنَ النّادِرِ أَن نَجِدَ عِندَ ابنِ العَرَبِيِّ تَحديدًا جامِعًا مانِعًا لِمَفهومِ السُّنِّيَّةِ. فَهُوَ يَتَحَدَّثُ عَن "أُصولِ السُّنَّة"[7]، دونَ أَن نَجِدَ تَحديدًا قاطِعًا لِهاذا المَفهوم. وَهاذِهِ الأُصولُ هِيَ عِندَهُ في أَحسَنِ الأَحوال: "هُوَ ما كانَ عَلَيهِ السَّلَفُ المُتَقَدِّمونَ مِنَ الصَّحابَةِ وَالتّابِعين".[8] وَيَظهَرُ أَنَّ السّنِيَّةَ عِندَهُ دَرَجاتٌ أَو فِئات. فَفيها العالِمَة، وَفيها الجاهِلَة، إذ هُوَ يُشيرُ إلى "اَلجَهَلَةِ المُنتَمينَ إلى أَهلِ السُّنَّة".[9]

وَيَستَعمِلُ لِوَصفِ غَيرَ السُّنِّيّينَ مُصطَلَحاتٍ جامِعَةً تَشمَلُهُم. وَهِيَ ما يَلي:

1. "أَهلُ البِدَعِ مِنَ المِلَّة"[10]، 2. "المُبتدِعون"[11]، "أَهلُ البِدعَة"[12]، "المُبتَدِعَة"[13]، "أَهلُ الأَهواء".[14]

وَمَعَ عَدَمِ تَحديدِ مَفهومِ السُّنِّيَّةِ وَالبِدعِيَّةِ وَالإلحادِ عَلى العُموم، فَإنَّهُ يَسعى، مَعَ هذا، إلى تَحديدِ مَفاهيمَ أُخرى، مِنها: "الإيمان"، وَ"الإسلام"، وَ"الكُفر"، وَ"الشِّرك"، وَ"النِّفاق"، وَ"الفِسق"،[15] وَالزَّندَقَة.[16] وَهِيَ مَفاهيمُ أَساسِيَّةٌ في عِلمِ الكَلام، فَـ"اَلزِّنديق: هُوَ الَّذي يُسِرُّ الكُفرَ وَيُظهِرُ الإيمان".[17]، وَنَقَلَ ذلك عَن إمامِ

---

2 اَلكِتابُ المُتَوَسِّط: 328.

3 اَلكِتابُ المُتَوَسِّط: 328، 330.

4 اَلكِتابُ المُتَوَسِّط: 405.

5 اَلكِتابُ المُتَوَسِّط: 405.

6 اَلعَواصِم: 103، 118.

7 اَلكِتابُ المُتَوَسِّط: 400.

8 قانونُ التَّأويل: 105.

9 اَلكِتابُ المُتَوَسِّط: 228.

10 اَلكِتابُ المُتَوَسِّط: 104.

11 قانونُ التَّأويل: 120.

12 قانونُ التَّأويل: 120.

13 أَحكامُ القُرآن: 1/294.

14 أَحكامُ القُرآن: 1/294.

15 اَلكِتابُ المُتَوَسِّط: 455.

16 أَحكامُ القُرآن: 1/12.

17 أَحكامُ القُرآن: 1/12.

المَذهب، فَقال: "وَسُئِلَ مالِكٌ عَنِ الزَّندَقَة، فَقال: ما كانَ عَلَيهِ المُنافِقونَ في عَهدِ رَسولِ الله، (ص) مِن إظهارِ الإيمان، وَكِتمانِ الكُفر، هُوَ الزَّندَقَةُ عِندَنا اليَوم."[18]

وَرَأى في الزَّندَقَةِ عَينَ "النِّفاقِ" الدّينِيِّ في عَهدِ النُّبُوَّة، تَبَعًا لِلإمامِ مالِك. وَإنَّما يَكمُنُ الفَرقُ عِندَهُ في التَّخصيصِ اللُّغَوِيِّ لِلمُصطَلَحَين، أَمّا مَضمونُهُما فَواحِد. قال: اَلنِّفاق: "فَأَمّا النِّفاق...وَهُوَ اعتِقادُ خِلافِ ما يُظهِرُ مِنَ الاِنقِياد. فَهاذِهِ التَّسمِيَةُ تُختَصُّ بِمَن كانَ يَعتَقِدُ ذلك عَلى عَهدِ النَّبِيّ، (ص). وَمَن كانَ بِهاذِهِ الصِّفَةِ بَعدَه، أُختُصَّ بِاسمِ الزِّنديق. وَذلك مِمّا اتَّفَقَ عَلَيهِ العُلَماء."[19]

وَقَد زادَ الأَمرَ تَخصيصا، فَخَصَّصَ مَفهومَ الزَّندَقَةِ بِالدَّهرِيَّة، وَبِإنكارِ الأُلوهِيَّة، أَو إنكارِ تَدَخُّلِها في العالَم، فَقال: "وَلا يُقالُ لِكُلِّ مَن وَحَّدَ اللّهَ تَعالى وَأَقَرَّ بِالصّانِع، زِنديق، وَإنَّما يَقَعُ اسمُ الزِّنديقِ عَلى المُعَطِّلِ الَّذي لا يُقِرُّ بِالصّانِع، وَيَزعُمُ أَنَّ النّاسَ يَتَكَوَّنونَ مِن غَيرِ خالِقٍ لَهُم، وَلا مُبدِئٍ لَهُم وَلا مُعيد، وَأَنَّ البارِئَ تَعالى قَد فَرَغَ مِن جَميعِ الأَشياء، لا يُحدِثُ شَيئا، وَإنَّما تَصدُرُ عَنهُ الكائنات..."[20]

وَأَورَدَ تَصنيفًا لِفِرَقِ الزَّنادِقَة،[21] نَقلاً عَن أَبي عاصِمِ النَّسائِيّ، (253هـ\867م) خُشَيشِ بنِ أَصرَم، في كِتابِه: "اَلتَّأكيدِ في لُزومِ السُّنَّة، وَحُبِّ خِيارِ هاذِهِ الأُمَّة"، فَقال: "وَمِنهُمُ الرّوحانِيَّة. وَهُم أَصناف. وَإنَّما سُمّوا الرّوحانِيَّة، لِأَنَّهُم زَعَموا أَنَّ أَرواحَهُم تَنظُرُ إلى المَلَكوتِ الأَعلى، وَتُعايِنُ الحور، وَتُشاهِدُ الجَنَّة، وَتَتَنَعَّمُ بِرُؤيَةِ البارِئِ تَعالى. وَسُمّوا أَيضًا "الفِكرِيَّة"، لِأَنَّهُم زَعَموا أَنَّهُم يَتَفَكَّرونَ في هذا حَتّى يَعرِفوا الله، فَجَعَلوا الفِكرَةَ غايَةَ العِبادَة، وَهُوَ مَذهَبُ الحَلاّجِ السّاحِرِ الكافِر. وَمِنهُم صِنفٌ مِنَ الرّوحانِيَّة، زَعَموا أَنَّ حُبَّ اللهِ تَعالى يَغلِبُ عَلى قُلوبِهِم وَأَهوائِهِم، حَتّى يَكونَ حُبُّهُ أَغلَبَ الأَشياءِ عَلَيهِم. فَإذا كانَ كَذلك، كانوا عِندَ اللهِ بِهاذِهِ المَنزِلَة، وَإذا كانوا عِندَ اللهِ بِهاذِهِ المَنزِلَة، وَجَبَت لَهُمُ الجَنَّة، وَارتَفَعَ عَنهُمُ التَّكليف، وَأَحَلَّ لَهُمُ المَحظور، فَهُوَ عِندَهُم حَلال. وَمِنهُم صِنفٌ زَعَموا أَنَّ تَركَ الدُّنيا اشتِغالُ القُلوبِ. وَرابِعَتُهُم رابِعَة."[22]

وَلَمّا كانَ ابنُ العَرَبِيِّ مالِكِيّا، وَالمالِكِيَّةُ لا يُسامِحونَ في أَمرِ الزَّندَقَة، أَنكَرَ تَجويزَ الشّافِعِيَّةِ لِاستِتابَةِ الزِّنديق، فَقال: "فَهاذا المُتَأَخِّرُ مِن أَصحابِ الشّافِعِيِّ الَّذي قال: إنَّ استِتابَةَ الزِّنديقِ جائزَة، قالَ ما لَم يَصِحّ، قَولاً واحِدا."[23] وَكَأَنَّهُ يُذَكِّرُهُ بِأَنَّهُ خالَفَ كَلامَ إمامِهِ الشّافِعِيّ.

---

18 اَلمَسالِك: 216/3.

19 اَلكِتابُ المُتَوَسِّط: 461–462.

20 اَلمَسالِك: 404/3.

21 اَلمَسالِك: 400/3–404.

22 اَلمَسالِك: 403/3.

23 أَحكامُ القُرآن: 12/1.

وَيَظهَرُ أَنَّ القاضِيَ أَبا بَكرٍ بَدَأَ مَشروعَهُ الكَلامِيَّ في أَوَّلَ شَبابِه، وَهُوَ في الأَندَلُس، وَأَنَّهُ فيهِ هَيَّأَ نَفسَهُ لِلنِّضالِ عَن مَذهَبِ أَهلِ السُّنَّة، وَتَفنيدِ مَذاهِبِ المُخالِفين، عَلى اختِلافِهِم. يَقول: "...وَأَنا بِغَرارَةِ الشَّباب...وَالقَدَرُ يُحَبِّبُها عِندي لِلانتِفاعِ بِها في الرَّدِّ عَلى المُلحِدين، وَالتَّمهيدِ لِأُصولِ الدّين"[24]، وَلا سِيَما بَعدَ اختِراقِهِ لِمِصر، حَيثُ استَبانَ لَهُ خُمولُ أَهلِ السُّنَّةِ بِسَبَبِ قَهرِ الفاطِمِيّين.[25]

2 مَفهومُ "الإلحادِ" مَفهومًا سُنِّيًّا أَشعَرِيًّا

إذا تَتَبَّعنا مَفهومَ الإلحادِ في أَصلِه، فَإِنَّهُ يَقَرُّ في الذِّهنِ أَنَّهُ مَفهومٌ سُنِّيٌّ قَبلَ كُلِّ شَيء. فَالكَلِمَةُ وارِدَةٌ في القُرآنِ الكَريم. وَقَدِ التَفَتَ الخَليلُ بنُ أَحمَد، (170هـ\786م) وَهُوَ أَحَدُ أَئِمَّةِ السُّنَّة، إلى تَفسيرِها، فَقال: "أَلحَدَ إِلَيه، وَلَحَدَ إِلَيهِ بِلِسانِه: أَي مال. وَيُقرَأ: "لِسانُ الَّذي يُلحِدونَ" وَ"يَلحَدون". وَأَلحَدَ في الحَرَم...إذا تَرَكَ القَصد، وَمالَ إِلى الظُّلم. وَمِنهُ قَولُهُ تَعالى: "وَمَن يُرِد فيهِ بِإِلحادٍ"."[26] وإذا تَتَبَّعنا كِتابَ "فِرَقِ الشّيعَةِ" لِلنَّوبَختِيّ، (310هـ\922م) فَإِنَّ هذا المَفهومَ غائِبٌ فيه.

لَن يَسَعَنا تَتَبُّعُ هذا المَفهومِ في كُتُبِ الفِرَقِ في هذا المَقام، وَسَوفَ نَكتَفي بِفِرقَةِ أَهلِ السُّنَّةِ الأَشاعِرَة، وَنَرى هَل حَدَّدَ أَبو الحَسَنِ الأَشعَرِيُّ مَفهومَ الإلحادِ وَالبِدعَةِ فيما وَصَلَنا مِن كُتُبِه؟

فَفي كِتابِه: "مَقالاتِ الإسلامِيّين، وَاختِلافِ المُصَلّين"، لا نَجِدُ هذا المَفهومَ وارِدًا عِندَه. وَإِنَّما نَجِدُهُ يَستَعمِلُ مَفاهيمَ "أَهلِ الاِستِقامَة"[27]، وَ"أَهلِ القِبلَةِ المُوَحِّدين"[28]، وَفي مُقابِلِها يَستَعمِلُ مَفاهيمَ "أَهلِ البِدَع"[29]، وَ"أَهلِ الأَهواء".[30] وَالظّاهِرُ أَنَّ "البِدعِيَّة" لا تَكونُ عِندَ الإمامِ أَبي الحَسَنِ الأَشعَرِيِّ بِغَيرِ تَأويلٍ لا تَقبَلُهُ "أُصولُ السُّنَّة"، فَهُوَ يُسَمّيهِمُ "المُتَأَوِّلين"[31]، وَ"أَهلَ التَّأويل".[32]

فَهَل استَعمَلَ تَلامِذَةُ أَبي الحَسَنِ الأَشعَرِيِّ مَفهومَ "الإلحادِ" وَوَظَّفوهُ في نَسَقِهِمُ الفِكرِيّ؟

24 قانونُ التَّأويل: 74.
25 قانونُ التَّأويل: 89.
26 كِتابُ العَين: 4/73–74.
27 مَقالاتُ الإسلامِيّين: 2/147.
28 مَقالاتُ الإسلامِيّين: 2/248.
29 مَقالاتُ الإسلامِيّين: 2/146.
30 مَقالاتُ الإسلامِيّين: 2/147.
31 مَقالاتُ الإسلامِيّين: 2/151.
32 مَقالاتُ الإسلامِيّين: 2/151.

في "كِتابِ التَّنبيه"، وَالرَّدِّ عَلى أَهلِ الأَهواءِ وَالبِدَع، لِأَبي الحُسَين، مُحَمَّدِ بنِ أَحمَدَ المَلَطِيّ. (377هـ\987م) وَهُوَ مِن أَقدَمِ كُتُبِ الأَشاعِرَة، نَجِدُهُ يَتَحَدَّثُ عَنِ الشّيعَة، أَو طائفَةٍ مِنهُم، فَيُسَمّيهِم: "أَهلَ الضَّلالَةِ الرّافِضَة".[33] وَعِندَهُ أَنَّهُم "هُم مُلحِدون، لِأَنَّ هِشامًا [هِشامَ بنَ الحَكَم] كانَ مُلحِدًا دَهرِيًا، ثُمَّ انتقَلَ إلى الثَّنوِيَّةِ وَالمَنّانِيَّة، ثُمَّ الإسلام، فَكانَ قَولُهُ في الإِسلامِ بِالتَّشبيهِ وَالرَّفض."[34]

وَفي "كِتابِ الحُدودِ في الأُصول"، (أَوِ الحُدودُ وَالمُواضَعات) لأَبي بَكر، مُحَمَّدِ بنِ الحَسَنِ ابنِ فورَكَ الأَصبَهانِيّ، (406هـ\1015م) نَجِدُ صاحِبَهُ يَذهَبُ إلى ما هُوَ أَبعَد، فَيُحَدِّدُ مَفهومَ الإلحادِ بِقَولِه: "حَدُّ الإلحاد: هُوَ الخُروجُ عَنِ الحَقّ، وَكَذلك المُروق"،[35] دونَ أَن يَربِطَهُ بِشيعَةٍ رافِضَةٍ أَو قَدَرِيَّة، أَو إنكارٍ لِلذّاتِ الإلاهِيَّة، فَمُطلَقُ الخُروجِ عَنِ الحَقِّ إلحادٌ فيه.

أَمّا أَبو إسحاقَ الإسفَرايينِيّ: (418هـ\1027م) فَقَدِ اهتَمَّ بِظاهِرَةِ الإلحاد، وَصَنَّفَ فيهِ كِتابًا سَمّاه: "جامِعَ الجَلِيِّ والخفي في أُصولِ الدّين، وَالرَّدِّ عَلى المُلحِدين"[36] ذَكَرَ عَنهُ أَبو المُظَفَّرِ الإسفَرايينِيّ: (471هـ\1078م) صاحِبُ "التَّبصير"، أَنَّهُ "لَم تَخفَ عَلَيهِ كَيفِيَّةُ الرَّدِّ عَلى شَيءٍ مِن مَقالاتِ المُلحِدينَ وَالمُبتَدِعين".[37]

وَفي الجيلِ اللاّحِقِ مِنَ الأَشاعِرَة، نَرى أَبا المُظَفَّرِ الإسفَرايينِيَّ المَذكور، (471هـ\1078م) يُقَسِّمُ المُسلِمينَ إلى قِسمَينِ هُما: "أَهلُ التَّحقيق"،[38] وَهُم "أَهلُ الحَقِّ وَمِلَّةِ الإسلام"[39]، وَهُم أَهلُ السُّنَّة، وَغَيرُ أَهلِ التَّحقيق، وَهُم "أَهلُ الأَهواء"[40]، وَ"أَهلُ البِدَع"[41]، وَ"أَهلُ الأهواءِ وَالبِدَع"[42]، وَقَد يُسَمّيهِم "أَهلَ الزَّيغِ وَالمُبتَدِعين".[43] وَ"أَهلُ البِدَعِ" عِندَهُ يَفتَرِقونَ إلى فِرَق، بَيدَ أَنَّ عَدَمَ سُنِّيَّتِهِم تَقتَضي عَدَمَ إدراجِهِم ضِمنَ مَفهومِ الإسلامِ وَالمُسلِمين، فَقال: "في بَيانِ فِرَقِ أَهلِ البِدَعِ الَّذينَ يَنتَسِبونَ إلى دينِ الإسلام، وَلا يُعَدّونَ في زُمرَةِ المُسلِمين".[44]

---

33 اَلتَّنبيه: 14.

34 اَلتَّنبيه: 19.

35 كِتابُ الحُدود: 112.

36 سِيَرُ أَعلامِ النُّبَلاء: 353/17. ع. 220.

37 اَلتَّبصير: 427.

38 اَلتَّبصير: 169.

39 اَلتَّبصير: 373.

40 اَلتَّبصير: 168.

41 اَلتَّبصير: 170.

42 اَلتَّبصير: 418.

43 اَلتَّبصير: 170.

44 اَلتَّبصير: 326.

وَتَقتَرِنُ "البِدعَةُ" عِندَهُ بِالإلحاد، بَل بِالكُفرِ الصُّراحِ أَيضا، كَما اقتَرَنَت بالأَهواءِ وَالزَّيغ، فَتَراهُ يُشيرُ إلى "أَهلِ الإلحادِ وَالبِدعَة"[45]، وَ"شَوائبِ البِدَعِ وَالإلحادِ وَأَنواعِ الكُفر".[46]

وَهاذا الإلحادُ لَيسَ طارِئًا في مِلَّةِ الإسلام، ناتِجًا عَنِ افتِراقِ فِرَقِ المُسلِمينَ بِالضَّرورة، بَل هُوَ أَقدَمُ حَتّى مِن دينِ الإسلامِ نَفسِه، فَقَد كَتَبَ فَصلاً "في بَيانِ مقالاتِ أَقوامٍ مِنَ المُلحِدين، كانوا قَبلَ ظُهورِ دَولَةِ الإسلام..."[47] وَ"في بَيانِ مقالاتِ قَومٍ كانوا قَبلَ ظُهورِ دَولَةِ الإسلام"[48]، هُوَ في الواقِعِ عَرضٌ لِمَقالاتِ الفَلاسِفَةِ اليونانِيّين، مِمَّن لَم يَعرِفوا الأَديانَ السَّماوِيَّةَ أَصلا.[49]

وَقَدِ افتَخَرَ أَبو المُظَفَّرِ بِاعتِناءِ أَهلِ السُّنَّةِ بِـ"الرَّدِّ عَلى المُلحِدين"، فَقال: "لِأَهلِ السُّنَّةِ وَالجَماعَةِ التَّفَرُّدُ بِأَكثَرَ مِن أَلفِ تَصنيفٍ في أُصولِ الدّين...في نُصرَةِ الدّين، وَالرَّدِّ عَلى المُلحِدين، وَالكَشفِ عَن أَسرارِ بِدَعِ المُبتَدِعين. وَلَم يَكُن لِواحِدٍ مِن مُتَقَدِّمي القَدَرِيَّةِ وَالرَّوافِضِ وَالخَوارِجِ تَصنيفٌ في هذا النَّوعِ يَظهَرُ وَيُتَداوَل."[50] ذلك أَنَّ مِن تَوَجُّهاتِ عِلمِ الكَلامِ السُّنِّيّ، وَأَصنافِ التَّآليفِ فيه، صِنفًا يَعمَلُ عَلى "الرَّدِّ" عَلى المُخالِفين، "يَتَعَرَّضُ فيهِ لِمَذاهِبِ الضّالّين، وَتَقريرِ شُبَهِهِم وَتَشكيكاتِهِم، وَرَدِّها وَحَلِّها، وَمُناظَرَتِهِم وَإِبطالِ دَعاويهِم"، كَما ذَكَرَ الشَّيخُ القاضي، أَبو عَبدِ الله، مُحَمَّدٌ الطّالِبُ ابنُ الحاجّ. (1273هـ\1856م).[51]

لاكِنَّ المَفاهيمَ الَّتي وَظَّفَها أَبو المُظَفَّرِ دونَ أَن يُحَدِّدَها لَيسَت بِالضَّرورَةِ تَلزَمُ كُلَّ الأَشاعِرَة. فَهاذا مُعاصِرُهُ عَبدِ القاهِرِ البَغدادِيّ، (429هـ\1037م) صاحِبُ "المِلَلِ وَالنِّحَل"، لا يُوَظِّفُ هذا المَفهوم. وَإِنَّما كَلامُهُ عَنِ "الضُّلاّل"[52]، وَ"البِدعَة".[53]

وَكَذلك لا يُعَرِّجُ عَلَيها أَبو الفَتح، مُحَمَّدُ بنُ عَبدِ الكَريمِ الشَّهرِستانِيّ، (548هـ\1153م) صاحِبُ "المِلَلِ وَالنِّحَل"، وَمُعاصِرُ أَبي بَكرِ ابنِ العَرَبِيّ. وَإِنَّما يَكتَفي بِأَن يَقولَ عَنِ الإسماعيلِيَّة: "وَأَشهَرُ أَلقابِهِم الباطِنِيَّة...وَبِخُراسانَ التَّعليمِيَّةَ وَالمُلحِدَة"[54]، دونَ أَن يَكونَ هذا المَفهومُ مَفهومًا رَءيسًا عِندَه.

---

45 اَلتَّبصير: 428.

46 اَلتَّبصير: 169.

47 اَلتَّبصير: 172.

48 اَلتَّبصير: 366.

49 اَلتَّبصير: 366.

50 اَلتَّبصير: 426.

51 اَلأَزهارُ الطَّيِّبَةُ النَّشر: 2/219–220.

52 اَلمِلَلُ وَالنِّحَل: 149.

53 اَلمِلَلُ وَالنِّحَل: 141.

54 اَلمِلَلُ وَالنِّحَل: 192.

3 مَفهومُ الإلحادِ في الأَندَلُسِ بَينَ المالِكِيَّةِ وَابنِ حَزم

لا يَظهَرُ مِن قَرائنِ الحالِ أَنَّ مَفهومَ الإلحادِ قَديمٌ في الأَندَلُس. فَهاذا مُحَمَّدِ بنِ وَضّاحٍ القُرطُبِيّ، (287هـ\900م) يَكتُبُ كِتابَهُ في "اَلبِدَعِ وَالنَّهيِ عَنها". أَمّا الفَقيهُ الكَبير، أَبو عَبدِ اللّٰهِ ابنُ زَربٍ القُرطُبِيّ، (381هـ\991م) فَقَد رَأى أَن "يُستَتابَ الخَوارِجُ الإباضِيَّةُ وَالحَرورِيَّة، وَأَهلُ الأَهواءِ كُلُّهُم"[55]، دونَ يُجرِيَ ذِكرَ الإلحادِ وَالمُلحِدين.

وَقَد حاوَلَ مُعاصِرُه، ابنُ أَبي زَمَنين، (399هـ\1008م) في كِتابِهِ في "أُصولِ السُّنَّة"، أَن يُشيرَ إلى بَعضِ خَصائصِ السُّنِّيَّةِ أَوَّلا، دونَ أَن يُحَدِّدَ مَفهومَها بِدِقَّة، فَقال: "اِعلَم، رَحِمَكَ اللّٰه، أَنَّ السُّنَّةَ دَليلُ القُرآن، وَأَنَّها لا تُدرَكُ بِالقِياس، وَلا تُؤخَذُ بِالعُقول. وَإِنَّما هِيَ في الاِتِّباعِ لِلأَئمَّة، وَلِما مَشى عَلَيهِ جُمهورُ هاذِهِ الأُمَّة".[56]، "وَهُوَ الَّذي عَلَيهِ أَهلُ السُّنَّة، وَجُمهورُ هاذِهِ الأُمَّة".[57] وَقَد سَمّى أَهلَ السُّنَّةِ بِـ"أَهلِ التَّوحيد"[58]، وَ"أَهلِ القِبلَة"[59]، وَ"أَهلَ العِلمِ".[60]

مُقابِلَ السُّنِّيَّة، أَشارَ إلى "أَهلِ الزَّيغِ وَالأَهواءِ المُضِلَّة"[61]، وَ"أَهلِ البِدَع"، وَأَدرَجَ ضِمنَهُم الإباضِيَّةَ وَالمُعتَزِلَة، قائلا: "...سُئِلَ سُحنونُ عَن قَولِ مالِكٍ في أَهلِ البِدَع: اَلإباضِيَّةِ وَالقَدَرِيَّةِ وَجَميعِ أَهلِ الأَهواء..."[62]

وَقَرَّرَ مَذهَبَ العُلَماءِ المالِكِيّينَ بِالأَندَلُس، اَلَّذي لا يُكَفِّرُهُم قائلا: "اِختَلَفَ أَهلُ العِلمِ في تَكفيرِ أَهلِ الأَهواء. فَمِنهُم مَن قال: إنَّهُم كُفّارٌ مُخَلَّدونَ في النّار، وَمِنهُم مَن لا يَبلُغُ بِهِمُ الكُفر، وَلا يُخرِجُهُم عَنِ الإسلام...وَهاذا مَذهَبُ مَشايِخِنا بِالأَندَلُس، وَالَّذي يَعتَقِدونَهُ فيهِم."[63]

55 كِتابُ الخِصال: 296.
56 أُصولُ السُّنَّة: 8.
57 أُصولُ السُّنَّة: 59.
58 أُصولُ السُّنَّة: 43.
59 أُصولُ السُّنَّة: 58.
60 أُصولُ السُّنَّة: 90.
61 أُصولُ السُّنَّة: 70.
62 أُصولُ السُّنَّة: 59.
63 أُصولُ السُّنَّة: 90.

وَهُوَ في كُلِّ هذا لا يَستَعمِلُ مَفهومَ الإلحاد. وَإنَّما استَعمَلَ مَفهومَ "الزَّندَقَةِ" لِلدَّلالَةِ عَلى مُنكِري البَعث، فَقالَ ناقِلاً عَن عَبدِ المَلِكِ بنِ حَبيب، (238هـ\852م) أَنَّهُ قال: "وَفِتنَةُ القَبرِ وَعَذابُه، عِندَ أَهلِ السُّنَّة. وَإنَّما يُكَذِّبُ بِهِ الزَّنادِقَةُ الَّذينَ لا يُؤمِنونَ بِالبَعث."[64]

وَقَد فَسَّرَ ابنُ الحاجِّ التُّجيبِيّ، (529هـ\1135م) الزَّندَقَة، لا بِإنكارِ البَعث، وَلاكِن بِالإنكارِ السِّرِّيِّ لِلإسلامِ نَفسِه، مَعَ إظهارِه، فَهُوَ ازدِواجِيَّةٌ دينِيَّة، أَي إسلامٌ وَلاإسلامٌ في الآنِ عَينِه، فَقال: "اَلنِّفاقُ عَلى عَهدِ رَسولِ اللّه، (ص) هُوَ الزَّندَقَةُ بَعدَه. ذَكَرَهُ ابنُ قُتَيبَة".[65]

وَنُلحِقُ بِمالِكِيَّةِ الأَندَلُس، اَلَّذينَ لَم يَجعَلوا مَفهومَ الإلحادِ مَفهومًا أَساسِيًّا عِندَه، اَلقاضِيَ عِياضًا اليَحصُبِيَّ السَّبتِيّ، (544هـ\1149م) إذ قالَ عَن أَئِمَّةِ المُسلِمين: "وَلا حَكَّموا الآراءَ المُضِلَّةَ في الدّين، وَلا انهَمَلوا انهِمالَ المُلحِدين، وَلا تَنَطَّعوا تَنَطُّعَ المُعتَدين".[66]

وَعَلى دَربِ مالِكِيَّةِ الأَندَلُسِ سارَ ابنُ حَزم، (458هـ\1066م) فَقَرَّرَ أَنَّ "أَهلَ السُّنَّة...أَهلُ الحَقّ، وَمَن عَداهُم فَأَهلُ البِدعَة".[67] بَيدَ أَنَّهُ تَوَسَّعَ في استِعمالِ المَفهوم. فَأَشارَ إلى "المُلحِدين"[68]، وَ"بَعضِ أَهلِ الإلحاد"[69]، وَ"مُلحِدي أَهلِ زَمانِنا"[70]، وَ"سائر المُلحِدين"[71]، دونَ أَن يُحَدِّدَ مَفهومًا مَضبوطًا لِلإلحاد. وَقَد سَمّى بَعضَ مَن لَقِيَ بِالأَندَلُسِ مُلحِدا، فَقال: "وَقَد أَلزَمتُ بَعضَ المُلحِدين، وَهُوَ ثابِتُ بنُ مُحَمَّدٍ الجُرجانِيّ، في هذا البُرهان"[72]، وَ"بَعضَ المُلحِدين، وَهُوَ عَبدُ اللّهِ بنُ عَبدِ اللّهِ بنِ شَنيف".[73]

وَلَمّا كانَ المَفهومُ غَيرَ مُحَدَّدٍ عِندَه، كانَ لِزامًا عَلَينا البَحثُ عَمّا يُمَثِّلُهُ مِنَ الفِرَق، فَوَجَدنا عِندَهُ أَنَّ فِرَقَ التَّشَيُّعِ الباطِنِيّ، اَلَّتي دَعَت إلى القَولِ بِالإمامَة، مَرفوقَةً بِالدَّعوَةِ إلى تَعطيلِ الشَّريعَةِ بِوَجهٍ مِنَ الوُجوه، هِيَ مِن "أَهلِ الإلحاد". قال: "اَلإسماعيلِيَّةَ وَالقَرامِطَة، وَهُما طائِفَتانِ مُجاهِرَتانِ بِتَركِ الإسلامِ جُملَة...فَاللّهَ اللّهَ، عِبادَ اللّه، في أَنفُسِكُم، وَلا يَغُرَّنَّكُم أَهلُ الإلحاد"[74]، فَالإسلامُ هُوَ مَذهَبُ أَهلِ السُّنَّةِ عِندَه.

64 أُصولُ السُّنَّة: 37.
65 نَوازِلُ ابنِ الحاجّ: 579/3.
66 تَرتيبُ المَدارِك: 5/1.
67 اَلفَصل: 271/2.
68 اَلفَصل: 167/1.
69 اَلفَصل: 64/1.
70 اَلفَصل: 73/1.
71 اَلفَصل: 122/1.
72 اَلفَصل: 61/1.
73 اَلفَصل: 63/1.
74 اَلفَصل: 274/2.

وَقرَنَ المُعتَزِلَةَ بِالإلحاد، فَقال: "فَاعتَقَدَ أَكثَرُهُمُ الإلحادَ وَالتَّعطيل"[75]، كَما قَرَنَ الفَلاسِفَةَ بِه، فَقال: "ذِكرُ فُصولٍ يَعتَرِضُ بِها جُهّالُ المُلحِدين، عَلى ضَعَفَةِ المُسلِمين"[76]، أَي أَهلِ الحَديث.

وَلَمّا كانَ ابنُ حَزمٍ مُنحَرِفًا عَنِ الأَشاعِرَة، مُتَعَصِّبًا عَلَيهِم، أَدرَجَ اَلباقِلاّنِيَّ (403هـ\1012م) ضِمنَ المُلحِدين، فَقال: "فَلاحَ كُفرُ هذا المُلحِد، وَبانَ كَيدُهُ لِلإسلام".[77] فَاجتَمَعَ مِن كُلِّ هذا أَنَّ مَفهومَ الإلحادِ عِندَهُ يَشمَلُ طَوائفَ مُتَبايِنَةً مِنَ المُسلِمين، مِنهُمُ اَلإسماعيلِيَّةُ وَالقَرامِطَة، وَالفَلاسِفَة، وَالمُعتَزِلَة، وَحَتّى الأَشاعِرَة.

4 الإلحادُ مَفهومًا مَركَزِيًا في فِكرِ ابنِ العَرَبِيّ

وَظَّفَ ابنُ العَرَبِيِّ مَفهومَي البِدعَةِ وَالزَّندَقَة، كَما استَعمَلَهُما المالِكِيَّةُ وَابنُ حَزمٍ قَبلَه، وسارَعَ إلى تَكفيرِ المُبتَدِعَة، وَلَم يَتَوَقَّف فيهِم، كَما صَنَعَ المالِكِيَّة، وَلا سِيَما المُعتَزِلَة، فَقال: "إِنَّ المُبتَدِعَةَ وَأَهلَ الأَهواءِ كُفّار...أَمّا القَدَرِيَّة، فَلا شَكَّ في كُفرِهِم. وَأَمّا مَن عَداهُم، فَنَستَقرِئُ فيهِمُ الأَدِلَّة."[78] وَأَدانَ مِنهُم القاضِيَ الكَبير، مُنذِرَ بنَ سَعيدٍ البَلّوطِيَّ اَلمُعتَزِلِيَّ الظّاهِرِيّ، فَقالَ عَنه: "عَجَبًا لِهاذا القائلِ البائس، شَدَّ الرِّحال، وَلَقِيَ الرِّجال، وَكُلُّ ذلك عَلَيهِ وَبال، لَم يَلقَن عَنهُم تَحقيقَ الأَدِلَّة، وَلا تَعَلَّمَ مِنهُم أُصولَ السُّنَّة."[79] هاذا وَالاِبتِداعُ عِندَهُ دَرَجات، فَمُنذِرُ بنُ سَعيدٍ البَلّوطِيّ: "رَضِيَ مِن قَولِ الأُمَّة، بِقَولِ المُبتَدِعَة، وَمِن قَولِ المُبتَدِعَة، بِقَولِ رُذالَتِهِم."[80]

وَعَلى الرَّغمِ مِنَ التَّبايُنِ الشَّديدِ بَينَ المُعتَزِلَةِ وَبَينَ الظّاهِرِيَّةِ في الأُصول، فَقَد رَأى أَنَّ داوودَ الظّاهِرِيَّ الفَقيه، (-270هـ) كانَ "مِنَ المُبتَدِعَة".[81] وَقَد يُسَمّي الخَوارِجَ وَالمُعتَزِلَةَ هاؤُلاءِ "أَهلَ البَغي".[82]

---

75 اَلفَصل: 235/2.

76 اَلفَصل: 233/2.

77 النَّصائحُ المُنجِيَة: 235.

78 أَحكامُ القُرآن: 294/1.

79 اَلكِتابُ المُتَوَسِّط: 400.

80 اَلكِتابُ المُتَوَسِّط: 400.

81 اَلمَسالِك: 538/5.

82 تَهذيبُ المَسالِك: 589.

أَمّا مَفهومُ الزَّندَقَة، فَهُوَ أَحسَنُ حالا، فَـ"اَلزِّنديق: وَهُوَ الَّذي يُسِرُّ الكُفرَ وَيُظهِرُ الإيمان"[83]، وَانتَقَدَ الشّافِعِيَّةَ في هذا الأَمر.[84] وَقَسَّمَ الزَّنادِقَةَ إلى أَقسامٍ خَمسَة، فَقال: "وَهُم خَمسُ فِرَق...وَافتَرَقَت مِنها فِرقَةٌ عَلى سِتِّ فِرَق. فَمِنهُمُ المُعَطِّلَة، اَلّذينَ يَزعُمونَ أَنَّ الأَشياءَ كائِنَةٌ مِن غَيرِ تَكوين، وَأَنَّهُ لَيسَ لَها مُكَوِّنٌ وَلا مُدَبِّر"[85]، وَهُمُ اَلدَّهرِيَّة.[86] فَهَل كانَت هاذِهِ الفِرَقُ جَميعًا مِمَّن يُسِرُّ الكُفرَ وَيُظهِرُ الإيمان؟

بَيدَ أَنَّهُ تَوَسَّعَ في استِعمالِ مَفهومِ الإلحاد، كَما تَوَسَّعَ قَبلَهُ ابنُ حَزم، دونَ أَن يُحَدِّدَه، حَتّى وَهُوَ يَستَعمِلُهُ في تَسمِيَةِ بَعضِ كُتُبِه، وَإن كانَ قَدِ انتَبَهَ إلى تَحديدِ بَعضِ المُصطَلَحاتِ تَحديدًا لُغَوِيّا.[87]

يَقرِنُ ابنُ العَرَبِيِّ مَفهومَ الإلحادِ بِمَفهومِ الكُفر، فَيَقول: "اَلكُفرُ وَالإلحاد".[88] وَقَد يَقرِنُ طَوائفَ مُفتَرِقَةً شَديدَةَ التَّبايُنِ الفِكرِيِّ في قَرَنٍ واحِد، هُمُ "المُلحِدُ وَصاحِبُ الطَّبيعَة، وَاليَهودِيُّ وَالنَّصرانِيّ"[89]، إيحاءً بِأَنَّ قاسِمًا مُشتَرَكًا يَجمَعُ بَينَها، وَيَرى أَنَّ "اَلكُفرَ لا يَكونُ إلاّ بِمُخالَفَةِ النَّصِّ وَالإجماعِ في أَبوابِ العَقائد"[90]، وَهُوَ ما يَعني أَنَّ الإلحادَ كَذلك، إن كانا مُتَرادِفَينِ عِندَه.

لَقَد رَفَضَت سُنِّيَّةُ القاضي أَبي بَكرٍ كُلَّ تَفكيرٍ كَلامِيٍّ مُخالِف، وَضَمَّنَت ذلك مَفهومَ الإلحاد، كَما قَد رَأَينا، فَتَعَدَّدَت مَناحي الإلحادِ عِندَهُ وَتَوَجُّهاتُه، حَتّى تَضَمَّنَ كَثيرًا مِنَ الفِرَقِ الإسلامِيَّةِ وَالفَلسَفِيَّةِ وَغَيرِها. وَعَلى الرَّغمِ مِن أَنَّ القاضِيَ أَبا بَكرٍ لَم يَكتُب كِتابًا مَخصوصًا لِلمِلَلِ وَالنِّحَل، أَو لِلفِرَقِ الإسلامِيَّة، فَإنَّ تُراثَهُ مَعَ ذلِكَ كَفيلٌ بِتَشكيلِ تَصَوُّرِهِ لَها، وَما كانَ يَراهُ فيها. وَإضافَةً إلى ما سَبَق، لَم يَنسَ القاضي أَن يَتَّهِمَ المُخالِفينَ بِوَضعِ الأَحاديثِ وَالأَخبارِ لِتَأييدِ مَذهَبِهِم أَو مَذاهِبِهِم، دونَ أَن يُحَدِّدَ فاعِلاً بِعَينِه، أَو طائفَةً بِعَينِها. وَلَعَلَّهُ يَقصِدُ في هاذِهِ الحالَةَ بَعضَ المُعتَزِلَةِ وَالشّيعَة. لاكِنَّ هذا الوَضعَ لِلأَخبارِ لا يُؤَثِّرُ كَثيرًا في نَظَرِهِ عَلى العالِمِ السُّنِّيِّ وَأَهلِ السُّنَّة، فَاستِعمالُ التَّأويلِ يُفرِغُ الأَخبارَ المَوضوعَةَ مِن مَضمونِها وَغايَتِها مَعَ ذلِك. يَقول: "وَذلِكَ أَنَّ كُلَّ خَبَرٍ تَضَعُهُ المُلحِدَةُ في هاذِهِ الأَبواب، لا يَخرُجُ عَن طَريقِ التَّأويل".[91] فَلنَتَتَبَّع هاذِهِ الفِرَقَ عِندَ القاضي أَبي بَكرٍ ابنِ العَرَبِيِّ عَلى قَدرِ الإمكان.

---

83 تَهذيبُ المَسالِك: 551.
84 أَحكامُ القُرآن: 12/1.
85 اَلتَّنبيه: 72.
86 اَلكِتابُ المُتَوَسِّط: 122.
87 اَلكِتابُ المُتَوَسِّط: 446.
88 اَلكِتابُ المُتَوَسِّط: 443.
89 اَلعَواصِم: 113.
90 أَحكامُ القُرآن: 3/1.
91 اَلكِتابُ المُتَوَسِّط: 161.

5 اَلإلحادُ وَالفَلسَفَةُ المادِّيَّة

لَم تَكُنِ المَقالاتُ الفَلسَفِيَّةُ شائعَةً في الأَندَلُس، بَعدَ تَحريقِ كُتُبِها في القَرنِ الرّابِعِ عَلى يَدِ المَنصورِ ابنِ أَبي عامِر.[92] وَيَظهَرُ أَنَّ ابنَ العَرَبِيِّ لَم يَصطَدِم بِها اصطِدامًا قَوِيًّا إلاّ في المَشرِق، وَلا سِيَما أَنَّهُ قَد وَجَدَ الفَلسَفَةَ قَد داخَلَت عِلمَ الكَلامِ المُعتَزِلِيّ، كَما داخَلَتِ الإسماعيلِيَّة، مُمَثَّلَةً في إخوانِ الصَّفا. كَما يَظهَرُ أَنَّهُ لَم يَقرَإِ الفَلسَفَةَ اليونانِيَّةَ في كُتُبِها مُباشَرَة، فَخَلَطَ بَينَ الأَطِبّاءِ وَبَينَ الفَلاسِفَةِ المادِّيّين. يَقول: "يا لَلّهِ وَلَذَهابِ العُقول، إلى ذَهابِ الأَديان! يُترجِمُ اليَهودِيُّ وَالنَّصرانِيُّ وَالمُلحِدُ عَن رَجُلٍ يُسَمّى جالينوس، لا نَدري مَن هو، وَلا عَلى أَيِّ مِلَّةٍ كان، إلاّ ما حَكَوْا عَنهُ مِن أَنفُسِهِم، أَو تَرجَموهُ بِاختِيارِهِم، فَيُجعَلُ أَصلاً ما تَرجَموهُ في الاِعتِقادِ وَالعَمَل."[93] وَقَد جَعَلَهُ هذا التَّوَجُّهُ يَبتَهِجُ بِرَدِّ شَيخِهِ الإمامِ الغَزالِيِّ عَلى الفَلاسِفَةِ في كِتابِه، "تَهافُتِ الفَلاسِفَة"، وَيُشيدُ إشادَةً بِه.[94]

وَقَد حَمَلَهُ العَداءُ لِلفَلسَفَة، وَهِيَ الغَريبَةُ القادِمَةُ مِنَ اليونان، عَلى تَتَبُّعِ أَصلِها في تاريخِ دارِ الإسلام، فَاكتَشَفَ أَنَّ المَسؤولِيَّةَ راجِعَةٌ إلى البَرامِكَةِ أَوَّلاً وَأَخيرا، وَإلى دَسِّ التَّراجِمَةِ في كُتُبِ الطِّبِّ ما هُوَ مِنَ الإلحادِ في الدّين. وَقَد بَيَّنَ السَّبَبَ في ذلك، وَهُوَ عِلمِيّ وَتاريخِيّ، فَصَبَّ جامَّ غَضَبِهِ عَلَيهِم، دونَ الخُلَفاءِ العَبّاسِيّينَ أَصحابِ السُّلطَةِ العُليا، فَمَقامُ الخِلافَةِ أَعظَمُ مِن أَن يَتَلَوَّثَ بِهذا الأَمر. يَقول: "ثُمَّ أَنشَأَتِ البَرامِكَةُ طامَّةً عَظيمَة، إذ كَلَّفوا الأَخباثَ أَيضًا تَرجَمَةَ كُتُبِهِم، طِبّاً وَطَبيعَة، بِالعَرَبِيَّة. فَتَوَلّى ذلك يَهودِيٌّ أَو نَصرانِيٌّ أَو مُلحِد، لا رَأسَ مالٍ لَهُ في الإسلام، فَمَزَجَ ما نَقَلَ مِنَ الطِّبِّ بِأَلفاظٍ وَعَقائِدَ تَتَعَلَّقُ بِالإلحاد، وَتُعارِضُ الشَّريعَة، في فُروعِها وَأُصولِها، لِيَتَوَهَّمَ مَن تُرجِمَت لَهُ أَنَّ هاذِهِ الأُمَمَ الفاضِلَة، الَّتي تَوَلَّت هاذِهِ العُلومَ الغَريبَة، كانَت عَلى هاذِهِ النِّحَل...وَلَم يُقَدِّموا قاضِيًا في البِلادِ إلاّ أَن يَكونَ عَلى هاذِهِ العَقيدَة، وَلا أَميرًا وَلا كاتِباً إلاّ وَهُوَ فيها، وَلا يُنظَمُ في سِلكِ الخاصَّةِ إلاّ مَن كانَ قائماً بِها...وَعَمَّ الباطِل، وَظَهَرَتِ الزَّندَقَة، وَثارَتِ البِدَع...وَكانَتِ المِلَّةُ عَلى الذَّهاب."[95]

وَرَأى أَنَّ البَرامِكَةَ أَعَدّوا الرِّجالَ المُفَكِّرينَ لِأَجلِ نَشرِ الفَلسَفَةِ المادِّيَّة، حَتّى مَن كانَ قَد ماتَ قَبلَ أَن يَظهَرَ شَأنُهُم في الدَّولَةِ العَبّاسِيَّة، كَابنِ المُقَفَّع، (142هـ\759م) فَقَد "رَبَّت في حِجرِها طَوائفَ كَابنِ المُقَفَّع، وَابنِ الرّاوَندِيّ، وَالجاحِظِ المُعتَزِلِيّ...قَدِ استَسَنّوا في البَشَر، أَنَّهُ لا مَدرَكَ إلاّ العُقول، وَأَنَّها

92 أُنظُر ما كَتَبناهُ في هذا الأَمر، في مُقَدِّمَةِ تَحقيقِنا لِكِتاب: اَلجَوهَرَةِ الفَريدَة: 58–59.

93 اَلعَواصِم: 75.

94 اَلعَواصِم: 78.

95 اَلعَواصِم: 70.

تُغني عَنِ الرُّسُل..."[96] فالمُهِمُّ عِندَه، أَنَّ كُلَّ عَقلانيَّة، مُؤمِنةٍ أَو غَيرِ مُؤمِنَة، فَلسَفيَّةٍ أَو كَلامِيَّة، هِيَ مِمّا اقتَرَفَتهُ يَدا البَرامِكَة.

وَكانَت نَتيجَةُ مَشروعِهِمُ الكَبيرِ سَلبيَّةً جِدّاً في نَظَرِ ابنِ العَرَبيّ، فَقَد "مَلَأوا الأَرضَ مِنَ الباطِل، وَاستَخلَفوا شَياطينَ الإنسانِ عَلى إِضلالِ الخَلق، مِن فَيلَسوفٍ وَأَديب، حَتّى لَم يَبقَ بَيتٌ إلاّ وَفيهِ مِن كُتُبِ الأَباطيل، ما بَينَ ناظِرٍ فيها، حَتّى يَعلَمَ المُرادَ مِنها، أَو مُختارٍ ما يَصلُحُ مِنها، وَتارِكٍ ما سِواه، أَو رادٍّ عَلَيه، لِئَلاّ يَعتَلِقَ بِمُوَحِّدٍ فَيَختَلَّ عَقدُه، أَو يَتزَلزَل...لِيَتَحَقَّقَ الوَعدُ الصّادِقُ في فَسادِ الزّمان، وَذَهابِ الأَديان."[97] فَكانَ عَلى ابنِ العَرَبيِّ أَن يَملأَ الأَرضَ حَقّا، بَعدَما مَلَأَها البَرامِكَةُ باطِلا، فَسَعى إِلى التَّحذيرِ مِنَ الفَلسَفَةِ وَأَهلِها وَحُلَفائها مِنَ المُعتَزِلَة، وَجَعَلَ البَديلَ لَها هُوَ الكِتابَ وَالسُّنَّة، حَتّى لَو كانَت هاذِهِ الفَلسَفَةُ كِتابًا أَدَبيًّا مِنَ الخُرافاتِ الرَّمزيَّة، بَعيدًا كُلَّ البُعدِ عَنِ الحَديثِ في اللّهِ وَأَصلِ الوُجودِ وَالنُّبُوّات، كَكِتابِ "كَليلَةَ وَدِمنَة". يَقول: "وَإِذا أَرَدتُم أَن تَكونوا حُكَماء، فَعَلَيكُم بِكِتابِ اللّهِ وَسُنَّةِ رَسولِه، فَهِيَ الغايَة. وَإِيّاكُم وَهاذِهِ الكُتُبَ الَّتي تُنسَبُ إِلى الحِكمَة، فَإِنَّها مَجموعَةٌ مِن أَقوالِ الفَلاسِفَةِ وَالمُعَطِّلَة، كَكِتابِ "كَليلَةَ وَدِمنَة"، اَلَّذي تَرجَمَهُ المُلحِدون، وَالكُتُبِ الَّتي جَمَعَتها المُلحِدَة، كَالجاحِظِ وَغَيرِه، لِيُشغَلَ بِها الخَلقُ عَن كَلامِ اللّه، وَكَلامِ نَبيِّه، وَدُسَّ فيها مِن كَلامِ اللّهِ تَعالى وَكَلامِ نَبيِّهِ ما لَم يَصِحّ..."[98]

لا يُسَمّي ابنُ العَرَبيِّ الفَلسَفَةَ المادِّيَّةَ أَوِ "الدَّهرِيَّة" بِاسمِها في العادَة. لاكِنِ استِقراءَ إِشاراتِهِ يَدُلُّ عَلى أَنَّ خِطابَهُ يُهاجِمُ هؤلاء الفَلاسِفَةَ الدَّهرِيّينَ مُباشَرَة، وَيُسَمّيهِم بِأَهلِ "الطَّبيعَة"، دونَ أَن يُسَمِّيَهُم بِأَسمائهِم. قَسَّمَ ابنُ العَرَبيِّ الفَلاسِفَةَ إِلى أَقسام[99]، وَكَفَّرَ الفَلاسِفَةَ المادِّيّين، اَلقائلينَ بِالطَّبيعَةِ دونَ سِواها، وَرَأى أَنَّ اللّهَ "رَدَّ عَلى الكُفّار...مِن مُلحِدَةٍ وَعَبَدَةِ أَوثان، وَأَهلِ كِتاب، وَطَبيعَة، وَصابئَة، وَشِركَة، وَيَهودِيَّة، بِكَلامِه."[100] وَلَم يَنسَ أَن يَدعُوَ عَلَيهِم كَذلك.[101] ذلك أَنَّ الفَلاسِفَة، "أَصحابَ الطَّبيعَة"، يَتَعَمَّدونَ تَضليلَ خَلقِ اللّه، وَصَرفَهُم عَن طَريقِه. وَلاكِن، هَيهاتَ هَيهات: "فَإِنَّما دارَت هاذِهِ الطّائفَةُ عَلى تَضليلِ الخَلقِ عَنِ الطَّريقِ إِلى الباري تَعالى، وَيَأبى اللّهُ إلاّ أَن يُتِمَّ نورَه."[102]، وَذلك بِالتَّعاوُنِ مَعَ القَدَرِيَّةِ

96 اَلعَواصِم: 73.

97 اَلعَواصِم: 71.

98 سِراجُ المُريدين: 2/340–341.

99 قانونُ التَّأويل: 105.

100 اَلعَواصِم: 80.

101 قانونُ التَّأويل: 109.

102 اَلكِتابُ المُتَوَسِّط: 433.

اَلمُتَستِّرَةِ بِالدّين، "لِتَغتالَ الدّين".[103] وَرأى أَنَّ الفَلسَفَةَ المادِّيَّة مُغرِقَةٌ في الإلحاد، فَ"أَصلُ هذا المَذهَب، وَهُوَ القَولُ بِالطَّبائع، مَردودٌ بِما قَدَّمنا مِنَ الأَدِلَّة. وَفيهِ أَربَعَةُ أُصولٍ مِنَ الإلحاد."[104]

إنَّ مِن أَوَّلِ ما يَرفُضُهُ ابنُ العَرَبِيِّ في الفَلسَفَةِ المادِّيَّة، هُوَ القَولُ بِالسَّبَبِيَّةِ في الفيزِياء، وَهُوَ قَولٌ تَلَقَّفَهُ المُعتَزِلَةُ في نَظَرِهِ مِنَ الفَلاسِفَة، فَقَد "نَبَغَت في الإسلامِ عِندَ حُدوثِ الأَهواءِ فِرقَةٌ زَعَمَت أَنَّ الإحراقَ فِعلُ النّار، وَالإشباعَ فِعلُ الخُبز...وَمَوضِعُ الرَّدِّ عَلى هؤلاء، اَلقَولُ عَلى مَذاهِبِ الطَّبائعِيّينَ مِنَ المُلحِدين، فَإنَّهُ شُعبَةٌ مِن شُعَبِهِم...وَأَوَّلُ مَن يُؤثَرُ عَنهُ هذا المَذهَب، مُعَمَّرٌ القَدَرِيّ، وَالجاحِظُ المُفتَري".[105] وَهُوَ كَذلك الاِحتِجاجُ الدّائمُ بِالحِسِّ وَالمُعايَنَة، طَرِيقًا لِإثباتِ العَقائدِ أَو نَفيِها، فَعَذابُ القَبر: "إن قيل: هُوَ خِلافُ المُشاهَدَة...قُلنا: تَاللّٰهِ لَقَد ظَهَرَت مَخايِلُ الإلحاد."[106]

وَمِثلَما رَفَضَ ابنُ العَرَبِيِّ نَظَرِيَّةَ السَّبَبِيَّةِ الفيزِيائِيَّةِ أَساسًا لِلتَّفكيرِ الفَلسَفِيّ، رَفَضَ كَذلك نَظَرِيَّةَ "مَراتِبِ الوُجود"، وَما فيها مِنَ القَولِ بِالبَسيطِ وَالمُرَكَّب، دونَ أَن يَتَكَلَّفَ عَناءَ فَحصِ هاذِهِ الأُطروحاتِ قَبلَ رَفضِها. يَقولُ مُناقِشًا لِأُطروحاتِ الفَلاسِفَة: "فَإن قيل: جَهِلتُم قَولَ الفَلاسِفَة...وَإنَّما يَعنونَ بِهِ أَنَّ المَوجوداتِ عِندَهُم عَلى قِسمَين: بَسيطٍ وَمُرَكَّب...فَالجَوابُ عَن ذلك أَن نَقول: حَلَّقتُم عَلى الإلحاد، وَأَشفَفتُم عَلى القَولِ بِالطَّبائع، وَأَنَّ البَسيطَ غيرَ المُرَكَّبِ لا يَفنى..."[107] فَمِن شَأنِ الفَلاسِفَة، عِندَه، أَو بَعضِهِم، إنكارُ الإلاهِيَّة، فَلِمَ تَكَلُّفُ كُلِّ هذا العَناءِ في الإدحاض؟ يَقولُ ذاكِرًا مَذهَبَهُم وَشُبهَتَهُم: "زَلَّتِ المُلحِدَةُ حَتّى قالوا: لَيسَ الباري بِمَوجود، وَلا عالِم، لِأَنَّ في ذلك تَشبيهًا لَهُ بِخَلقِه."[108]

وَبَعدَ إنكارِ الإلاهِيَّة، يَأتي إنكارُ النُّبُوَّة، لاكِن، مَعَ تَعليلِ هذا الإنكار. يَقول: "وَزَعَمَ أَهلُ الإلحادِ أَنَّ إرسالَ الرُّسُلِ غَيرُ جائز...فَمِنهُم مَن قال: ذلك عَبَثٌ لا فائدَةَ فيه، لِأَنَّهُ لا يَخلو مِن أَن يَأتِيَ بِما يُوافِقُ العُقولَ أَو يُخالِفُها. فَإن جاءَ بِما يُوافِقُ العُقول، لَم يُفِد شَيئا، وَإن كانَ جاء بِما يُخالِف، فَذلك مُحال، لِأَنَّ قَلبَ الأَدِلَّةِ وَنَقضَها وَإبطالَ الحَقائقِ مُمتَنِع."[109]

103 اَلعَواصِم: 81.
104 اَلكِتابُ المُتَوَسِّط: 433.
105 اَلكِتابُ المُتَوَسِّط: 261.
106 اَلكِتابُ المُتَوَسِّط: 391.
107 اَلكِتابُ المُتَوَسِّط: 433.
108 اَلكِتابُ المُتَوَسِّط: 134.
109 اَلكِتابُ المُتَوَسِّط: 323.

ثُمَّ يَأتي إنكارُ الغَيبيّاتِ الدّينيَّة، وَأَوَّلُها وُجودُ المَلائكَة، "وَمَن حامَ عَلى ذلك، أَسَفَّ عَلى أَرضِ الإلحاد."[110]، ثُمَّ إنكارُ البَعثِ بَعدَ المَوت: "وَظَنَّهُ بَعضُ المُلحِدَةِ عَدَمًا مَحضا، وَفَناءً كُلّيًّا لا وُجودَ بَعدَه. وَأَشكَلَ أَيضًا عَلى طَوائفَ مِن أَهلِ السُّنَّة.".[111]

وَلَمّا كانَ المَوتُ وَالقَولُ بِالبَعثِ يَدعُوانِ في الإسلامِ إلى الاِعتِقادِ في وُجودِ الآخِرَةِ بَعدَ دارِ الدُّنيا، كانَ هذا الاِعتِقادُ في نَظَرِ ابنِ العَرَبِيِّ عَقَبَةً في سَبيلِ انتِشارِ الأَفكارِ الفَلسَفِيَّةِ الطَّبيعِيَّة، اَلَّتي لا تَرى الإيمانَ إلاّ بِالمَحسوس، فَقَد أَوَّلَتِ الفَلاسِفَةُ الآخِرَةَ تَأويلاً رَمزِيًّا: "وَهُوَ رَأيٌ فَلسَفِيّ، مَبناهُ عَلى أَنَّ الدّارَ الآخِرَةَ لا حَقائقَ لَها، وَإنَّما هِيَ أَمثال.".[112]

وَقَدَ وَجَدَ ابنُ العَرَبِيِّ هذا التَّأويلَ الرَّمزِيَّ مُتَهافِتا: "وَإذا ذَكَرَتِ الفَلاسِفَةُ مَعنى هاذِهِ الأَمثال، فَفيها حُجَّةٌ عَلَيها.".[113] وَقَرَنَ بَينَ الفَلسَفَةِ وَالدّينِ المَسيحِيّ، وَجَعَلَهُما مُشتَرِكَينِ في القَولِ بِالتَّخييلِ وَالتَّمثيل، دونَ فَحصٍ لِدينِ النَّصارى، أَو تَحديدٍ لِلمَذهَبِ النَّصرانِيِّ القائلِ بِه. قال: "وَأَمّا مَنِ اعتَقَدَ في الدّارِ الآخِرَةِ أَنَّها خَيالاتٌ وَتَمثيلات...فَهاذا مَذهَبُ النَّصارى وَالفَلاسِفَة."[114] وَإنكارُ هذا التَّأويلِ قائمٌ عِندَهُ عَلى مَبدَإٍ طَبَّقَهُ عَلى التَّأويلِ الفَلسَفِيِّ وَالتَّأويلِ الباطِنِيِّ لِلقُرآن. ذلك أَنَّ "كَلامَ المَولى لَم يَكُن رُموزًا وَلا كِنايَة".[115]

وَبَعدَ هذا، سَجَّلَ ابنُ العَرَبِيِّ عَلى الفَلاسِفَةِ المادِّيّينَ إنكارَهُم وُجودَ الجِنِّ وَالشَّياطين، لِيَكونَ إنكارُهُم تَمهيدًا لِإنكارِ وُجودِ المَلائكَة: "وَصَرَّحَت طَوائفُ كَثيرَةٌ مِنَ المُبتَدِعَةِ بِإنكارِ الجِنِّ وَالشَّياطين: قَصَدَت بِذلك إلى التَّلبيسِ عَلى العِباد، وَنَصَبَتهُ لِإنكارِ المَلائكَةِ جَهالَةَ إلحاد."[116]، وَجَعْلَهُم إيّاهُم "صِنفًا روحانِيّا. وَهاذِهِ كَلِمَةٌ هائلَة...وَأَكثَرُ تَرَدُّدِها في كُتُبِ الفَلاسِفَة."[117] كَما سَجَّلَ عَلَيهِم إنكارَهُم عَذابَ القَبر، "فَإن قيل: هُوَ خِلافُ المُشاهَدَة...قُلنا: تَاللّٰهِ لَقَد ظَهَرَت مَخايِلُ الإلحاد."[118]

110 اَلكِتابُ المُتَوَسِّط: 363.

111 اَلكِتابُ المُتَوَسِّط: 379.

112 قانونُ التَّأويل: 357.

113 قانونُ التَّأويل: 169.

114 قانونُ التَّأويل: 250–251.

115 اَلعَواصِم: 110.

116 اَلكِتابُ المُتَوَسِّط: 430.

117 اَلكِتابُ المُتَوَسِّط: 431.

118 اَلكِتابُ المُتَوَسِّط: 391.

وَيُلحَقُ بِكُلِّ هذا إنكارُ مَولِدِ آدَمَ وَعيسى مِن غَيرِ أَبَوَين: يَقول: "اَلرَّدُّ عَلى المُلحِدَةِ الَّذينَ يَقولون: لا إنسانَ إلاّ مِن نُطفَة".[119]، وَإنكارُ مَسخِ الإنسانِ حَيَوانًا بِأَمرٍ مِنَ اللّه.[120] وَفي جَميعِ الأَحوال، فَمِن شَأنِ الإلحادِ الفَلسَفِيِّ إنكارُ الحَقيقَةِ الدّينِيَّةِ وَالقَولِ بِوُجودِها. يَقول: "وَإنَّما قَصَدَتِ المُلحِدَةُ بِذِكرِ هذا التَّلاعُبِ بِالعالَم، لِتَستَرسِلَ العامَّة".[121]

وَلَمّا كانَ عِلمُ المَنطِقِ مَدخَلاً إلى الفَلسَفَة، وَتُرجِمَ عِندَما تُرجِمَت، وَ"المَدخَلُ إلى الشَّرِّ شَرّ"، وَكانَ شَيخُهُ الغَزاليُّ مُشتَغِلاً بِهذا العِلم، فَقَد تَوَزَّعَ مَوقِفُ ابنِ العَرَبيِّ مِن هذا العِلمِ اليونانِيِّ بَينَ الإشادَةِ بِمَنطِقِ الغَزاليِّ في "مِعيارِ العِلم"،[122] وَبَينَ التَّحذيرِ مِنَ "الغَرَر" في مَنطِقِه.[123] بَيدَ أَنَّهُ لَم يَنسَ أَن يُدينَ إدانَةً واضِحَةً شَديدَةً مَنطِقَ ابنِ حَزم، وَابنَ حَزمٍ أَيضا.[124]

لَم يَكُن ظُهورُ الفَلسَفَةِ بِمُبادَرَةٍ مِنَ البَرامِكَةِ عِندَهُ سَبَبًا في ظُهورِ فَلاسِفَةٍ مادِّيّينَ فَقَط. بَل كانَ ذلك سَبَبًا فيما هُوَ أَشَدُّ اعتِياصا، وَهُوَ الجَمعُ بَينَ الفَلسَفَةِ وَالدّين، وَالقَولُ بِأَنَّ السَّبيلَينِ يَقودانِ إلى غايَةٍ واحِدَة، فَكانَ هذا مِمّا جَرَّتهُ البرامِكَةُ عَلى الإسلامِ كَذلك، فَقَد "نَبَغَت طائفَةٌ تَستَّرَت بِالإسلام، وَهِيَ تُبطِنُ عَقائدَ الأَوائل"[125]، فَحاوَلَتِ جَمعًا تَلفيقِيًّا بَينَهُما: فَ"أَرادوا أَن يُلَفِّقوا بَينَ مَوارِدِ الشَّرع، وَأَغراضِ الفَلاسِفَة، وَادَّعَوْا أَنَّها مُتلائمَة، وَأَنَّ كَلامَ النَّبِيِّ مَعَ كَلامِ الفَيلَسوف، يَخرُجُ عَن مِشكاةٍ واحِدَة."[126]، وَرَأى ابنُ العَرَبِيِّ في هذا الجَمعِ "قاصِمَة".[127] ذلكَ أَنَّ الإيمانَ وَالقَولَ بِوُجودِ إلاهٍ فَلسَفِيٍّ لا يُؤَثِّرُ في الكَونِ لا يَجتَمِعانِ أَبَدا، بَل هُوَ أَصلٌ مِن أُصولِ الإلحادِ الفَلسَفِيّ. ذلك "أَنَّ الإسلامِيّينَ مِنَ الفَلاسِفَة، قَد حَكَموا عَلى أَفلاطونَ وَأَرسطوطاليسَ بِاستِحالَةِ الإيثار =[التَّأثير]، وَأَنَّ صانِعًا مُؤَثِّرًا لا يُتَصَوَّر. وَهاذا أَحَدُ أُصولِ الإلحادِ الأَربَعَة."[128]

---

119 اَلكِتابُ المُتَوَسِّط: 142.

120 اَلمَسالِك: 523/7.

121 اَلعَواصِم: 12.

122 اَلعَواصِم: 78.

123 اَلعَواصِم: 80.

124 اَلعَواصِم: 78.

125 اَلعَواصِم: 158.

126 قانونُ التَّأويل: 353.

127 قانونُ التَّأويل: 109.

128 اَلعَواصِم: 96.

6 اَلإلحادُ وَالباطِنِيَّةُ وَالرَّفض: اَلشِّيعَةُ الإسماعيلِيَّة

كانَ المَذهَبُ الشّيعِيّ الإسماعيلِيّ، اَلمُرتَبَطُ بِالباطِنِيَّةِ ارتِباطًا يَزيدُ أَو يَنقُصُ حَسَبَ العُصورِ وَالظُّروف، وَالمُبايِنُ لِمَذهَبِ أَهلِ السُّنَّةِ في تَأويلِهِ لِنُصوصِ الشَّرع، قَد تَمَكَّنَ في مِصرَ وَبِلادِ الشّامِ بِتَمَكُّنِ الفاطِمِيّين مِن إرساءِ إمامَتِهِم فيها. فَفي بَيتِ المَقدِس، "رُؤوسُ العُلَماء، وَرُؤوسُ المُبتَدِعَة"[129]، وَفيهِم تَلَطُّفٌ في الدَّعوَة، وَصَبرٌ في المُناظَرَة، وَ"فيهِم، لَعَمرُ الله، وَإن كانوا عَلى مَذهَبٍ باطِل، اِنطِباعٌ وَإنصاف، وَإقرارٌ لِلرَّجُلِ بِفَضلِهِ، إذا ظَهَر، وَاعتِراف".[130] وَساحِلُ الشّامِ هُوَ بَلَدُ الإسماعيلِيَّة: "وَكانَ السّاحِلُ المَذكورُ مَملوءاً مِن هاذِهِ النِّحَلِ المُلحِدِيَّة، وَالمَذاهِبِ الباطِنِيَّةِ وَالإمامِيَّة".[131] وَعَلى الرَّغمِ مِن هذا الإنصاف، وَلَمّا كانوا مِنَ "اَلرّافِضَة"، فَهُم بِالضَرورَةِ "أَرذَلُ الأُمَم"؛[132] فَلا يَشفَعُ لَهُم شَيء، وَلَم يَستَنكِف عَنِ الرَّدِّ على "الهِشامِيَّةِ مِنَ الرَّوافِض".[133]

أَقامَ القاضي أَبو بَكرٍ في مِصر، عاصِمَةِ الإمامَةِ الإسماعيلِيَّة، وَ"دارِ الدَّعوَة"، حَيثُ "بَيتُ الحِكمَة"، وَمَقَرُّ داعي الدُّعاة، مُدَّةً كافِيَةً مَكَّنَتهُ مِن تَحصيلِ المَذهَبِ الإسماعيلِيّ، حَتّى إنَّهُ ناظَرَ دُعاةَ الإسماعيلِيَّة. يَقول: "وَناظَرتُ الشّيعَةَ وَالقَدَرِيَّة، وَتَدَرَّبتُ في جُملَةٍ مِنَ الجَدَل، وَنَظَرتُ في نُبَذٍ مِن عِلمِ الكَلام، وَتَفَطَّنتُ مِن سَخافَةِ هاذِهِ الطّائِفَةِ بِنَفسي...أُمَّةٌ غَلَبَ عَلَيها سوءُ الاِعتِقاد، وَنُشِّئَت مِن غَيرِ فَطمٍ بِلَبَنِ العِناد، وَاستَولى اليَأسُ مِنهُم بِما هُم فيهِ مِنَ الفَساد."[134]

وَيَظهَرُ أَنَّ قِراءَتَهُ لِرَسائِلِ "إخوانِ الصَّفا"،[135] أَوِ "الصَّفَوِيَّة"[136]، كَما سَمّاها كَذلك، قَد أَعانَتهُ عَلى فَهمِ أَبعادِ المَشروعِ الكَلامِيِّ الشّيعِيِّ الباطِنِيِّ الإسماعيلِيّ، وَأَبعادِهِ الفَلسَفِيَّة، وَمَدى مُبايَنَتِهِ لِمَذهَبِ أَهلِ السُّنَّة. وَقَد جَعَلَتهُ الإقامَةُ وَالمُناظَرَةُ في مِصرَ يَرى المَذهَبَ الإسماعيلِيَّ أَسوَأَ مِن دينِ النَّصارى نَفسِه، وَقَد خَلَّفَهُ في بِلادِهِ مُخالِطًا لِأَهلِ الأَندَلُس، وَما فيهِ مِن قَولٍ بِالأُبُوَّةِ وَالحُلول. يَقول: "فَلَبِثتُ فيهِم، ثَمانِيَةَ أَشهُر، لَم يَبقَ باطِلٌ إلاّ سَمِعتُه، وَلا كُفرٌ إلاّ شوفِهتُ بِهِ وَوَعَيتُه...وَهُم لَم يَدَّعوا لِلرَّحمانِ وَلَدا، وَلكِنَّهُم

129 اَلعَواصِم: 45.

130 اَلعَواصِم: 46.

131 اَلعَواصِم: 46.

132 اَلكِتابُ المُتَوَسِّط: 470.

133 اَلكِتابُ المُتَوَسِّط: 142.

134 قانونُ التَّأويل: 90.

135 اَلعَواصِم: 109.

136 اَلعَواصِم: 128.

جاءوا بِأَعظَمَ مِن ذلك كُفرًا وَعِنادا، مَعَ انهِماكٍ في الكُفرِ وَاستِهتار، وَانحِلالٍ عَن رَبقَةِ الدِّيانَة، وَالمُروءَةِ وَالحِشمَة، وَخَلعِ عِذار".[137]

وَيَظهَرُ أَنَّ انفِصالَ إفريقِيَةَ عَن الإمامَةِ الإسماعيلِيَّةِ في القَرنِ الخامِسِ جَعَلَت مَذهَبَهُم يَنحَسِرُ فيها. فَالقاضي أَبو بَكرٍ ابنُ العَرَبِيِّ يُسَجِّلُ اصطِدامَهُ بِمَقالاتِ هاذِهِ الطّائفَة، وَلا سِيَما القَولَ بِالإمامَةِ وَعِصمَتِها، عِندَما خَرَجَ مِنَ الأَندَلُسِ وَبَلَغَ تُخومَ مِصرَ قائلا: "وَهاذِهِ أَوَّلُ بِدعَةٍ لَقيتُ في طَريقي...حَتّى بَلَغتُ بِلادَ هاذِهِ الطّائفَة".[138] وَرَأى في عَقيدَةِ الإمامَةِ "كَلِماتٍ غَرّارَة، خاتِمَتُها نَبذُ الحَقيقَةِ وَالشَّريعَة، وَالِاستِرسالُ عَلى الإباحَة"[139]، أَي عَقيدَةً يَلزَمُ مِنَ القَولِ بِها مُمارَسَةُ الإباحِيَّةِ الجِنسِيَّة.

وَرَأى في عَقيدَةِ الإمامَة، وَإن لَم يَكُن فيها تَصريحٌ بِأُبُوَّةٍ لِلَّهِ أَو بُنُوَّة، ضَربًا مِنَ القَولِ بِحُلولِ اللاّهوتِ في النّاسوت: "وَهاذا مِمّا يَنبَغي أَن تَعلَموا، أَنَّهُ راجِعٌ إلى القَولِ بِالحُلول. وَإنَّما عَرَّجوا عَلَيه، لِيَبعُدوا مِنه، وَهُم عَلَيهِ مُحَلِّقون، وَإلَيهِ راجِعون".[140]

وَلَمّا كانَت عَقيدَةُ الإمامَة، بَل كُلُّ المَذهَبِ الإسماعيلِيّ، مُرتَكِزَةً عَلى التَّأويلِ الرَّمزِيِّ وَالمَجازِيِّ لِلقُرآنِ الكَريم، وَهُوَ تَأويلٌ باطِنِيّ، فَقَد عَمِلَ ابنُ العَرَبِيِّ عَلى نَقدِهِ[141] مِن مَنظورِ أَهلِ الظّاهِر، وَصَرَّحَ أَنَّ "كَلامَ المَولى لَم يَكُن رُموزًا وَلا كِنايَة".[142]

وَضَرَبَ مَثَلاً عَلى ذلك تَأويلَ الإسماعيلِيَّةِ بَيتَ الله، أَي الكَعبَة، بِالقَلب، فَقال: "وَلَو هُدِيَت لِهاذِهِ الفِرقَةِ الضّالَّة، مِنَ الشّيعَةِ وَالباطِنِيَّة، لَما كانَت عَنِ الحَقِّ ناكِبين، وَقالَت إنَّ المُرادَ بِقَولِه: "وَطَهِّر بيتي"، القَلب، وَلا حَظَّ لِلكَعبَةِ فيه. وَلكِنَّهُ كَما أَخبَرَ تَعالى عَنه: "يُضِلُّ بِهِ كَثيرًا وَيَهدي بِهِ كَثيراً. وَما يُضِلُّ بِهِ إلاّ الفاسِقينَ."[143]

وَرَأى في التَّأويلِ الباطِنِيِّ لِلقُرآنِ مَشروعًا لِإبطالِ الشَّريعَةِ وَتَعطيلِها، يَقودُ إلى الإباحِيَّة، بِاعتِبارِها مِن لَوازِمِ الِاعتِقادِ بِه: "وَهاذا كُلُّهُ مُجاهَدَةٌ عَلى الدّين، وَحِيَلٌ في هَدمِ قَواعِدِ الشَّرائع، مِنَ الإباحِيَّةِ وَالتَّعطيلِيَّة."[144]

---

137 اَلعَواصِم: 45.
138 اَلعَواصِم: 44.
139 اَلعَواصِم: 44.
140 اَلعَواصِم: 46.
141 اَلعَواصِم: 114.
142 اَلعَواصِم: 110.
143 قانونُ التَّأويل: 226.
144 اَلعَواصِم: 43.

وَيَظهَرُ عَلى العُمومِ أَنَّ القاضِيَ أَبا بَكرٍ ابنَ العَرَبِيِّ كانَ يَتَمَثَّلُ الشّيعَةَ الإسماعيلِيَّةَ في الباطِنِيَّةِ قَبلَ غَيرِهِم مِن طَوائفِ الباطِنِيَّة، لَمّا اختَرَقَ وَأَقامَ في بِلادِ "الإمامَةِ الإسماعيلِيَّة"، وَلِما كانَ يُحِسُّهُ مِن خَطَرٍ يُشَكِّلونَهُ عَلى "أَهلِ السُّنَّة"، بِما طَوَّروا مِن وَسائلِ التَّأثيرِ وَالدَّعوَة. بَيدَ أَنَّ كِتاباتِهِ لَم تَخلُ مِن إشارَةٍ إلى طَوائفَ باطِنِيَّةٍ مُنقَرِضَةٍ في زَمانِه، هُم مِنَ "اَلمُلحِدَةِ مِنَ الخُرَّمِيَّةِ وَالباطِنِيَّة"، لَم يُعَرِّج عَلَيها كَثيرا.[145]

7 اَلإلحادُ وَالاِعتِزال

يُسَمّي ابنُ العَرَبِيِّ المُعتَزِلَةَ كَثيرًا بِغَيرِ اسمِهِم، وَلاكِن بِما يَلزَمُ مِن أُطروحاتِهِمُ الفِكرِيَّة. فَهُم عِندَهُ "قَدَرِيَّة"[146]، أَي نُفاةَ القَدَر، وَنُفاةُ الصِّفاتِ الإلاهِيَّة، إذ "زَلَّتِ المُعتَزِلَةُ في الطَّرَفِ الآخَرِ مَعَ المُلحِدَةِ في نَفيِ الصِّفات"،[147] وَ"جَبرِيَّة"،[148] وَ"مُعَطِّلَة".[149] هاذا وَالقَولُ بِالقَدَرِ وَإثباتِ الصِّفاتِ مِن أَركانِ الإيمانِ عِندَ أَهلِ السُّنَّة، وَمَذهَبُ المُعتَزِلَةِ "مَذهَبُ المُؤَوِّلَة"،[150] وَشَأنُهُم "اَلبِدعَةُ وَالتَّعطيل"،[151] وَ"الإلحادُ وَالتَّعطيل".[152] وَقَد يُسَمّيهِمُ بِاسمَينِ اثنَين، في آنٍ واحِد، فَيَصيرونَ عِندَهُ "اَلمُعتَزِلَةَ وَالقَدَرِيَّة"،[153] وَقَد يُكَنّي عَنهُم بِقَولِه: "اَلقَدَرِيَّةُ وَالمُبتَدِعَةُ مِن إخوانِهِم".[154] وَلِكَي يَكونَ لِكَلامِهِ صَدى طَيِّب، وَيَكونَ حُجَّةً في أُطروحاتِ المُعتَزِلَة، أَكَّدَ أَنَّهُ عَرَفَهُم حَقَّ المَعرِفَة، إذِ التَقاهُم في بَيتِ المَقدِسِ وَناظَرَهُم، وَقالَ في شَأنِهِم: "وَكُنّا نُفاوِضُ الكَرّامِيَّةَ وَالمُعتَزِلَةَ وَالمُشَبِّهَةَ وَاليَهود...وَخاصَمنا النَّصارى".[155]

وَكَذلك ناظَرَهُم في مِصر: "وَناظَرتُ الشّيعَةَ وَالقَدَرِيَّة، وَتَدَرَّبتُ في جُملَةٍ مِنَ الجَدَل، وَنَظَرتُ في نُبَذٍ مِن عِلمِ الكَلام، وَتَفَطَّنتُ مِن سَخافَةِ هاذِهِ الطّائفَةِ بِنَفسي...أُمَّةٌ غَلَبَ عَلَيها سوءُ الاِعتِقاد، وَنُشِّئَت مِن غَيرِ فَطمٍ بِلَبَنِ العِناد، وَاستَولى اليَأسُ مِنهُم بِما هُم فيهِ مِنَ الفَساد."[156]

---

145 اَلعَواصِم: 39.
146 قانونُ التَّأويل: 293–295.
147 اَلكِتابُ المُتَوَسِّط: 135.
148 اَلكِتابُ المُتَوَسِّط: 270.
149 سِراجُ المُريدين: 2/340–341.
150 قانونُ التَّأويل: 185.
151 قانونُ التَّأويل: 350.
152 اَلكِتابُ المُتَوَسِّط: 455.
153 اَلكِتابُ المُتَوَسِّط: 270. 306. 310. 359.
154 اَلكِتابُ المُتَوَسِّط: 377.
155 قانونُ التَّأويل: 95.
156 قانونُ التَّأويل: 90.

صَنَّفَ ابنُ العَرَبِيِّ المُعتَزِلَةَ ضِمنَ المُبتَدِعَةِ وَالمُلحِدَةِ في آنٍ واحِد، وَقَرَنَهُم في قِرانٍ واحِد: فَهُم عِندَهُ "المُبتَدِعَةُ وَالمُلحِدَة"،[157] وَ"المُبتَدِعَةُ مِنَ المُعتَزِلَةِ وَشِبهِهِم"،[158] وَ"المُلحِدَةُ وَالمُعتَزِلَة"[159]، وَ"المُعتَزِلَةُ وَسائرُ المُبتدِعَة".[160]، "وَقَرَنَهُم في قِرانٍ واحِدٍ مَعَ الفَلاسِفَةِ الإسلامِيّينَ قائلا: "وَهاذِهِ الكُتُبَ الَّتي تُنسَبُ إلى الحِكمَة، فَإنَّها مَجموعَةٌ مِن أَقوالِ الفَلاسِفَةِ وَالمُعَطِّلَة..."[161]

وَلَمّا كانَ المُعتَزِلَةُ مُبتَدِعَةً وَمُلحِدينَ زائغينَ عَنِ الحَقِّ الكَلامِيِّ في رَأيِ ابنِ العَرَبِيّ، فَقَد أَنذَرَهُم وَمَن يَنوي اتِّباعَ طَريقِهِم بِسوءِ الخاتِمَة، فَهُم "فَعَلى مَهواة"،[162] وَفَسَّرَ قَولَهُ تَعالى: "وَما أَضَلَّنا إلاّ المُجرِمون"، وَنَزَّلَهُ عَلى المُعتَزِلَة، بَل جَعَلَهُم رُؤَساءَ المُجرِمينَ قائلا: "وَمِن أَئمَّةِ المُجرِمينَ القَدَرِيَّة..."[163]

وَقَد كَفَّرَهُمُ ابنُ العَرَبِيِّ وَضَلَّلَهُم بِخَرقِ الإجماع: "وَذلك كُلُّهُ خَرقُ الإجماعِ السّابِق، وَتَعَلُّقٌ بِأَذيالِ الكُفر، وَانغِماسٌ في غَمرَةِ الضَّلالَة."[164] وَكانَ هذا كافِيًا في نَظَرِهِ لِتَكفيرِهِم، وَكانَ رَدُّ عَمرِو بنِ عُبَيدٍ لِحَديثِ الخَلقِ كافِيًا لِتَكفيرِه: "وَهاذا هُوَ الكُفرُ الصُّراح"،[165] كَما رَأى في كِتاباتِ الجاحِظِ مَشروعًا لِصَرفِ المُسلِمينَ عَن كَلامِ اللّهِ وَرَسولِه، وَاتَّهَمَهُ بِالإستِدلالِ السَّيِّئِ في شُؤونِ الدّين، فَقال، وَهُوَ يُقَدِّمُ لِأَهلِ السُّنَّةِ البَديلَ عَنِ الفِكرِ الإعتِزالِيِّ وَالفَلسَفِيّ: "وَإذا أَرَدتُم أَن تَكونوا حُكَماء، فَعَلَيكُم بِكِتابِ اللّهِ وَسُنَّةِ رَسولِه، فَهِيَ الغايَة. وَإيّاكُم وَهاذِهِ الكُتُبَ الَّتي تُنسَبُ إلى الحِكمَة، فَإنَّها مَجموعَةٌ مِن أَقوالِ الفَلاسِفَةِ وَالمُعَطِّلَة، كَكِتابِ كَليلَةَ وَدِمنَة، اَلَّذي تَرجَمَهُ المُلحِدون، وَالكُتُبَ الَّتي جَمَعَتها المُلحِدَة، كَالجاحِظِ وَغَيرِه، لِيَشغَلَ بِها الخَلقَ عَن كَلامِ اللّه، وَكَلامِ نَبِيِّه، وَدَسَّ فيها مِن كَلامِ اللّهِ تَعالى وَكَلامِ نَبِيِّهِ ما لَم يَصِحّ..."[166]

أَعلَنَ ابنُ العَرَبِيِّ بِكُفرِ المُعتَزِلَةِ وَبِتَكفيرِهِم دونَ قَيدٍ أَو شَرط، مَعَ تَرَيُّثِهِ في شَأنِ غَيرِهِم، فَكُفرُ المُعتَزِلَةِ ثابِتٌ في نَظَرِه، وَلا شَكَّ فيه، وَكُفرُ غَيرِهِم مَحَلُّ نَظَر، فَ"إنَّ المُبتَدِعَةَ وَأَهلَ الأَهواءِ كُفّار...أَمّا القَدَرِيَّة،

157 اَلكِتابُ المُتَوَسِّط: 122.
158 اَلكِتابُ المُتَوَسِّط: 313.
159 اَلكِتابُ المُتَوَسِّط: 298.
160 اَلكِتابُ المُتَوَسِّط: 131.
161 سِراجُ المُريدين: 2/340–341.
162 اَلكِتابُ المُتَوَسِّط: 322.
163 قانونُ التَّأويل: 127.
164 اَلكِتابُ المُتَوَسِّط: 284.
165 اَلكِتابُ المُتَوَسِّط: 386.
166 سِراجُ المُريدين: 2/340–341.

فَلا شَكَّ في كُفرِهِم، وَأَمّا مَن عَداهُم، فَنستَقرِئُ فيهِمُ الأَدِلَّة."[167] بَل ذَهَبَ إلى ما هُوَ أَبعَدُ مِن ذلك، فَرَأى فيهِم "مُتَستِّرين" بِالإسلام،[168] تَمامًا كَشَأنِ الزَّنادِقَةِ أَو مُنافِقي عَهدِ النُّبُوَّة.

وَقَد أَنكَرَ عَلَيهِم مِنَ الأُطروحاتِ الكَلامِيَّةِ ما يَلي:

أ. نَفيَ القَدَرِ وَوَضعَ الحَديثِ في إنكارِه، فَـ"كُلُّ حَديثٍ في النَّهيِ عَنِ القَدَرِ لا أَصلَ لَه. وَإنَّما أَحدَثَ النَّهيَ عَنهُ أَقوام، مِثلَ مَن أَحدَثَ القَولَ فيه، كَأَنَّهُم قَصَدوا حِمايَةَ الشَّريعَةِ بِما لَيسَ مِنها."[169] وَفي وَضعِ الحَديثِ في نَفيِه، وافَقوا بَعضَ أَهلِ السُّنَّةِ في وَضعِ الحَديثِ في إثباتِه.

ب. تَأويلَ الِاستِواءِ بِالِاستيلاء: "اَلرَّحمانُ عَلى العَرشِ استَوى...وَالكُلُّ عِندي في حُسنِ الِاعتِقاد، وَسَلامَتِهِ عَن مَذهَبِ أَهلِ الإلحادِ سَواء".[170]

ت. القَولَ بِالتَّوَلُّدِ أَوِ السَّبَبِيَّة، فَقال: "وَنَبَغَت في الإسلامِ عِندَ حُدوثِ الأَهواءِ فِرقَةٌ زَعَمَت أَنَّ الإحراقَ فِعلُ النّار، وَالإشباعَ فِعلُ الخُبز...وَمَوضِعُ الرَّدِّ عَلى هؤلاء، اَلقَولُ عَلى مَذاهِبِ الطَّبائِعِيّينَ مِنَ المُلحِدين، فَإنَّهُ شُعبَةٌ مِن شُعَبِهِم...وَأَوَّلُ مَن يُؤثَرُ عَنهُ هذا المَذهَب، مُعَمَّرٌ القَدَرِيّ، وَالجاحِظُ المُفتَري".[171]

ث. إنكارَ الكَرامات، فَقالَ عَنهُم: "وَأَنكَرَتِ القَدَرِيَّةُ وَالمُعتَزِلَةُ وَطائِفَةٌ ضَعيفَةٌ مِن أَهلِ الحَديثِ الكَرامات"[172]، وَالحالُ أَنَّ الكَرامات، وَهِيَ مَجالٌ تَوارَدَ عَلى إنكارِهِ أَوِ الغُلُوِّ في الِاعتِقادِ بِها بَعضُ أَهلِ السُّنَّة، كَابنِ أَبي زَيدٍ القَيرَوانِيّ، "هِيَ أَصلُ الدّين، وَعُمدَةٌ مِن عُمَدِ المُسلِمين، لا يُنكِرُها إلاّ جاهِل؛ اِتَّفَقَ عَلَيها العُلَماء..."[173] فَقَد خافَ القاضي أَبو بَكرٍ مِن أَن يَكونَ إنكارُ الكَراماتِ مَدخَلاً إلى إنكارِ المُعجِزات، وَحاوَلَ تَفسيرَها.[174] وَقَد رَأى في بَعضِ أُطروحاتِهِم أَصلاً فَلسَفِيًّا: "وَهاذا سُؤالٌ حادٌّ وَجَّهَتهُ المُعتَزِلَة، إذ كَرِهَت هاذِهِ الدَّلالَة، لِأَنَّها قُرآنِيَّة، وَعَدَلَت إلى دَلالَةِ الفَلاسِفَة، وَهِيَ استِحالَةُ الكَثرَةِ في العِلَّة."[175]

وَلَمّا كانَ أَهلُ السُّنَّةِ الأَشاعِرَةُ قَد وافَقوا المُعتَزِلَةَ في نَصبِ الأَدِلَّةِ العَقلِيَّة، لِلدِّفاعِ عَنِ الأُطروحاتِ السُّنِّيَّة، فَقَد اضطُرَّ اضطِرارًا إلى عَقدِ فَصلٍ يُبَيِّنُ فيهِ أَنَّ هاذِهِ المُوافَقَةِ الَّتي يَفتَرِضُ فيها خُروجًا عَنِ

167 أَحكامُ القُرآن: 294/1.

168 اَلعَواصِم: 81.

169 اَلعَواصِم: 140.

170 اَلكِتابُ المُتَوَسِّط: 257.

171 اَلكِتابُ المُتَوَسِّط: 261.

172 اَلكِتابُ المُتَوَسِّط: 359.

173 اَلعَواصِم: 37.

174 اَلعَواصِم: 37.

175 قانونُ التَّأويل: 305.

المَنهَجِ السُّنِّيِّ الأَصلِيّ، مِمّا يُمكِنُ الاِعتِذارُ لَهُ وَتَسويغُه: "ذِكرِ الاِعتِذارِ عَن عُدولِ العُلَماءِ عَنِ الكِتابِ إلى أَدِلَّةِ العُقول"[176]

وَفَسَّرَ هاذِهِ المُوافَقَةَ بِرَغبَةِ أَهلِ السُّنَّةِ في مُواجَهَةِ الخَصمِ بِسِلاحِه، بَل بِالتَّفَوُّقِ عَلَيهِ في استِعمالِه، فَقال: "أَرادوا أَن يُبَصِّروا المُلحِدَة، وَيُعَرِّفوا المُبتَدِعَة، أَنَّ مُجَرَّدَ العُقولِ الَّتي يَدَّعونَها لِأَنفُسِهِم، وَيَعتَقِدونَ أَنَّها مِعيارُهُم، لا حَظَّ لَهُم فيها...حَتّى يَتَبَيَّنَ لَهُم أَنَّهُ كَيفَ دارَتِ الحالُ مَعَهُم مِن كَلامِهِم، بِمَنقولٍ أَو مَعقول، فَإِنَّهُم فيهِ عَلى غَيرِ تَحصيل."[177]

وَمَعَ قَولِهِ بِبُطلانِ مَذهَبِ المُعتَزِلَةِ في نَظَرِه، فَقَد أَشارَ إلى مُوافَقَةِ بَعضِ أَهلِ السُّنَّةِ لِبَعضِ أُطروحاتِهِم، عَلى قِلَّة، فَقال: "وَعِندَ القَدَرِيَّةِ وَابنِ الرّاوَندِيّ، وَيُؤثَرُ ذلك عَن أَبي حَنيفَة، وَابنِ سُرَيج، أَنَّ الاِستِطاعَةَ قَبلَ الفِعل."[178]

وَلَمّا كانَ العَداءُ مُستَحكِمًا بَينَ المُعتَزِلَةِ وَأَهلِ السُّنَّة، أَشاعِرَةً وَغَيرَ أَشاعِرَة، ساغَ لَهُ أَن يَتَّهِمَ الصّاحِبَ ابنَ عَبّادٍ وَالمُعتَزِلَةَ عُمومًا بِتَدبيرِ إحراقِ كِتابِ "المُختَزِن"، لِأَبي الحَسَنِ الأَشعَرِيّ،[179] وَهُوَ كِتابُهُ الكَبيرُ في التَّفسير، وَإعدامِهِ كَي لا يَنتَفِعَ بِهِ أَهلُ السُّنَّةِ في مُعارَضَةِ بِدعَتِهِم.

8 اَلإلحادُ وَالتَّشبيه

في مُقابِلِ الأُطروحاتِ الاِعتِزالِيَّة، بِما هِيَ مَنظومَةٌ "تَعطيلِيَّةٌ" تَبني تَنزيهَها عَلى "تَعطيل" الصِّفاتِ الإلاهِيَّة، وَتَجعَلُ الذّاتَ عَينَ الصِّفات، بِناءً عَلى تَنزيهٍ عَقلانِيٍّ مُفرِط، قَرَّرَ ابنُ العَرَبِيِّ مَذهَبَ أَهلِ السُّنَّةِ الأَشاعِرَةِ في الأَمر، فَقال: "وَلا يَجوزُ أَن يَكونَ لَهُ شَبَهٌ في ذاتِه، وَلا في صِفاتِه، وَلا في أَفعالِه. وَكُلُّ مَن أَضافَ ذلك إلَيهِ فَهُوَ مُشَبِّه. وَلِذلك كانَ جَميعُ مَن يَخرُجُ عَن رَسمِ المُوَحِّدينَ مُشَبِّها."[180] لكِنَّهُ لَم يَستَفرِغ طاقَتَهُ في الرَّدِّ عَلى المُشَبِّهَة، وَلا أَعلَنَ بِكُفرِهِم، كَما فَعَلَ في شَأنِ المَذاهِبِ الأَشَدِّ إقلاقًا لَهُ وَلِأَهلِ السُّنَّةِ الأَشاعِرَةِ في زَمَنِه. وَاكتَفى بِنَقدِ ابنِ أَبي زَيدٍ القَيرَوانِيّ، (386هـ\996م) في بَعضِ المَسائلِ الكَلامِيَّةِ الخِلافِيَّةِ بَينَ أَهلِ السُّنَّة، الَّتي تَقتَرِبُ مِن مَذهَبِ التَّشبيه.[181]

---

176 قانونُ التَّأويل: 176.

177 قانونُ التَّأويل: 177.

178 اَلكِتابُ المُتَوَسِّط: 287.

179 اَلكِتابُ المُتَوَسِّط: 439.

180 قانونُ التَّأويل: 127.

181 اَلعَواصِم: 215.

9 ٱلإلحادُ وَالتَّصوُّفُ الفَلسَفِيّ

لا يَخلو تُراثُ ابنِ العَرَبِيِّ مِن تَنويهٍ بِالتَّصَوُّف، أَو قُل بِضَربٍ مِنه، يُفارِقُ التَّأويلَ وَالإشارَة، وَيُجانِبُ الباطِنِيَّةَ وَالفَلسَفَة، فَالصّوفِيَّةُ عِندَهُ هُم "عُلَماءُ القُلوب".[182] بَل كانَ لَهُ اعتِزازٌ بِمَرحَلَةٍ مِن عُمُرِهِ اشتَغَلَ هُوَ فيها بِالتَّصَوُّفِ مُمارَسَةً وَرِياضَة، فقالَ عَنها: "فَاستَفَدتُ مِن أَهلِ السُّنَّة، وَجادَلتُ بِالَّتي هِيَ أَحسَنُ أَهلَ البِدعَة، وَأَفنَيتُ عَظيمَ الزَّمانِ في طَريقَةِ الصّوفِيّين".[183] فكَأَنَّهُ يَرى التَّصَوُّفَ مَنزِلَةً بَينَ المَنزِلَتَين، فَلا هُوَ مِن شَأنِ أَهلِ السُّنَّة، وَلا هُوَ مِن شَأنِ "أَهلِ البِدعَة".

وَلا يَخلو كَلامُ ابنِ العَرَبِيِّ مِنِ استِطرادٍ صوفِيِّ المَنزَع. يَقول: "وَباطِنُه: قَلبُ عَبدي المُؤمِنِ الَّذي كَرَّمتُهُ بِأَن جَعَلتُهُ مَحَلَّ مَعرِفَتي، وَشَرَحتُهُ بِنورِ هِدايَتي، وَمَلَأتُهُ حِكمَةً مِن عِلمي، وَخَصَصتُهُ بِأَن أَحيَيتُهُ بِروحي."[184]

"اَلمُتَفَكِّر: وَحَقيقَتُه: تَرَدُّدُ العُلومِ في القَلب، وَتَرتيبُها حَتّى تُثمِرَ أَمثالَها في أَمثالِها".[185]

وَلَمّا كانَ ابنُ العَرَبِيِّ مُوافِقًا لِلصّوفِيَّةِ في جانِبٍ مِن فِكرِهِم، وَهُوَ القَولُ بِالكَرامات، قَرَّرَ أَنَّها "أَصلُ الدّين، وَعُمدَةٌ مِن عُمَدِ المُسلِمين، لا يُنكِرُها إلاّ جاهِل؛ اِتَّفَقَ عَلَيها العُلَماء."[186]، وَحاوَلَ تَفسيرَها، كَما قَد رَأَينا.[187]

وَعَلى الرَّغمِ مِن صُحبَةِ القاضي أَبي بَكرٍ ابنِ العَرَبِيِّ لِحُجَّةِ الإسلامِ الغَزاليّ، (505هـ\1111) وَاعتِزازِهِ بِها، وَاعتِزالِهِ مَعَهُ في خَلوَةٍ صوفِيَّة، فَقَد كانَ لَهُ مَوقِفٌ حَذِرٌ مِنَ التَّصَوُّفِ ذي البُعدِ الباطِنِيِّ أَوِ الفَلسَفِيِّ أَوِ الإشارِيّ، إذ كانَ يَخشى مِنِ انزِلاقِ الصّوفِيَّةِ إلى الإلحادِ وَالبِدعَة، فَانتَقَدَ الصّوفِيَّةَ وَشَيخَهُ الغَزالِيَّ كَذلك.[188] كَما انتَقَدَ القُشَيرِيّ، (465هـ\1072م) قائِلاً عَن كَلامِه، دونَ تَصريحٍ بِذلك: "كَلامٌ غَريبٌ عَلى طَريقَتِهِم، لَستُ لَه، وَلا أَنتُم".[189]

---

182 قانونُ التَّأويل: 381.

183 قانونُ التَّأويل: 120.

184 قانونُ التَّأويل: 224.

185 سِراجُ المُريدين: 2/459.

186 اَلعَواصِم: 37.

187 اَلعَواصِم: 37.

188 اَلعَواصِم: 12–13.

189 قانونُ التَّأويل: 265.

يَرى ابنُ العَرَبيِّ في الصّوفيَّةِ "أَصحابَ الإشارات"[190]، أَي قومًا يَستَعمِلونَ مَنطِقًا في تَفسيرِ القُرآنِ الكَريمِ لا يَنبَني عَلى نَقلٍ أَو عقلٍ أَو لُغة. وَقَدِ انصَبَّ نَقدُ القاضي أَبي بَكرٍ ابنِ العَرَبيِّ عَلى "الإشارَةِ الصّوفيَّةِ" وَتَفسيرِها لِلقُرآنِ الكَريم تَفسيرًا إشاريًّا وَرَمزيًّا[191]، حَتّى لَو كانَت هاذِهِ الإشارَةُ صادِرَةً عَن صوفيَّةٍ مَحسوبينَ عَلى مَذهَبِ أَهلِ السُّنَّة. وَقَد سَجَّلَ انتِشارَ ظاهِرَةِ التَّفسيرِ الإشاريّ، بِما يوحي أَنَّ المَغرِبَ وَالأَندَلُسَ لَم يَكونا يعرفان هذا التَّفسيرَ الإشارِيّ، فَقال: "وَهاذا فَنٌّ مِن تَوابِعِ التَّفسيرِ اعتَنَت بِهِ الصّوفيَّة، حَتّى غَلَبَ عَلى بِلادِ المَشرِقِ مِن تُركِستانَ إلى الشّام، فَلا يَقدِرُ أَحَدٌ عَلى دَفعِه."[192] ولا يَظهَرُ مِن قَرائِنِ الحالِ أَنَّهُ كانَ مُرتاحًا لِهاذا التَّفسيرِ الإشارِيّ.

لَم يُكَفِّرِ ابنُ العَرَبيِّ الصّوفيَّةَ المُتَفَلسِفَةَ تَكفيرًا واضِحا، بَل لَم يَنسِبهُم إلى الإلحادِ أَو يَصِفهُم بِالبِدعَة، مِن حَيثُ المَبدَإ، وَعَلى وَجهِ التَّصريح، كَما فَعَلَ بِغَيرِهِم مِن طَوائفِ المُسلِمين، فَكانَ بِهِم رَحيماً رَحمَةً مَرموقَة، بَلِ اكتَفى بِالقَولِ إنَّهُم مُتَشَبِّهونَ بِالمُبتَدِعَة. لكِنَّهُ قَرَنَهُم بِـ"دُعاةِ الباطِنيَّة"، وَاتَّهَمَهُم بِوَضعِ الحَديث، وَاستِعمالِ الحَديثِ المَوضوعِ وَالاِحتِجاجِ بِه، فَقال: "إنَّ غُلاةَ الصّوفيَّة، وَدُعاةَ الباطِنيَّة، يَتَشَبَّهونَ بِالمُبتَدِعَة، في تَعَلُّقِهِم بِمُتَشابِهاتِ الآياتِ وَالآثارِ عَلى مُحكَماتِها، فَيَختَرِعونَ أَحاديث، أَو تُختَرَعُ لَهُم عَلى أَغراضِهِم، يَنسِبونَها إلى النَّبِيّ، وَيَتَعَلَّقونَ بِها عَلينا، فَمِنها حَديث..."[193] وَنَبَّهَ إلى أَنَّ بَعضَ هاذِهِ الأَحاديثِ المَوضوعَة، مُسنَدٌ إلى الإمامِ مالِك، وَرَأى في ذلك حيلَةً مِن غُلاةِ الصّوفيَّةِ لِلاِلتِفافِ عَلى أَهلِ السُّنَّة، وَلا سِيَما المالِكِيّينَ مِنهُم، فَقال: "وَأَثَرُوا ذلك عَن مالِك... إسكاتًا لَنا، وَاعتِضادًا بِإمامَتِهِ عَلَينا".[194]

وَقَد نَبَّهَ كَذلك إلى اقتِباسِ "غُلاةِ الصّوفيَّة" بَعضَ أُطروحاتٍ فَلسَفيَّة، فَقال: "وَتابَعَتِ الصّوفيَّةُ الفَلاسِفَةَ في شَيءٍ مِن أَغراضِها فيها، [الرّوح] قالوا: اَلعالَمُ عالَمان: عالَمُ الخَلق، وَعالَمُ الأَمر".[195] وَرَأى في ذلك استِنساخًا لِلخِطابِ الفَلسَفِيّ، فَقال: "وَهاذا كَلامٌ يَحومُ عَلى مَقاصِدِ الفَلاسِفَة...هذا القَولُ الَّذي أَملَتِ الفَلاسِفَة، وَانخَدَعَ بِهِ زُمرَةٌ مِنَ الصّوفيَّة".[196] كَما تَناوَلَ أَمرَ أَهلِ "اَلوَحدَةِ المَحضَة".[197]

190 اَلعَواصِم: 193.

191 اَلعَواصِم: 109.

192 قانونُ التَّأويل: 264.

193 اَلعَواصِم: 14.

194 اَلعَواصِم: 16.

195 قانونُ التَّأويل: 172–173.

196 قانونُ التَّأويل: 246.

197 اَلعَواصِم: 132.

وَقد تَفاوَتَ كَلامُهُ في شَأنِ المَقالاتِ الصّوفِيَّةِ بَينَ الحِدَّةِ وَاللُّطف، وَالوَصفِ وَالإدانَة. قال: "وَزَعَمَت طائفَةٌ مِنَ الصّوفِيَّة، أنَّ الباري، تَعالى، إنَّما ضَرَبَ الأَمثالَ في المَنامِ لِانتِقاشِ العُلومِ في اللَّوحِ المَحفوظ...وَهاذِهِ نَزعَةٌ فَلسَفِيَّة، وَأَغراضٌ عَنِ الحَقِّ قَصِيَّة."[198] وَرَأى في هذا الاِقتِباسِ مِنَ الفَلسَفَةِ ابتِداعاً وَزَيغا، دونَ أن يَصِفَهُ بِالإلحاد، فَقال: "وَمِن هاهُنا زاغَت طائفَةٌ كَثيرَةٌ مِنَ الصّوفِيَّة، فَاستَرسَلَت في ضَربِ الأَمثال، وَاستِعارَةِ الأَلفاظ، وَنَقلِها مِنَ المَخلوقِ إلى الخالِق. وَذلك ابتِداع."[199]

بَيدَ أَنَّهُ لَم يَخلُ مَعَ كُلٍّ هذا مِن حَطٍّ عَلى الصّوفِيَّةِ المُتَفَلسِفَة، وَوَصفِهِم بِالإلحاد، دونَ تَصريحٍ مِنهُ بِذلك، فَقال: "وَأَرادَت طائفَةُ التَّشغيب، أَن تُفرِدَ الرّوحَ بِبَيان، وَتَخُصَّهُ بِنَوعٍ مِنَ البُرهان، حَتّى انتَهى بِهِمُ القَولُ إلى أَن يَقولوا: وَما الإنسان؟...فَإنَّ المُلحِدَةَ أَدخَلَت هاذِهِ الأَلفاظَ في بابِ الإشكال، تشغيبًا وَتَلبيسا".[200]

بَل بَلَغَ بِهِ الأَمرُ أَنَّهُ اتَّهَمَ "غُلاةَ الصّوفِيَّةِ" بِالتَّستُّرِ بِالإسلام، فَقالَ في مَعرِضِ حَديثِهِ عَن رُؤيَةِ اللّٰه: "وَهاذا ما يَمنَعونَهُ سِرّا، وَلا يَقدِرونَ عَلَيهِ جَهرا، لِأَنَّهُم **يَتَظاهَرونَ بِالإسلام**، فَأَمّا الفَلاسِفَةُ فَيَمنَعونَه."[201]

وَأَورَدَ تَصنيفًا لِفِرَقِ الزَّنادِقَة،[202] نَقلاً عَن أَبي عاصِمٍ النَّسائِـــــيّ، (253هـ\867م) في كِتابِه: "اَلتَّأكيدِ في لُزومِ السُّنَّة، وَحُبِّ خِيارِ هاذِهِ الأُمَّة"، فَقال: "وَمِنهُمُ الرّوحانِيَّة. وَهُم أَصناف. وَإنَّما سُمّوا الرّوحانِيَّة، لِأَنَّهُم زَعَموا أَنَّ أَرواحَهُم تَنظُرُ إلى المَلَكوتِ الأَعلى، وَتُعايِنُ الحور، وَتُشاهِدُ الجَنَّة، وَتَتَنَعَّمُ بِرُؤيَةِ الباريءِ تَعالى. وَسُمّوا أَيضًا "الفِكرِيَّة"، لِأَنَّهُم زَعَموا أَنَّهُم يَتَفَكَّرونَ في هذا حَتّى يَعرِفوا اللّٰه، فَجَعَلوا الفِكرَةَ غايَةَ العِبادَة، وَهُوَ مَذهَبُ الحَلاّجِ السّاحِرِ الكافِر. وَمِنهُم صِنفٌ مِنَ الرّوحانِيَّة، زَعَموا أَنَّ حُبَّ اللّٰهِ تَعالى يَغلِبُ عَلى قُلوبِهِم وَأَهوائِهِم، حَتّى يَكونَ حُبُّهُ أَغلَبَ الأَشياءِ عَلَيهِم. فَإذا كانَ كَذلك، كانوا عِندَ اللّٰهِ بِهاذِهِ المَنزِلَة، وَإذا كانوا عِندَ اللّٰهِ بِهاذِهِ المَنزِلَة، وَجَبَت لَهُمُ الجَنَّة، وَارتَفَعَ عَنهُمُ التَّكليف، وَأَحَلَّ لَهُمُ المَحظور، فَهُوَ عِندَهُم حَلال...وَمِنهُم صِنفٌ زَعَموا أَنَّ تَركَ الدُّنيا اشتِغالُ القُلوبِ...وَرابِعَتُهُم رابِعَة."[203]

198 قانونُ التَّأويل: 262.

199 قانونُ التَّأويل: 292.

200 اَلعَواصِم: 27–28.

201 اَلعَواصِم: 35.

202 اَلمَسالِك: 3/400–404.

203 اَلمَسالِك: 3/403.

10 اَلظّاهِرِيَّة

عَلى الرَّغمِ مِن كَونِ الظّاهِرِيَّةِ مَحسوبينَ عَلى أَهلِ السُّنَّةِ في كَلامِهِم وَفِقهِهِم، مِن حَيثُ المَبدَإ، فَإِنَّ ابنَ العَرَبِيِّ وَجَّهَ لَهُم نَقدًا شَديدا،[204] وَلا سِيَما ابنِ حَزمٍ الأَندَلُسِيّ. (456هـ\1064م)[205] وَهاذِهِ الحالَةُ نادِرَة، إذ كانَت عادَةُ ابنِ العَرَبِيِّ أَن يُجَرِّحَ طَوائفَ "أَهلِ الإلحادِ وَالبِدعَةِ" مِنَ المَشارِقَة، دونَ الأَندَلُسِيّينَ وَالمَغارِبَة. وَلَم يَخلُ كَلامُهُ مِن تَضليلٍ وَتَبديعٍ لِلظّاهِرِيَّة، دونَ إسرافٍ في الأَمر، أَوِ استِعمالِ مُصطَلَحِ الإلحاد، فَقالَ عَنهُم: "وَلِهاذا ضَلَّتِ المُبتَدِعَة، [الظّاهِرِيَّة] وَأَنكَرَتِ القِياسَ بِالرَّأي، فَأَبطَلَت رُكنًا مِن أَركانِ الشَّريعَة."[206]

11 مِن عِلمِ الكَلامِ إلى التّاريخِ وَالقَفزِ عَلَيه: اَلقَفزُ عَلى التّاريخ، لِخِدمَةِ التّاريخ

كانَ مُصطَلَحُ الإلحادِ مَفهومًا واسِعًا فَضفاضًا عِندَ ابنِ العَرَبِيّ، يَتَضَمَّنُ كُلَّ ما هُوَ غَيرُ سُنِّيّ، كَما فَهِمَ ابنُ العَرَبِيِّ السُّنِّيَّةَ وَالتَّسَنُّن، نازِلَ بِهِ طَوائفَ مُتَنَوِّعَةً مِنَ المُسلِمين، مِن أَكثَرِهِم عَقلانِيَّة، إلى أَكثَرِهِم نَقلانِيَّة، إلى الباطِنِيَّة. فَهَل تَعَمَّدَ ابنُ العَرَبِيِّ استِعمالَ مُصطَلَحٍ فَضفاضٍ لِحاجَةٍ في نَفسِه؟ إنَّ شَيئًا جَديرًا بِالتَّنبيهِ عَلَيهِ في هذا الصَّدَد، وَهُوَ أَنَّ ابنَ العَرَبِيِّ قَدِ اعتَنى بِالكَلامِ عَلى طَوائفِ المُسلِمين، مِن مُعتَزِلَةٍ وَباطِنِيَّةٍ وَإسماعيلِيَّةٍ وَغَيرِهِم، مِنَ المُعاصِرينَ لَه، وَمَنِ احتَكَّ بِهِم في المَشرِقِ في أَثناءِ رِحلَتِهِ الشَّهيرَة. لكِنَّهُ لَم يَعتَنِ إطلاقًا بِالحَديثِ عَن طَوائفِ المُسلِمينَ مِن أَهلِ المَغرِبِ وَالأَندَلُس، مِنَ المُعاصِرينَ لَه، أَو حَتّى مِنَ السّابِقينَ عَلَيه، فَلا حَديثَ إطلاقًا فيما وَقَفنا عَلَيهِ مِن تُراثِه، عَن أَهَمِّ فِرقَةٍ ظَهَرَت في القَرنِ السّادِسِ الهِجرِيِّ في كُلِّ "دارِ الإسلام"، وَهِيَ "الطّائفَةُ المَهدِيَّة"، أَي فِرقَةُ المُوَحِّدين، وَعَن إمامِها المَهدِيِّ ابنِ تومَرت (515–524هـ/1121–1130م) وَقَد كانَ مُعاصِرًا لَها، بَلِ انتَهَت أَيّامُهُ هُوَ أَيّامَ إقبالِها وَعِزِّها، فَضلاً عَن أَن يَتَحَدَّثَ عَنِ المَسَرِّيَّةِ وَصوفِيَّةِ الأَندَلُس، أَو عَن إحراقِ كِتابِ "إحياءِ عُلومِ الدّين"، فَضلاً عَن أَن يَتَحَدَّثَ عَن نِحلَةِ بَرغَواطَة، أَو الشّيعَةِ البَجَلِيَّة.

فَهَل تَعَمَّدَ ابنُ العَرَبِيِّ السُّكوتَ عَنِ المُوَحِّدينَ وَمَذهَبِهِم خَوفًا مِنهُم، لَمّا كانوا بِحَسَبِ مَقاييسِهِ مِنَ "أَهلِ الإلحاد"؟ وَهَل تَعَمَّدَ ابنُ العَرَبِيِّ مُهاجَمَةَ المُوَحِّدينَ بِالسُّكوتِ عَنهُم، وَبِالاِلتِفافِ عَلى أَمرِهِم،

---

204 اَلعَواصِم: 208.

205 اَلعَواصِم: 249.

206 اَلمَسالِك: 405/3.

وَذلك بِمُهاجَمَةِ الفاطِمِيّينَ الإسماعيلِيّين، لِتَشابُهِ العَقيدَتَين: اَلفاطِمِيَّةِ الإسماعيلِيَّةِ وَالمُوَحِّدِيَّةِ في قَولِهِما بِالإمامَةِ وَالعِصمَةِ وَالزَّمَنِ الدَّورِيّ، وَأَشياءَ عَقَدِيَّةٍ أُخرى مُشتَرَكَة؟

لَقَد رَأَينا فيما سَبَقَ أَنَّهُ أَورَدَ تَصنيفًا لِفِرَقِ الزَّنادِقَة،[207] نَقلاً عَن أَبي عاصِمٍ النَّسائِيّ، (253هـ\867م) في كِتابِه: "اَلتَّأكيدِ في لُزومِ السُّنَّة"، فَقال: "وَمِنهُمُ الرّوحانِيَّة...وَسُمّوا أَيضًا "الفِكرِيَّة"، لِأَنَّهُم زَعَموا أَنَّهُم يَتَفَكَّرونَ في هذا حَتّى يَعرِفوا الله، فَجَعَلوا الفِكرَةَ غايَةَ العِبادَة، وَهُوَ مَذهَبُ الحَلاّجِ السّاحِرِ الكافِر..."[208]

فَهَل كانَ هذا التَّصنيفُ لِفِرَقِ الزَّنادِقَة، وَتَضمينُهُ لِلطّائفَةِ الرّوحانِيَّةِ وَالفِكرِيَّة، وَهُم صوفِيَة، وَرُبَّما كانوا مِن غُلاتِهِم، تَعريضًا بِالمُوَحِّدين، وَقَد نُسِبَ إِلى أَوائِلِهِم أَنَّهُم عَلى هذا المَذهَب؟

لَيسَ الجَوابُ عَلى هذا السُّؤالِ بِالأَمرِ السَّهل، فَالمَفروضُ أَن نَدرُسَ ما قالَ ابنُ العَرَبِيّ صَراحَة، وَأَن نَستَنتِجَ مِنه، لا ما لَم يَقُل صَراحَة. لاكِنَّ هذا الغِيابَ المُطلَقَ لِلواقِعِ التّاريخِيِّ "الإلحادِيّ" بِاصطِلاحِه، لِلمَغرِبِ وَالأَندَلُس، ما عَدا إشاراتِهِ الطَّفيفَةَ إِلى إباضِيَّةِ طَرابُلُس[209]، مُثيرٌ لِلانتِباهِ حَقًّا، وَيَدعو إِلى التَّأَمُّل، وَلا سِيَما أَنَّهُ كانَ يَكتُبُ كُتُبَهُ الأولى وَهُوَ في كَنَفِ المُرابِطينَ السُّنِّيّين، فَقَد كانَ في إمكانِهِ أَن يَخدِمَهُم بِالطَّعنِ في عَدُوِّهِم مِنَ المُوَحِّدينَ وَرَميِهِم بِالإلحاد، وَتَوسيعِ هذا المَفهومِ لِيَشمَلَهُم صَراحَة. لكِنَّهُ اختارَ السُّكوتَ المُطلَق، وَالقَفزَ عَلى التّاريخِ وَالواقِعِ الأَندَلُسِيِّ المَغرِبِيّ، إلاّ في النّادِر، وَفي هذا الأَمرِ دَلالَةٌ بَليغَة. فَيظهَرُ أَنَّهُ كانَ يَعلَمُ عِلمَ يَقينٍ أَنَّ مَصيرَ المَذهَبِ السُّنِّيِّ بَعدَ المُرابِطينَ إِلى سُقوطٍ أَو خُمول، عاجِلاً أَو آجِلا. وَلَم تَخلُ كِتاباتُهُ مِن تَسجيلِ امتِعاضِهِ مِن أُمَراءِ المُرابِطين، إذ قالَ مُتَوَجِّعا: "...أَم عَلى أَميرِهِ الَّذي لا يَرعى فيهِ إلاّ وَلا ذِمَّة؟..."[210]، وَإذ كَنّى عَنِ الأَميرِ سير بنِ أَبي بَكرٍ المُرابِطِيّ بِالأَميرِ الأَعجَمِيّ.[211]

وَلَرُبَّما كانَت مَخائِلُ سُقوطِهِم واضِحَةً عِندَه، وَمَخائِلُ انتِشارِ المَذهَبِ المُوَحِّدِيِّ واضِحَةً عِندَهُ كَذلك، كَما انتَشَرَ المَذهَبُ الفاطِمِيُّ مِن قَبل، فَاستَكثَرَ مِن تَرويجِ مَفهومِ "الإلحاد"، لِيَكونَ مَفهومًا واسِعًا شامِلاً لِكُلِّ مَن لا يَرضى عَنهُ فِكرُه، مِن طَوائِفِ الماضينَ وَالمُعاصِرينَ لَه، فَصَرَّحَ بِأَمرِ مَنِ استَطاع، كَالإسماعيلِيّينَ الفاطِمِيّين، وَسَكَتَ عَنِ المُوَحِّدين، لِيَضمَنَ عَدَمَ إحراقِ كُتُبِهِ في حَياتِهِ أَو بَعدَ مَوتِه.

---

207 اَلمَسالِك: 3/400–404.

208 اَلمَسالِك: 3/403.

209 اَلمَسالِك: 7/107.

210 سِراجُ المُريدين: 2/102.

211 سِراجُ المُريدين: 4/262.

وَمهما يَكُن، فَإِنَّ مُعاصِرَه ابنَ القَلانِسِيّ، (555هـ\1160م) وَهُوَ مُؤَرِّخُ دِمَشق، وَأَقدَمُ مَن كَتَبَ عَنِ المُوَحِّدينَ مِنَ المُؤَرِّخين، يَنقُلُ بَعدَ شِدَّةِ التَّحَرّي، فيما يَذكُر،[212] عَنِ ابنِ عَبدِ الجَبّارِ الصِّقِلِّيّ، مَصدَرِهِ الرئيس، أَنَّ ابنَ تومَرت، كانَ قَد سَبَقَ لَهُ أَن دَعا النّاسَ في مَدينَةِ المَهدِيَّةِ بِإِفريقِيَة، في عامِ 512هـ\1118م، إِلى مَذهَبِ "الفِكرَةِ" الصّوفِيّ. قال: "وَأَمَرَ أَهلَها أَن يَبنوا قَصرًا عَلى نِيَّةِ الفِكرَة، وَأَن يَعبُدوا اللّٰهَ فيهِ بِالفِكرَة"[213]، وَكادَ هاذا الأَمرُ أَن يَتِمّ.

وَمَهما يَكُن مِن ضَعفِ القَرائِنِ أَو قُوَّتِها في هذا الصَّدَد، فَلَيسَ مِنَ المُستبعَدِ في نَظَرِنا أَن يَكونَ نَقلُ ابنِ العَرَبِيّ لِهاذا التَّصنيفِ لِلزَّنادِقَةِ عَن أَبي عاصِمٍ النَّسائِيّ، (253هـ\867م) اَلمُتَضَمِّنِ لِمَذهَبِ "الفِكرَة"، أَمرًا مَقصودا، اَلغايَةُ مِنهُ التَّعريضُ بِالمُوَحِّدينَ وَمَذهَبِهِم، وَنِسبَتُهُم ضِمنًا إِلى الزَّندَقَة، بِناءً عَلى ما كانوا عَلَيهِ في بَدءِ أَمرِهِم، أَو نَسَبَهُ أَعداؤُهُم إِلَيهِم بِالحَقِّ أَو بِالباطِلِ في هذا البَدء.

فَهرَسُ المَصادِرِ وَالمَراجِع

أُصولُ السُّنَّة، لأَبي عَبدِ اللّٰهِ ابنِ أَبي زَمَنَين. (399هـ) دارُ الفُرقان. اَلقاهِرَة. 1428هـ\2007م.

أَحكامُ القُرءان، لِأَبي بَكرٍ ابنِ العَرَبِيّ. (543هـ) تَحقيقُ عَلِيٍّ مُحَمَّدٍ البِجاوِيّ. عيسى البابي الحَلَبِيّ وَشُرَكاه. اَلقاهِرَة. 1388 هـ 1968م. 4 أَجزاء.

اَلأَزهارُ الطَّيِّبَةُ النَّشر، فيما يَتَعَلَّقُ بِبَعضِ العُلومِ مِنَ المَبادِئِ العَشر، لِأَبي عَبدِ اللّه، مُحَمَّدٍ الطّالِبِ ابنِ الحاجِّ السُّلَمِيّ. (1273هـ) تَحقيقُ جَعفَرِ ابنِ الحاجِّ السُّلَمِيّ. مَنشوراتُ جَمعِيَةِ تِطّاوُنَ أَسمير. سِلسِلَةُ تُراث. 12. ج.1: تِطوان. 1431 هـ 2010م. ج.2. 1428 هـ 2007م. ج. 3. 1433 هـ 2012م.

تاريخُ أَبي يَعلى القَلانِسِيّ. (555هـ) (ذَيلُ تاريخِ دِمَشق) بِاعتِناءِ هـ ف أَمِدروز. مَطبَعَةُ الآباءِ اليَسوعِيّين. بَيروت. 1908م. أَعادَت نَشرَهُ مَكتَبَةُ المُثَنّى. بَغداد. د.ت.

اَلتَّبصيرُ في الدين، وَتَمييزِ الفِرقَةِ النّاجِيَةِ عَنِ الفِرَقِ الهالِكين، لِأَبي المُظَفَّرِ الأَسفَرايينِيّ. (471هـ) تَحقيقُ مَجيدٍ الخَليفَة. دارُ ابنِ حَزم. بَيروت. 1429 هـ 2008م.

تَرتيبُ المَدارِك، وَتَقريبُ المَسالِك، لِمَعرِفَةِ أَعلامِ مَذهَبِ مالِك، لِلقاضي عِياض. (544هـ) تحقيقُ مُحَمَّدِ ابنِ تاويتَ الطَّنجِيّ، وَعَبدِ القادِرِ الصَّحراوِيّ، وَمُحَمَّدِ ابنِ شَريفَة، وَسَعيدٍ أَعراب. مَنشوراتُ وِزارَةِ الأَوقافِ بِالرِّباط. 1383-1403هـ 1965-1983م. 8 أَجزاء.

212 تاريخُ ابنِ القَلانِسِيّ: 291.

213 تاريخُ ابنِ القَلانِسِيّ: 291.

اَلجَوهَرَةُ الفَريدَة، في حَلِّ رُموزِ الخَريدَة، لِأَبي عَبدِ الله، مُحَمَّدٍ المُحَدِّثِ ابنِ الحاجِّ السُّلَمِيِّ المِرداسِيِّ الفاسِيّ. (ـ1274هـ) تَحقيقُ جَعفَرِ ابنِ الحاجِّ السُّلَمِيّ. تَقديمُ مُحَمَّدٍ العَدلونِيِّ الإدريسِيّ. مَنشوراتُ جَمعِيَّةِ تِطّاوُنَ أَسمير. سِلسِلَةُ تُراث. 19. مَنشوراتُ الجَمعِيَّةِ المَغرِبِيَّةِ لِلدِّراساتِ الأَندَلُسِيَّة. تِطوان. 1439 هـ 2018م.

سِراجُ المُريدين، في سَبيلِ الدّين، لِأَبي بَكرٍ ابنِ العَرَبِيّ. (-542هـ) تَحقيقُ عَبدِ اللهِ التَّوراتِيّ. أَعلاقٌ أَندَلُسِيَّةٌ إِشبيلِيَّة. 4. سِلسِلَةُ مُؤَلَّفاتِ الإمامِ أَبي بَكرٍ ابنِ العَرَبِيّ. 4. دارُ الحَديثِ الكَتّانِيَّة. طَنجَة. بَيروت. 1438 هـ 2017 م. 6 أَجزاء.

سِيَرُ أَعلامِ النُّبَلاء، لِشَمسِ الدّينِ الذَهَبِيّ. (ـ748هـ) تَحقيقُ جَماعَةٍ مِنَ العُلَماء، بِإشرافِ شُعَيبٍ الأَرناءوط. مُؤَسَّسَةُ الرِّسالَة. بَيروت. 1419هـ 1998م. 28 جُزءا.

اَلعَواصِمُ مِنَ القَواصِم، لِأَبي بَكرٍ ابنِ العَرَبِيِّ المَعافِرِيِّ الإشبيلِيّ. (ـ543هـ) (اَلنَّصُّ الكامِل) تَحقيقُ عَمّارٍ الطّالِبِيّ. دارُ التُّراث. اَلقاهِرَة. 1417هـ 1997م.

اَلفَصل، في المِلَلِ وَالأَهواءِ وَالنِّحَل، لِابنِ حَزمٍ الأَندَلُسِيّ. (ـ456هـ) تَحقيقُ مُحَمَّدٍ إبراهيمَ نَصر، وَعَبدِ الرَّحمانِ عُمَيرَة. دارُ الجيل. بَيروت. 1416 هـ 1996م. 5 أَجزاء.

فِرَقُ الشّيعَة، لِأَبي مُحَمَّد، اَلحَسَنِ بنِ موسى النَّوبَختِيّ. (ـ310هـ) تَحقيقُ هـ. ريتر. بِدايات. جَبَلَة. 2007م.

قانونُ التَّأويل، لِأَبي بَكرٍ ابنِ العَرَبِيِّ المَعافِرِيِّ الإشبيلِيّ. (ـ543هـ) تَحقيقُ مُحَمَّدٍ السُّلَيمانِيّ. دارُ الغَربِ الإسلامِيّ. بَيروت. 1990م.

كِتابُ التَّنبيه، وَالرَّدِّ عَلى أَهلِ الأَهواءِ وَالبِدَع، لِأَبي الحُسَين، مُحَمَّدِ بنِ أَحمَدَ المَلَطِيّ. (377–هـ) تَحقيقُ س. دبدرينغ. اَلمَعهَدُ الأَلمانِيُّ لِلأَبحاثِ الشَّرقِيَّة. بَيروت. 1420هـ–2009م.

كِتابُ الحُدودِ في الأُصول، (اَلحُدودُ وَالمُواضَعات) لأَبي بَكر، مُحَمَّدِ بنِ الحَسَنِ ابنِ فورَكَ الأَصبَهانِيّ. (406–هـ) بِاعتِناءِ مُحَمَّدٍ السُّلَيمانِيّ. دارُ الغَربِ الإسلامِيّ. بَيروت. 1999م.

كِتابُ الخِصال، لأَبي بَكر، مُحَمَّدِ بنِ يَبقى ابنِ زَربٍ القُرطُبِيّ. (ـ381هـ) بِاعتِناءِ عَبدِ الحَميدِ العَلَمِيّ. مَنشوراتُ وِزارَةِ الأَوقاف. اَلرِّباط. 1426 هـ 2005م.

اَلكِتابُ المُتَوَسِّطُ في الاِعتِقاد، وَالرَّدُّ عَلى مَن خالَفَ السُّنَّةَ مِن ذَوي البِدَعِ وَالإلحاد، لِأَبي بَكر، مُحَمَّدِ بنِ عَبدِ اللهِ ابنِ العَرَبِيِّ المَعافِرِيِّ الإشبيلِيّ. (-542هـ) بِاعتِناءِ عَبدِ اللهِ التَّوراتِيّ. أَعلاقٌ أَندَلُسِيَّةٌ إِشبيلِيَّة. 2. سِلسِلَةُ مُؤَلَّفاتِ الإمامِ أَبي بَكرٍ ابنِ العَرَبِيّ. 2. دارُ الحَديثِ الكَتّانِيَّة. طَنجَة. بَيروت. 1436 هـ 2015م.

كِتابُ العَين، لِلخَليلِ بنِ أَحمَدَ الفَراهيدِيّ. (ـ170هـ) (مُرَتَّبًا عَلى حُروفِ المُعجَم) تَحقيقُ عَبدِ الحَميدِ هِنداوِيّ. دارُ الكُتُبِ العِلمِيَّة. بَيروت. 1424 هـ 2003م. 4 أَجزاء.

مَقالاتُ الإسلامِيّين، وَاختِلافُ المُصَلّين، لِأَبي الحَسَنِ الأَشعَرِيّ. (ـ324هـ) تَحقيقُ مُحَمَّدٍ مُحيي الدّينِ عَبدِ الحَميد. دونَ مَطبَعَة. د. م. 1405 هـ 1985م. جُزءان.

المَسالِك، في شَرحِ مُوَطَّإِ مالِك، لِلقاضي أبي بَكر، مُحَمَّدِ بنِ عَبدِ اللَّهِ ابنِ العَرَبِيِّ المَعافِرِيّ. (ـ543هـ) تَحقيق مُحَمَّدِ بنِ الحُسَينِ السُّلَيمانِيّ، وَعائِشَةَ بِنتِ الحُسَينِ السُّلَيمانِيّ. دارُ الغَربِ الإسلامِيّ. بَيروت. 1428 هـ 2007م. 8 أجزاء.

اَلمِلَلُ وَالنِّحَل، لِعَبدِ القاهِرِ البَغدادِيّ. (ـ429هـ) تَحقيقُ أَلبير نَصري نادِر. دارُ المَشرِق. بَيروت. 1970م.

اَلمِلَلُ وَالنِّحَل، لِأبي الفَتح، مُحَمَّدِ بنِ عَبدِ الكَريمِ الشَّهرِستانِيّ. (ـ548هـ) دارُ الفِكر. بَيروت. د.ت.

اَلنَّصائِحُ المُنجِيَة، مِنَ الفَضائحِ المُخزِيَة، وَالقَبائحِ المُردِيَة، مِن أَقوالِ أهلِ البِدَع، منَ الفِرَقِ الأَربَع، اَلمُعتَزِلَةِ وَالمُرجِئَةِ وَالخَوارِجِ وَالشِّيَع، لِأَبي مُحَمَّد، عَلِيِّ بنِ أَحمدَ بنِ سَعيدِ ابنِ حَزم. (-458هـ) تَحقيقُ الزُّبَير دَحّان. مَكتبَةُ سَلمى الثَّقافِيَّة. 139. تِطوان. 2018م.

نَوازِلُ ابنِ الحاجِّ التُّجيبِيّ، لِأَبي عَبدِ الله، مُحَمَّدِ بنِ أَحمَدَ ابنِ الحاجِّ التُّجيبِيِّ القُرطُبِيّ. (-529هـ) تَحقيقُ أَحمَدَ شُعَيبٍ اليوسُفِيّ. مَنشوراتُ الجَمعِيَّةِ المَغرِبِيَّةِ لِلدِّراساتِ الأَندلُسِيَّة. تِطوان. 1439 هـ 2018م. ٣ أَجزاء.

4

Ibn al-ʿArabī and al-Ṭurṭūshī

Are Their al-Ghazālī Critiques Related?

Kenneth Garden

Abū Bakr Ibn al-ʿArabī's (468–543/1076–1148) ten-year journey to the East to study religious sciences (485–495/1092–1102) is a topic he returned to again and again in his writings. It was the subject of an entire work, his lost *Tartīb al-riḥla*, and warranted extensive mention in numerous other works, especially his *Qānūn al-taʾwīl*. A major focus of his travel accounts was a catalogue of all the religious scholars he studied with. The two most consequential of these were Abū Ḥāmid al-Ghazālī (448–505/1056–1111)[1] and Abū Bakr al-Ṭurṭūshī (451–520/1059–1126). Al-Ghazālī gave him one-on-one instruction in his private quarters in Baghdad in 490/1097, and Ibn al-ʿArabī called attention to their connection repeatedly in his writings for the rest of his life. He met al-Ṭurṭūshī twice: in Jerusalem, where the two took vows to live as ascetics, and again in Alexandria, where Ibn al-ʿArabī stayed with al-Ṭurṭūshī during and after his father's death. It seems certain that the two of them discussed al-Ghazālī in Alexandria, and both later wrote critically about al-Ghazālī, although al-Ṭurṭūshī did so much more extensively, uncompromisingly, and scathingly than Ibn al-ʿArabī. This paper will explore these connections and ask whether Ibn al-ʿArabī and al-Ṭurṭūshī may have influenced one another's readings of al-Ghazālī.

1 Ibn al-ʿArabī and al-Ṭurṭūshī Meet

Ibn al-ʿArabī set out on his celebrated journey to the Mashriq (Egypt, the Levant, Iraq, and the Hijaz) in Rabīʿ I 485/April 1092 with his father ʿAbd Allāh (435–493/1043–1099). Wherever they went, Abū Bakr studied with local

1 Frank Griffel has arrived at a new date for al-Ghazālī's birth based on information found in his Persian letters. He finds 448/1056 to be the most likely date, though 447 or even 446 are also possible. In any case, the date is at least two years earlier than the commonly cited 450/1058 found in Ibn al-Jawzī. Griffel, Frank, *Al-Ghazālī's philosophical theology*, New York: Oxford University Press, 2009, 23–25.

 | DOI:10.1163/9789004706859_006

scholars, and ʿAbd Allāh engaged in commerce to finance their journey. But they had a more concrete agenda of going ultimately to Baghdad to request from the Abbasid caliph a certificate recognising the Almoravid sultan, Yūsuf b. Tāshfīn (d. 500/1106) as the caliph's deputy in the Maghrib and al-Andalus (the certificate was known as an *ʿaqd al-wilāya*).[2] ʿAbd Allāh had been a minister (*wazīr*) in the regional (*ṭāʾifa*) kingdom of Seville and had been stripped of his family property by the Almoravids when they conquered Seville. Bringing back a caliphal document which Yūsuf b. Tāshfīn wanted was an opportunity for him to ingratiate himself with the new regime and possibly regain his status and property. For his part, Abū Bakr set out to become a religious scholar, a group that had gained new status and influence under the Almoravids.[3]

They travelled via Bijāya, Tripolitania, and Alexandria to Jerusalem in 486/1093, where Ibn al-ʿArabī's studies began in earnest, in no small part because of his encounter with al-Ṭurṭūshī. It was here, Abū Bakr tells us, that "the full moon of knowledge" appeared to him.[4] At his insistence his father cancelled plans to go on hajj, and they spent the next three years in and around Jerusalem while he studied. Ibn al-ʿArabī sought out numerous scholars, attending eighteen study circles and two madrasas, one Shāfiʿī and one Ḥanafī.[5] Of all who had come to Jerusalem for study, he was most impressed with the learning of the Khorasanians.[6] The closest scholarly bond he formed, though, was with his fellow Andalusī, Abū Bakr al-Ṭurṭūshī.

It seems that he had been advised to look for al-Ṭurṭūshī, for Ibn al-ʿArabī writes that he and his father searched for him first in a particular spot in the Masjid al-ʿAqsā called al-Ghuwayra and then found him in another spot called al-Sikīna. Once ʿAbd Allāh had explained his son's intention and al-Ṭurṭūshī had seen Abū Bakr's seriousness, he agreed to teach him and opened to him "the doors to knowledge".[7] A bond of spiritual affinity formed between the two, and they took a mutual oath to draw their happiness from religious science and to live as ascetics.[8] In 488/1095, after more than two years in and around

2 Many scholars have given their opinion on whether or not this mission on behalf of Yūsuf b. Tāshfīn was official or not. I have made a case for my conclusion (likely official), in my article Garden, Kenneth, "The *riḥla* and self-reinvention of Abū Bakr ibn al-ʿArabī", in *Journal of the American Oriental Society* 135 (2015), 1–17, at pp. 10–12.

3 For this facet of the journey, see Garden, "The *riḥla* and self-reinvention" 5–6.

4 *Fa-lāḥa lī badr al-maʿrifa*. Ibn al-ʿArabī, Abū Bakr, *Qānūn al-taʾwīl*, ed. Muḥammad Slīmānī, Beirut: Dar al-Gharb al-Islami, 1990, 91.

5 Ṭālibī, ʿAmmār, *Ārāʾ Abī Bakr ibn al-ʿArabī al-kalāmiyya*, 2 vols., Algiers: al-Sharika al-Waṭaniyya li-l-nashr wa-l-tawzīʿ, 1974, i, 35, quoted from *al-ʿAwāṣim min al-qawāṣim*.

6 Ibn al-ʿArabī, *Qānūn al-taʾwīl* 97–98.

7 Ibn al-ʿArabī, *Qānūn al-taʾwīl* 92–93.

8 Ṭālibī, *Ārāʾ Abī Bakr* 58, drawn from *Aḥkām al-Qurʾān*.

Jerusalem, Abū Bakr realised it was time to leave and go to Baghdad.[9] Shortly thereafter, al-Ṭurṭūshī also left Jerusalem, in his case to practice asceticism in the mountains of Lebanon.[10]

2 Meeting between Ibn al-ʿArabī and al-Ghazālī

Abū Bakr and his father arrived in Baghdad in 489/1096 and, a short time later, went on hajj, where Abū Bakr glimpsed al-Ghazālī.[11] He met the great scholar in Jumāda II 490/May–June 1097,[12] shortly after al-Ghazālī returned to Baghdad from his travels through the Levant after his famous crisis described in his *al-Munqidh min al-ḍalāl*. From one of al-Ghazālī's Persian letters, we know that al-Ghazālī had returned to his hometown of Ṭūs by Dhū al-Ḥijja 490/November 1097.[13] Given the long overland journey from Iraq to Khorasan, Ibn al-ʿArabī could only have studied with al-Ghazālī for a couple of months before his departure for the East.

Nonetheless, Ibn al-ʿArabī describes their studies together in the most glowing of terms. Of his introduction to al-Ghazālī he writes, "I went to him and presented him with my aspirations. I said to him, 'You are the one I have been searching for, my *imām* to whom I turn for guidance.' Ours was a meeting of knowledge and through him I saw that which was beyond description."[14] In this short time, he received a thorough grounding in al-Ghazālī's thought. He writes, "He met exclusively with me for reasons I have made clear in the book *Tartīb al-riḥla*. I studied all of his books with him, and heard (*samiʿtu*) his book, which he titled *al-Iḥyāʾ li-ʿulūm al-dīn* [sic]."[15] In later writings, Ibn al-ʿArabī

9 Ibn al-ʿArabī, *Qānūn al-taʾwīl* 102–3.

10 Al-Ṭurṭūshī, Abū Bakr, *Kitāb al-Ḥawādiṯ wa-l-bidaʿ* (*el Libro de las novedades y las innovaciones*), trans. and ed. Maribel Fierro, Madrid: Consejo Superior de Investigaciones Cientificas, 1993, 42.

11 Griffel, *Al-Ghazālī's philosophical theology* 64.

12 Ibn al-ʿArabī, Abū Bakr, *al-ʿAwāṣim min al-qawāṣim*, ed. ʿAmmār Ṭālibī, Cairo: Maktabat Dār al-Turāth, 1997, 24.

13 Griffel, *Al-Ghazālī's philosophical theology* 49.

14 Ibn al-ʿArabī, *Qānūn al-taʾwīl* 111.

15 Ibn al-ʿArabī, *al-ʿAwāṣim* 24. As an aside, I have argued elsewhere that the distinction he draws here between studying al-Ghazālī's books and "hearing from" the *Iḥyāʾ* strongly suggests that al-Ghazālī had not finished writing that book at this point and completed it only after his return to Khorasan. Garden, Kenneth, *The first Islamic reviver: Abū Ḥāmid al-Ghazālī and his Revival of the Religious Sciences*, New York: Oxford University Press, 2013, 125 and 203n3.

frequently referred to al-Ghazālī proudly by the Persian word for professor, *dānishmand*, which must have carried a certain exotic prestige for his Andalusī audience. In fact, it seems that al-Ghazālī encouraged Ibn al-ʿArabī and other students in Baghdad to call him *dānishmand*. Al-Ghazālī honoured a friend and fellow student of his teacher al-Juwaynī, a companion in his travels to Iraq and the Levant, and his co-teacher in at least some of his classes, one Abū l-Qāsim b. ʿAbd al-Malik al-Ḥākimī al-Ṭūsī, by calling him the "greater *dānishmand*" (*dānishmand al-akbar*), while referring to himself as the "lesser *dānishmand*" (*dānishmand al-aṣghar*).[16]

The admiration between al-Ghazālī and Ibn al-ʿArabī seems to have been mutual. In a letter to the Almoravid sultan Yūsuf b. Tāshfīn that he sent back to the Maghrib with Ibn al-ʿArabī, al-Ghazālī wrote,

> The shaykh and *imām* Abū Bakr achieved a quantity of knowledge in the course of his frequent visits to me that others do not achieve in all their lives. This is because of the golden discernment and clever sensibility he possesses and the fire of his genius (*ittiqād al-qarīḥa*). When he leaves Iraq, he will certainly be capable of undertaking independent legal reasoning, distinguished among his peers.[17]

The letter that carried these words accompanied a fatwa al-Ghazālī wrote giving his sanction to the Almoravid's overthrow in 484/1091 of the "regional kings" (*mulūk al-ṭawāʾif*) who had ruled al-Andalus, one of whom, as we have seen, ʿAbd Allāh had served as a minister in Seville. So al-Ghazālī contributed to ʿAbd Allāh Ibn al-ʿArabī's efforts to ingratiate himself and his son with the new regime upon his return to al-Andalus.[18]

---

16 Aʿrāb, Saʿīd, *Maʿ al-Qāḍī Abī Bakr ibn al-ʿArabī*, Beirut: Dār al-Gharb al-Islāmī, 1987, 46. This is the second case of evidence of a companion who accompanied al-Ghazālī on his travels in the Levant and Hijaz after his famous departure from Baghdad in 488/1095. The other is a religious scholar originally from Jurjan, Ibrahīm al-Sabbāk (d. 513/1119), who al-Ghazālī mentions in two of his Persian letters. While al-Ghazālī's flight from Baghdad is often imagined as a solitary spiritual quest, we can say now that he was accompanied on these travels by an entourage of at least two. On Ibrahīm al-Sabbāk, see Garden, *The first Islamic reviver* 134–135.

17 Ibn al-ʿArabī, Abū Bakr, *Shawāhid al-jilla*, in Muḥammad Yaʿlà, *Tres textos ārabes sobre Beréberes en el occidente Islámico*, Fuentes Arábico-Hispanas, 20, Madrid: Consejo Superior de Investigaciones Cientificas, Agencia Española de Cooperacion Internacional, Madrid, 1996, 312.

18 For a more extensive discussion of this, see Garden, "The *riḥla* and self-reinvention" 13–14.

While al-Ghazālī departed from Baghdad at some point in the summer of 490/1097, Ibn al-ʿArabī and his father stayed until the end of 492/1098, spending over two years in Baghdad. They departed with the caliphal certificate that Yūsuf b. Tāshfīn wanted and began to make their way back to the west. If we accept al-Ghazālī's assessment of Ibn al-ʿArabī—that he would be a legal expert in his own right by the time he left Baghdad—then he had also achieved the goal of his journey. Six years later, it was time to go home.

3 Meeting between al-Ṭurṭūshī and al-Ghazālī

Abū Bakr al-Ṭurṭūshī also claims to have met al-Ghazālī. Our source for this is also our most complete and extensive source for his critique of al-Ghazālī, as we will see below. In this letter to one Ibn al-Muẓaffar, likely written in or around 503/1109, al-Ṭurṭūshī writes of having met al-Ghazālī: "I saw the man and spoke with him and found him to be an (honourable) man of the people of knowledge and I encouraged in him his virtues (*qad nahaḍtu bihi faḍāʾilahu*) and he gathered in himself reason and understanding and the practice of the sciences throughout his life and so he was for most of his days."[19] The meeting must have occurred in Jerusalem. In 489/1096, al-Ṭurṭūshī was still living in Jerusalem prior to leaving for the mountains of Lebanon. Al-Ghazālī arrived in Jerusalem in the late spring or early summer of 489/1096 and departed in the autumn of the same year.[20]

It is striking to compare al-Ṭurṭūshī's attitude toward al-Ghazālī with that of Ibn al-ʿArabī. While the younger Ibn al-ʿArabī held al-Ghazālī in reverence as the most brilliant and famous of his teachers, al-Ṭurṭūshī clearly saw him as an equal or even an inferior. It is he, al-Ṭurṭūshī, who presumes to encourage al-Ghazālī in his cultivation of the virtues, to approve of his scholarly achievement, until he later withdraws his approval, as we shall see.

---

19 See al-Wansharīsī, Aḥmad. *Al-Miʿyār al-muʿrib wa-l-jāmiʿ al-mughrib ʿan fatāwa ʿulamāʾ ifrīqiya wa-l-Andalus wa-l-Maghrib*, ed. Muḥammad Ḥajjī, 13 vols, Rabat: Wizāra al-Awqāf wa-l-Shu'ūn al-Mamlaka al-Maghribiyya, 1981, xii, 186. For a more critical edition of the letter with a very illuminating study of the context of the letter and al-Ṭurṭūshī, see Ghrāb, Saʿd, "Ḥawl iḥrāq al-murābiṭīn li-Iḥyāʾ al-Ghazālī", in *Actas del IV coloquio Hispano-Tunecino* (*Palma de Mallorca, 1979*), Madrid: Instituto Hispano-Arabe de Cultura, 1983, 133–163.

20 For a detailed account of al-Ghazālī's time in Jerusalem, see Griffel, *Al-Ghazālī's philosophical theology* 45–47.

4 The Second Meeting between Ibn al-ʿArabī and al-Ṭurṭūshī in 493/1099

In 493/1099, Abū Bakr and his father ʿAbd Allāh arrived in Alexandria. There they met al-Ṭurṭūshī again. While al-Ṭurṭūshī had come to Egypt to live on the coast as an ascetic, he was convinced to come to Alexandria to serve as a Mālikī *faqīh*. When he met him again, Ibn al-ʿArabī was disappointed to find him sleeping in a bed and eating a dish called *malawwaf*, which he saw as a violation of their oath together to live as ascetics. He asked him, "Was this what we pledged to do?" Al-Ṭurṭūshī replied, "I didn't request these things, but when they came to me, I accepted them."[21] Later, he married a wealthy widow and turned the lower level of her house into a madrasa.[22]

It was during this stay in Alexandria that Ibn al-ʿArabī's father ʿAbd Allāh became ill, and died in Muḥarram 493 (November–December 1099). To have his father and travelling companion die on the brink of return, the end of a long exile in sight, was no doubt a cruel blow to Abū Bakr. The father and mentor of his youth had become a comrade in the course of their travels together, someone of whom he could write that he was a "father in rank, but a brother in companionship" (*ab fī l-rutba wa-akh fī l-ṣuḥba*).[23]

When the time came to depart, Ibn al-ʿArabī tried to convince al-Ṭurṭūshī to come with him back to al-Andalus, telling him that Fatimid Egypt was a centre of blameworthy innovation (*bidʿa*). Al-Ṭurṭūshī rejected this proposal:

> As for the land of the Maghrib, all of them are on a single path, ignorance has taken possession of them, scholarly imitation has spread among them, and they have abstained from rationalism (*zahadū fī l-naẓar*). An Umayyad disposition and a traditional culture (*sīrat umawiyya wa-nasha taqlīdiyya*) has hindered them from possessing this and if you were to live among them, you would live as one lost along with them.[24]

The two of them discussed Ibn al-ʿArabī's staying on in Alexandria, but Ibn al-ʿArabī reluctantly decided to go back to al-Andalus, since returning to his mother—whose only child he was—was a religious duty. His stay there does seem to have been a long one. He did not arrive in Tunis until Dhū

21 Quoted in Ṭālibī, *Ārāʾ Abī Bakr* 58. The exchange was found in a manuscript of *Sirāj al-muridīn*.

22 Al-Ṭurṭushī, *Kitāb al-Ḥawādiṯ* 52–53.

23 Ibn al-ʿArabī, *Shawāhid al-jilla* 277.

24 Ṭālibī, *Ārāʾ Abī Bakr* 59, from *Sirāj al-muridīn*.

al-Ḥijja 494/September–October 1101, two years after his arrival in Alexandria. Presumably, he spent this time in al-Ṭurṭūshī's company.

Did the two of them discuss al-Ghazālī? We can assume Ibn al-ʿArabī shared his experiences in Baghdad with al-Ṭurṭūshī, including his study with al-Ghazālī, and he would likely have showed him the books he was bringing back with him. In his *Sirāj al-muridīn*, Ibn al-ʿArabī lists a number of al-Ghazālī's books that he brought back from his travels: *Miḥakk al-naẓar fī l-manṭiq*, *Miʿyār al-ʿilm fī fann al-manṭiq*, *Tahāfut al-falāsifa*, and *al-Iqtiṣād fī l-iʿtiqād*.[25] Surely, he must have shown them to al-Ṭurṭūshī. What seems most suggestive of such a discussion is the fact that al-Ṭurṭūshī contributed a letter of his own to Yūsuf b. Tāshfīn like the one which al-Ghazālī had written.[26] Whether Ibn al-ʿArabī solicited it, or whether al-Ṭurṭūshī volunteered, both must have been aware that it was one of two letters from religious scholars that Ibn al-ʿArabī bore to the Almoravid sultan and that the other had come from al-Ghazālī. It is hard to see how they could not have discussed him on the occasion.

5 The Context of Critiques of al-Ghazālī

Ibn al-ʿArabī and al-Ṭurṭūshī are often discussed in the context of the controversy over al-Ghazālī in al-Andalus and the Maghrib. It is well known that al-Ghazālī's *Iḥyāʾ ʿulūm al-dīn* was burned in Cordoba in 503/1109, and several related events, some real, some legendary, are found in the sources about the burning.[27]

Less well known is the fact that al-Ghazālī himself had previously faced a controversy over the *Iḥyāʾ* and other writings of his when he returned to teaching in Nishapur in 499/1106, about which we know from al-Ghazālī's letters in Persian.[28] A key text in the controversy is the allegory of God's action in

25 The list is taken from an excerpt of *Sirāj al-muridīn* published at the end of ʿAmmār Ṭālibī's edition of Ibn al-ʿArabī, *al-ʿAwāṣim* 377–379.

26 Ibn al-ʿArabī, *Shawāhid al-jilla*.

27 For the most detailed account of the burning of the *Iḥyāʾ* in al-Andalus, see Fierro, Maribel, "Opposition to Sufism in al-Andalus," in Frederick de Jong and Bernd Radke (eds.), *Islamic mysticism contested: Thirteen centuries of controversies and polemics*, Leiden: Brill, 1999, 174–206.

28 Al-Ghazālī, Abū Ḥāmid, *Makātib-i fārsī-i Ghazzālī ba-nām-i Faḍāʾil al-anām min rasāʾil Ḥujjat al-Islām*, ed. ʿAbbās Iqbāl, Tehran: Kitābforūshī Ibn Sīnā, 1373/1954. For a German translation and study of the letters, see Krawulsky, Dorothea, *Briefe und Reden des Abū Ḥāmid al-Ghazālī: Übersetzt und Erläutert*, Islamkundliche Untersuchungen, Freiburg im Breisgau: Klaus Schwarz Verlag, 1971. For my analysis of these letters and reconstruction of the controversy described here, see Garden, *The first Islamic reviver* 124–168 and Garden,

the world in terms of a pen writing on paper found in Book 35 of the *Iḥyāʾ*, *Kitāb al-Tawḥīd wa-l-tawakkul* (*The Book of Divine Unity and Trust in Divine Providence*). We know this from the response al-Ghazālī wrote to his critics, *al-Imlāʾ fī ishkālāt al-iḥyāʾ*, (*The Composition on the Critiques of the Revival*), which consists of a response to eleven criticisms of this passage of *Kitāb al-Tawḥīd*. The best known of these is criticism number ten, al-Ghazālī's claim that God has made this the best of all possible worlds (summarised by those who came after with the phrase *laysa fī l-imkān abdaʿ mimmā kān*, "it is not in the realm of possibility that anything should exist that is more wonderful than what is"). Also important to us here are critiques two and seven. Two is a critique of al-Ghazālī's claim that "revealing the secret of the Divine Lordship is unbelief" (*ifshāʾ sirr al-rububiyya kufr*). Point number seven questions how the protagonist in al-Ghazālī's allegory is able to hear the voice of God when he is not a prophet. In his discussion of the second charge, al-Ghazālī makes an off-hand rejection of the doctrine of *iktisāb al-nubuwwa*, the acquisition of prophethood.[29]

Further detail on the critiques of al-Ghazālī by his enemies come from the compiler of al-Ghazālī's collected letters, who writes that his enemies charged that al-Ghazālī "did not have any belief whatsoever in Islam, but rather that he held the creed of the philosophers and the heretics (*iʿtiqād-i falāsifa va mulḥidān*) and he mixed all of his books with their words. … He called God the true light and this is the belief of the Zoroastrians (*madhhab-i majūs*), who speak of light and darkness." The specific meaning of the word *mulḥid* in this period in the Islamic East was Ismāʿīlī.[30] These charges seem very odd when made against the author of *Tahāfut al-falāsifa* and several refutations of the Ismāʿīlīs (or *bāṭiniyya*, as al-Ghazālī calls them). Recent scholarship has shown that al-Ghazālī's writings bear a strong influence of philosophy, especially Ibn Sīnā.[31] The charge of Zoroastrian influence seems to be a reference

---

Kenneth, "Al-Māzarī al-Dhakī: al-Ghazālī's Maghribi adversary in Nishapur," in *Journal of Islamic Studies* 21 (2010), 89–107.

29 For my list of the critiques found in *al-Imlāʾ fī ishkālāt al-iḥyāʾ*, see my *The first Islamic reviver* 149–152. For the *Imlāʾ*, see *al-Imlāʾ fī ishkālāt al-iḥyāʾ*, in al-Ghazālī, Abū Ḥāmid, *Iḥyāʾ ʿulūm al-dīn*, 5 vols., Cairo: Muʾassasat al-Ḥalabī wa-Shurakāʾuhu, 1387/1967–68, v, 18–58.

30 For this quote from the compiler of al-Ghazālī's letters, see *Makātib-i fārsī-i* 3. For my discussion of this quote and argument that the Persian word *mulḥidān* refers to Ismāʿīlīs, see Garden, *The first Islamic reviver* 153–154.

31 For important studies of al-Ghazālī and his complicated engagement with the philosophical tradition, and especially Ibn Sīnā, see Griffel, *Al-Ghazālī's philosophical theology*; Treiger, Alexander, *Inspired knowledge in Islamic thought: Al-Ghazālī's theory of mystical cognition and its Avicennian foundation*, New York: Routledge, 2012; Janssens,

to al-Ghazālī's discussion of light and darkness in *Mishkāt al-anwār wa-miṣfāt al-asrār*. A Maghribī involved in the controversy in Nishapur, Abū ʿAbd Allāh al-Māzarī al-Dhakī, took to referring to al-Ghazālī as "al-Ghazālī, the Zoroastrian Ṭūsī cow" (*al-Ghazālī al-majūsī al-baqarṭūsī*).[32] As we will see below, we know that news of al-Māzarī al-Dhakī and his role in the campaign against al-Ghazālī filtered back to the Maghrib; ʿIyāḍ b. Mūsā's biography of him in his *Tartīb al-madārik* refers to it.[33] As we will see below, both al-Ṭurṭūshī and Ibn al-ʿArabī refer unmistakably to critiques raised in this controversy as well.

6 Al-Ṭurṭūshī's Critique of al-Ghazālī

The above cited letter from al-Ṭurṭūshī was almost certainly a response to the burning of the *Iḥyāʾ* in al-Andalus in 503/1109. It is the most extensive surviving critique of al-Ghazālī and the *Iḥyāʾ* of anyone involved in the controversy. But as we will see, al-Ṭurṭūshī drew much of his critique from accounts that must have reached him directly or indirectly from Nishapur.

The letter is al-Ṭurṭūshī's response to one Ibn al-Muẓaffar[34] who wrote to ask his opinion on al-Ghazālī and his *Iḥyāʾ*. As we have seen, al-Ṭurṭūshī's initial impression of al-Ghazālī was favorable. But later, he writes, al-Ghazālī departed the sciences and became a Sufi, though he did not understand Sufism

---

Jules, "Al-Ghazzālī's *Tahāfut*: It it really a rejection of Ibn Sīnā's philosophy?" in *Journal of Islamic Studies* 12 (2001), 1–17; and al-Akiti, Afifi, "The good, the bad, and the ugly of falsafa: Al-Ghazālī's *Maḍnūn*, *Tahāfut*, and *Maqāṣid*, with particular attention to their *falsafī* treatments of God's knowledge of temporal events," in Y Tzvi Langerman (ed.), *Avicenna and his legacy: A golden age of science and philosophy*, Turnhout, Belgium: Brepols, 2009, 51–100.

32 Garden, "Al-Māzarī al-Dhakī" 103.

33 Al-Qāḍī ʿIyāḍ, Abū l-Faḍl b. Mūsā b. ʿIyāḍ al-Sabtī, *Tartīb al-madārik wa taqrīb al-masālik fī maʿrifat aʿlām madhhab Mālik*, ed. Saʿīd Aḥmad Aʿrāb et al, 8 vols., Rabat: Wizārat al-Awqāf wa-l-Shuʾūn al-Islāmiyya, 1384–1403/1965–1983, viii, 103.

34 Saʿd Ghrāb found one Ibn Muẓaffar, who in many respects fits the profile of the recipient of the letter. He was Abū al-Ḥakam ʿAbd Allāh b. Muẓaffar b. ʿAbd Allāh b. Muḥammad al-Yāhilī al-Marīnī al-Andalusī al-Maghribī, an Andalusī of an age to plausibly have been al-Ṭurṭūshī's correspondent, and he travelled to the East and studied in Alexandria during al-Ṭurṭūshī's lifetime. He distinguished himself, however, as a doctor, and had no recorded study of or interest in *fiqh* or other religious sciences. The point that decisively disqualifies him as the recipient of the letter is that he is described as a debauched poet (*shāʿir khalīʿ*), a description seemingly confirmed by the attribution to him of a book entitled *Nahj al-waḍāʿa li-uwalī al-khilāʿa* (*The Humble Method of the Men of Debauchery*). He is therefore unlikely to have been a correspondent of the ascetic al-Ṭurṭūshī.

and had not experienced its divinely bestowed knowledge (*maʿrifa*). He was neither worthy to be considered among the scholars nor was he firmly established in the states of the ascetics. He adulterated Sufism with the aims of the philosophers and the symbols of the controversial and martyred Sufi al-Ḥusayn ibn Manṣūr al-Ḥallāj. He criticised the *fuqahāʾ* and the *mutakallimūn* and abandoned religion. When he wrote *Iḥyāʾ ʿulūm al-dīn*, he was insincere in his treatment of Sufism and filled his book with more lies about the Prophet than any other book on earth, which is to say that he cited weak *ḥadīth*, a common and justified critique of the *Iḥyāʾ*.[35]

The charge of philosophical borrowing already suggests that al-Ṭurṭūshī had read or heard the critiques of al-Ghazālī's Nishapurian critics. Following this, though, are two critiques that leave no doubt of this connection. The first is the charge that al-Ghazālī claimed that prophecy is an acquisition (*al-nubuwwa iktisāb*), which a prophet achieves through his own efforts at ethical self-mastery, rather than prophecy being bestowed exclusively by God upon those prophets He sends to humankind. Al-Ṭurṭūshī attributes this doctrine first to the Ikhwān al-Ṣafāʾ and then to the philosophers, and then charges that, "anyone who tries to fortify the religion of Islam with the methods of the philosophers and views of the logicians is like someone who would purify water with urine."[36] The acquisition of prophecy (*iktisāb al-nubuwwa*) is the seventh charge al-Ghazālī rebuts in the *Imlāʾ*, as we have seen above.

Al-Ṭurṭūshī charges that, to further seduce his reader, al-Ghazālī holds out the promise of esoteric knowledge, saying that the matter at hand is the science of praxis (*ʿilm al-muʿāmala*) and that behind this lies the science of unveiling (*ʿilm al-mukāshafa*), and this is an accurate account of these two disciplines al-Ghazālī creates in the *Iḥyāʾ*, these being subdisciplines of his science of the hereafter (*ʿilm al-ākhira*). Al-Ṭurṭushī claims, "[Al-Ghazālī tells his reader], 'This is not permitted to be written in books', or he says, 'this is the secret of destiny (*sirr al-qadar*), which we have been forbidden to reveal (*alladhī nuhaynā ʿan ifshāʾihi*)', and this is what the *bāṭiniyya* do."[37] The formulation here is very close to the charge that revealing the secret of the Lordship is unbelief (*ifshāʾ sirr al-rubūbiyya kufr*), the second charge al-Ghazālī rebuts in the *Imlāʾ* as discussed above. That two such specific charges levelled against al-Ghazālī's allegory of the pens in Book 35 of the *Iḥyāʾ* should be echoed in al-Turṭūshī's letter leaves no doubt that details of the campaign in Nishapur had reached him in Alexandria.

35 Ghrāb, "Ḥawl iḥrāq" 159–160.
36 Ghrāb, "Ḥawl iḥrāq" 161.
37 Ghrāb, "Ḥawl iḥrāq" 161.

How would al-Ṭurṭūshī have known about the critiques of al-Ghazālī in Nishapur? A likely conduit was Abū ʿAbd Allāh al-Māzarī al-Dhakī, the scholar from Mahdiyya who participated in the campaign against al-Ghazālī in Nishapur as discussed above. We know that word of his efforts against al-Ghazālī were relayed to Mālikīs in the Maghrib. The Maghribī ʿIyāḍ b. Mūsā (d. 544/1149) ends his entry on him in his biographical dictionary of Mālikī jurists, *Tartīb al-madārik*, by writing, "He died in Isfahan after 500, after he became involved in conflicts that year over the claims of al-Ghazālī, for he was one of those who opposed him there. But [al-Ghazālī] was protected from them and they did not reach them."[38] So although al-Māzarī al-Dhakī's exact death date was not known to ʿIyāḍ and his fellow Westerners, his participation in the campaign against al-Ghazālī in the East was. Doctrinal details of the controversy likely accompanied this news. Whether al-Māzarī al-Dhakī was the conduit or not, the fact that the critiques generated by the Nishapurian controversy spread to Alexandria and then to the Maghrib seems indisputable.

Despite this harsh critique of al-Ghazālī and his *Iḥyāʾ*, al-Ṭurṭūshī ends his letter by declaring his intention to write a revised version of the *Iḥyāʾ*, purging it of its errors and heresies, writing, "I have resolved to devote myself to removing all of the book's mistakes and errors and to clarify it letter by letter."[39]

Al-Ṭurṭūshī did write his amended version of the *Iḥyāʾ*. In a 1988 article, Muḥammad al-Manūnī reports that the first volume of a book by al-Ṭurṭūshī entitled *Kitāb al-Asrār wa-l-ʿibar* was found in a private library in Marrakech in 1983 that is almost certainly the reworking of the *Iḥyāʾ* which al-Ṭurṭushī announces in the letter.[40] In an appendix to the article, al-Manūnī reproduces three printed pages of what he describes as " … the most important passages of the introduction, in which the author defines the aims of the book."[41]

In these passages, al-Ṭurṭūshī presents both his own objectives and al-Ghazālī's in a different light than he does in his letter. He writes of being inspired to write a book on "piety, its reality, and its divisions (*taqwā wa-ḥaqīqatuhu wa-aqsāmuhu*), and whether it consists of actions of the heart

38 Al-Qāḍī ʿIyāḍ, *Tartīb al-madārik* viii, 103.

39 Ghrāb, "Ḥawl iḥrāq" 162.

40 Al-Manūnī, Muḥammad, "Iḥyāʾ ʿulūm al-dīn fī manẓūr al-gharb al-islāmī ayām al-murābiṭīn wa-l-muwaḥḥidīn", in *Abū Ḥāmid al-Ghazālī: dirāsāt fī fikrihi wa-ʿaṣrihi wa-taʾthīrihi*, Rabat: Jāmiʿat Muhammad al-Khāmis, Nasharāt Kuliyyat al-Ādāb wa-l-ʿUlūm al-Insāniyya bi-Rabāṭ, Silsilat Nadawāt wa Munāẓarāt, 1988, 125–137. Al-Manūnī describes the single, aged manuscript as containing 177 folios, 185 mm by 129 mm, with varying numbers of lines on the page in a Mashriqī script. I was unable to learn anything about the current whereabouts of this manuscript, which was privately held at the time of the publication of the article.

41 Al-Manūnī, "Iḥyāʾ ʿulūm al-dīn". For the appendix, see 135–137.

or the limbs or both" by an encounter with a pious old man in al-Masjid al-Aqṣā in Jerusalem. His choice of al-Ghazālī's book as a model was motived by the book's virtues rather than its vices. He writes that there are countless books on the subject, and he lists seven of them, mostly books considered to be on the subject of Sufism, ending with the *Iḥyāʾ*, whose author he describes as the most knowledgeable of them and the strongest authority (*aqwāhum ḥujjatan*).[42]

But the book has five flaws which al-Ṭarṭūshī dislikes. The first of his objections is the proliferation of weak *ḥadīth*s in the *Iḥyāʾ*, a very common critique of the book, which we saw al-Ṭurṭūshī make in his letter. The second is that when al-Ghazālī classified the sciences, he objected to the science of divine unity (i.e. *kalām*), which is a calamity, because it is the foundation of all of the sciences. The third is that he relates unspecified material from the Ikhwān al-Ṣafāʾ, whom he describes as heretics (*kuffār*) and *bāṭiniyya*. The fourth is his permissiveness (*ibāḥa*) regarding singing and listening to music. The fifth is his mention in the *Kitāb al-Aṭʿima* (sic: *The Book of Foods*; the relevant book would be the *Kitāb Ādāb al-akl*, *Book of the Manners Relating to Eating*) of stories that are inappropriate to a book of religion. After the list, al-Ṭurṭūshī continues with further minor objections concerning al-Ghazālī's division of the days and nights, his prescribing certain prayers and litanies at different times, and his not following customary treatment of prophetic and saintly miracles and of weeping.[43]

These criticisms are much milder and more partial than the vehement and categorical attack on al-Ghazālī and the *Iḥyāʾ* in the letter. The critique of the use of weak *ḥadīth*s is common and serious, but al-Ṭurṭūshī excuses al-Ghazālī of malicious intent by writing that he was simply weak in his knowledge of Ḥadīth, even quoting al-Ghazālī himself to this effect. The critique of not giving *kalām* the centrality it deserves is rather mild. There is no mention of al-Ghazālī's demotion of *kalām*—and *fiqh*—as "sciences of the world", which is alluded to in the letter. The third point, al-Ghazālī's borrowings from the Ikhwān al-Ṣafāʾ, follows al-Ṭurṭūshī's critique in his letter, though instead of condemning the Ikhwān for holding the doctrine of the acquisition of prophecy, he charges them with inappropriate interpretation of the Qurʾān (*yataʾawwilūna al-Qurʾān ʿalā ghayr taʾwīlihi*) and doesn't specify which of their errors al-Ghazālī may have followed—at least in the passage al-Manūnī quotes in his excerpt.

Al-Ṭurṭūshī's overview of his rewriting of the *Iḥyāʾ* shows that *Kitāb al-Asrār wa-l-ʿibar* is significantly different from the *Iḥyāʾ*. Al-Ghazālī dedicates the

42 Al-Manūnī, "Iḥyāʾ ʿulūm al-dīn" 135.
43 Al-Manūnī, "Iḥyāʾ ʿulūm al-dīn" 135–136.

Iḥyā' to a discipline that he calls the science of the hereafter (*ʿilm al-ākhira*), which he divides into two parts: the science of praxis (*ʿilm al-muʿāmala*) and the science of unveiling (*ʿilm al-mukāshafa*). In the *Iḥyā'*, he says he will restrict himself to explicating the science of praxis, not the science of the unveiling, just as al-Ṭurṭūshī says. He divides the book into four sections. The first two are on exterior acts of the body, first actions of worship and then actions of daily life. The third and fourth sections are on interior states, the third on vices that must be eliminated, and the fourth on virtues that must be cultivated. Al-Ṭurṭūshī likewise divides his book into four sections, but these are realisations (*maʿārif*), acts of worship, (spiritual) practices (*muʿāmalāt*), and states and stations. But he then continues with "The Comprehensive Book" (*al-kitāb al-jāmiʿ*) that covers numerous topics not covered in the first four sections. The total number is one hundred books, while the *Iḥyā'* contains forty, ten in each section.[44] Al-Ṭurṭūshī may have taken inspiration from al-Ghazālī, but took great liberties in his version of the *Iḥyā'*, writing a book that seems from al-Manūnī's synopsis to be a much more Sufi-oriented work. A closer comparison will be possible only if the manuscript is some day published. It is worth noting that with this book, al-Ṭurṭūshī initiated a long tradition of creative re-writings of the *Iḥyā'* which has produced at least two dozen imitations over the centuries, including a Shiʿite version and even a Christian one. This tradition has persisted into the first years of the present century.[45]

Al-Ṭurṭūshī criticised al-Ghazālī in a third work as well, one that shows a more detailed knowledge of al-Ghazālī's broader writings. This we find in a fragment of a work entitled *Marāqī al-ʿārifīn*, quoted in Ibn al-Khaṭīb's *al-Iḥāṭa fī akhbār gharnāṭa*. There al-Ṭurṭūshī writes that al-Ghazālī conveys philosophical doctrines but passes them off as Sufi ones, leading his readers to error. He claims, correctly, that al-Ghazālī imitated Ibn Sīnā in his philosophy, which he transmitted in the *Maqāṣid al-falāsifa*, and in his logic, which he transmitted in *Miʿyār al-ʿilm*.[46]

---

44 Al-Manūnī, "Iḥyā' ʿulūm al-dīn" 137.

45 A point first made by ʿAbd al-Raḥmān Badawī in *Mu'allifāt al-Ghazālī*, Cairo: al-Majlis al-Aʿlā li-Riʿāyat al-funūn wa-l-Ādāb wa-l-ʿulūm al-ijtimaʿīya, 1961, 114–118 and further elaborated on by Michael Cook in *Commanding right and forbidding wrong in Islamic thought*, Cambridge: Cambridge University Press, 2001, 451–452 and 600–603. For my article on the latest of these efforts by Su ʿād al-Ḥakīm, see Garden, Kenneth, "*The Revival of the Religious Sciences* in the twenty-first century: Suʿād al-Ḥakīm's adaptation of al-Ghazālī's Revival", in Frank Griffel (ed.), *Islam and rationality: The impact of al-Ghazālī*, Leiden: Brill, 2015, ii, 310–331.

46 Ghrāb, "Ḥawl iḥrāq" 162n37. See also Schreiner, Martin, "Beiträge zur Geschichte der theologischen Bewegungen im Islam", in *Zeitschrift der deutschen morgenländischen Gesellschaft* 52 (1898), 463–512, at p. 503. Saʿd Ghrāb argues that *Marāqī al-ʿarifīn* is al-Ṭurṭūshī's rewrite of the *Iḥyā'* promised in the letter to Ibn Muẓaffar, but this is not

7 Ibn al-ʿArabī's Critique of al-Ghazālī

As we have seen, Ibn al-ʿArabī had none of the superior and critical attitude toward al-Ghazālī that al-Ṭurṭūshī displayed toward him. He refers to him frequently and reverently in his writings as his *dānishmand* and evidently hoped to gain prestige from his association from his most famous teacher. Ibn al-Qaṭṭān has written that at the time of the burning of *Iḥyāʾ ʿulūm al-dīn* under the Almoravids in 503/1109, Ibn al-ʿArabī was ordered to wash the pages of his copy that he had carried back from the Mashriq,[47] though we know this is not so. As we have seen, al-Ghazālī seems not to have completed the *Iḥyāʾ* at the time that Ibn al-ʿArabī studied with him; Ibn al-ʿArabī "heard" the book rather than reading it with al-Ghazālī, which is how he describes all the other books he studied with him. This is not to say that he couldn't have been ordered to erase a copy of the *Iḥyāʾ* that he had subsequently acquired, but more importantly, the campaign against al-Ghazālī's writings in the Almoravid Maghrib seems to have left no mark on Ibn al-ʿArabī's presentation of his relationship to him.

We further know that al-Ghazālī's writings deeply influenced Ibn al-ʿArabī's thought and writing. ʿAmmār Ṭālibī has documented extensive reliance on al-Ghazālī's writings in Ibn al-ʿArabī's *al-Awāṣim min al-qawāsim*: twenty-eight passages that draw from *Maqāṣid al-falāsifa*, and four that draw from *Tahāfut al-falāsifa*, several of which are more than a page long.[48] In several other passages in the book, he relates comments made to him by al-Ghazālī.[49] His *Qānūn al-taʾwīl* is closely connected to al-Ghazālī's *Fayṣal al-tafriqa bayna al-islām wa-l-zandaqa*, its title coming from a phrase coined by al-Ghazālī in that book. In it, he reproduces one of al-Ghazālī's main claims in the *Fayṣal*: that a text can only be read allegorically when it has been established that its literal meaning is false or impossible.[50] Frank Griffel has found further influence of al-Ghazālī in Ibn al-ʿArabī's main legal work *Aḥkām al-Qurʾān*.[51]

---

the case. *Kitāb al-Asrār wa-l-ʿibar*, excerpted by al-Manūnī and discussed just above, is plainly the work in question. Ibn al-Khaṭīb describes the *Marāqī* as "a long treatise", (*risāla ṭawīla*), while *Kitāb al-Asrār wa-l-ʿibar* is described as a work consisting of one hundred "books".

47 Ibn al-Qaṭṭān, *Naẓm al-jumān li-tartīb mā salafa min akhbār al-zamān*, ed. Muḥammad ʿAlī Makkī, Beirut: Dār al-Gharb al-Islāmī, 1990, 71.

48 Ṭālibi, *Ārāʾ Abī Bakr* 291–293. I was alerted to this passage by Griffel, Frank, *Apostasie und Toleranz im Islam*, Leiden: Brill, 2000, 383n58.

49 See Griffel, *Al-Ghazālī's philosophical theology* 69–70. For Griffel's fuller discussion of Ibn al-ʿArabī and his relation to al-Ghazālī, see pp. 62–71.

50 Griffel, *Apostasie und Toleranz* 383–385.

51 Griffel, *Apostasie und Toleranz* 385.

But Ibn al-ʿArabī did have his criticisms of al-Ghazālī, and they are related to al-Ṭurṭūshī's. Though he drew from al-Ghazālī's criticisms of philosophy in *al-ʿAwāṣim min al-qawāṣim*, he was aware of al-Ghazālī's debt to philosophy. Ibn Taymiyya quotes him as saying, "Our Shaykh Abū Ḥāmid entered deeply into the belly of philosophy, and when he wanted to come out, he was unable."[52]

Beyond this general criticism for al-Ghazālī's well attested engagement with philosophy, there is a specific criticism of al-Ghazālī's theodicy, the claim that God has created the best world possible. Ibn al-ʿArabī writes in *Sirāj al-murīdīn*:

> Our Shaykh Abū Ḥāmid al-Ghazālī has said something stupendous for which the Iraqis have criticized him. And, as God is my witness, this is indeed occasion for criticism! For he said, "There is nothing in the divine omnipotence more wonderful than this world in its perfection and its wisdom."[53]

This is a well-known critique of al-Ghazālī's statement of theodicy, which continued to be debated through the nineteenth century.[54] Like the critiques of al-Ghazālī advocating the doctrine of the acquisition of prophecy (the seventh point al-Ghazālī responds to in the *Imlāʾ*) and claiming that revealing the secret of the divine lordship is apostasy (the second point), this criticism had its origins in the controversy in Nishapur, and it is point ten to which al-Ghazālī responds in the *Imlāʾ*. Ibn al-ʿArabī here attributes the critique to "the Iraqis", which raises questions about the controversy in Nishapur that I cannot answer. How could the critiques to which al-Ghazālī responded in *al-Imlāʾ fī ishkālāt*

52 Ibn Taymiyya, *Dār taʿāruḍ al-ʿaql wa-l-naql*, ed. M. Rashād Sālim, 11 vols., Beirut: Dār al-Kunūz al-Adabiyya, 1980, i, 5, and Ibn Taymiyya, *Majmūʿ fatāwa*, ed. ʿAbd al-Raḥmān ibn Muḥammad ibn Qāsim al-ʿAṣimī, 35 vols., Riyadh, Maṭābiʿ al-Riyāḍ, 1961–1967, iv, 66. Cited from Griffel, *Al-Ghazālī's philosophical theology* 66.

53 Quoted from Ormsby, Eric, *Theodicy in Islamic thought: The dispute over al-Ghazālī's "best of all possible worlds"*, Princeton, Princeton University Press, 1984, 60. Ormsby in turn quotes from al-Zabīdī's commentary on the *Iḥyāʾ*. See al-Zabīdī, Muḥammad b. Muḥammad, *Itḥāf al-sāda al-muttaqīn bi-sharḥ Iḥyāʾ ʿulūm al-dīn*, 14 vols., Beirut: Dār al-Kutub al-ʿIlmiyya, 2002, xii, 107. I obtained a copy of Ibn al-ʿArabī's *Sirāj al-muridīn* late in the publication process, and on a first search, I was only able to find an abbreviated reference to the issue: "... do not restrict His power such that he can create the likes of this world, yes, but not more perfect than it, in distinction to the Sufis who say, 'Not more perfect than this', which is a philosophical direction." Ibn al-ʿArabī, Abū Bakr, *Sirāj al-murīdīn fī sabīl al-dīn*, ed. Abdallah Taourati, Tangiers: Dār al-Ḥadīth al-Katānīya, 2017, iii, 157. The editor, Abdallah Taourati, explains in a footnote that this is a reference to to al-Ghazālī.

54 See Ormsby, *Theodicy* for a full discussion of this controversy, its stakes, parameters, and history.

al-Iḥyāʾ, over which a controversy erupted that is described in al-Ghazālī's letters as occurring in Nishapur, be attributed to unspecified Iraqis? It is possible that the content of the controversy reached Ibn al-ʿArabī through connections in Baghdad. It is possible that a written critique originating in Baghdad was used against al-Ghazālī in Nishapur. What is important in the context of this article is that this point of critique had the same origin as the two points raised by al-Ṭurṭūshī.

So did Ibn al-ʿArabī learn about the critique of the allegory of the pens in Book 35 of the *Iḥyāʾ* from al-Ṭurṭūshī? It seems unlikely. Rather than focus on al-Ṭurṭūshī's two main objections, Ibn al-ʿArabī focuses instead on a third issue, one that would have a much longer afterlife than the issues al-Ṭurṭūshī focused on. But it does seem clear that both al-Ṭurṭūshī and Ibn al-ʿArabī were engaged in a controversy over al-Ghazālī and his *Iḥyāʾ* that ranged from Nishapur to Seville. Assuming that Ibn al-ʿArabī and his former teacher and fellow ascetic remained in contact, it seems likely that they would have corresponded about issues raised by the thought of this man whom they both had met and had surely discussed in Alexandria. But this is speculation.

8 Conclusion

Ibn al-ʿArabī and al-Ṭurṭūshī each met al-Ghazālī within a year of one another and had a chance to compare notes with one another afterwards. Al-Ṭurṭūshī seems to have seen al-Ghazālī as a junior and inferior scholar from the beginning, while Ibn al-ʿArabī saw him as a revered teacher who deeply influenced his thought and writing, and to whom he referred frequently in later writings. The critique of al-Ghazālī, specifically of the allegory of the pens found in Book 35 of the *Iḥyāʾ* that emerged from a campaign against him in Nishapur came to the attention of both men. Al-Ṭurṭūshī made it into a central scholarly concern, writing a letter savagely critiquing al-Ghazālī and calling for the burning of the *Iḥyāʾ*, critiquing him in a separate treatise, and writing his own version of the *Iḥyāʾ* in keeping with his scholarly outlook. Ibn al-ʿArabī, for his part, also criticised al-Ghazālī in terms that emerged from the Nishapurian controversy, but not on points identical to those of al-Ṭurṭūshī, and in both cases referring to al-Ghazālī as his shaykh. If he ever took a public stance on the burning of the *Iḥyāʾ* under the Almoravids, it has not come down to us. Clearly, both of their reactions must be understood within the same context, but despite their close relationship earlier in life, each man developed his own response to al-Ghazālī.

Bibliography

al-Akiti, Afifi, "The good, the bad, and the ugly of falsafa: Al-Ghazālī's Maḍnūn, Tahāfut, and Maqāsid, with particular attention to their falsafī treatments of God's knowledge of temporal events", in Y Tzvi Langerman (ed.), *Avicenna and his legacy: A golden age of science and philosophy*, Turnhout, Belgium: Brepols, 2009, 51–100.

Aʿrāb, Saʿīd, *Maʿ al-Qāḍī Abī Bakr ibn al-ʿArabī*, Beirut: Dār al-Gharb al-Islāmī, 1987.

Badawī, ʿAbd al-Raḥmān, *Muʾallifāt al-Ghazālī*, Cairo: al-Majlis al-Aʿlā li-Riʿāyat al-funūn wa-l-Ādāb wa-l-ʿulūm al-ijtimaʿīya, 1961.

Cook, Michael, *Commanding right and forbidding wrong in Islamic thought*, Cambridge: Cambridge University Press, 2001.

Fierro, Maribel, "Opposition to Sufism in al-Andalus", in Frederick de Jong and Bernd Radke (eds.), *Islamic mysticism contested: Thirteen centuries of controversies and polemics*, Leiden: Brill, 1999, 174–206.

Garden, Kenneth, "Al-Māzarī al-Dhakī: Al-Ghazālī's Maghribi adversary in Nishapur", in *Journal of Islamic Studies* 21 (2010), 89–107.

Garden, Kenneth, *The first Islamic reviver: Abū Ḥāmid al-Ghazālī and his* Revival of the Religious Sciences, New York: Oxford University Press, 2013.

Garden, Kenneth, "The riḥla and self-reinvention of Abū Bakr ibn al-ʿArabī", in *Journal of the American Oriental Society* 135 (2015), 1–17.

Garden, Kenneth, "The Revival of the Religious Sciences in the twenty-first century: Suʿād al-Ḥakīm's adaptation of al-Ghazālī's Revival", in Frank Griffel (ed.), *Islam and rationality: The impact of al-Ghazālī*, Leiden: Brill, 2015, ii, 310–331.

al-Ghazālī, Abū Ḥāmid, *Makātib-i fārsī-i Ghazzālī ba-nām-i Faḍāʾil al-anām min rasāʾil Ḥujjat al-Islām*, ed. ʿAbbās Iqbāl, Tehran: Kitābforūshī Ibn Sīnā, 1373/1954.

al-Ghazālī, Abū Ḥāmid, *Iḥyāʾ ʿulūm al-dīn*, 5 vols., Cairo: Muʾassasat al-Ḥalabī wa-Shurakāʾuhu, 1387/1967–1968.

Ghrāb, Saʿd, "Ḥawl iḥrāq al-murābiṭīn li-Iḥyāʾ al-Ghazālī", in *Actas del IV coloquio Hispano-Tunecino (Palma de Mallorca, 1979)*, Madrid: Instituto Hispano-Arabe de Cultura, 1983, 133–163.

Griffel, Frank, *Al-Ghazālī's philosophical theology*, New York: Oxford University Press, 2009.

Griffel, Frank, *Apostasie und Toleranz im Islam*, Leiden: Brill, 2000.

Ibn al-ʿArabī, Abū Bakr, *Qānūn al-taʾwīl*, ed. Muḥammad Slīmānī, Beirut: Dar al-Gharb al-Islami, 1990.

Ibn al-ʿArabī, Abū Bakr, *Shawāhid al-jilla*, in Muḥammad Yaʿlà, *Tres textos ārabes sobre Beréberes en el occidente Islámico*, Fuentes Arábico-Hispanas, 20, Madrid: Consejo Superior de Investigaciones Cientificas, Agencia Española de Cooperacion Internacional, Madrid, 1996.

Ibn al-ʿArabī, Abū Bakr, *al-ʿAwāṣim min al-qawāṣim*, ed. ʿAmmār Ṭālibī, Cairo: Maktabat Dār al-Turāth, 1997.

Ibn al-ʿArabī, Abū Bakr, *Sirāj al-murīdīn fī sabīl al-dīn*, ed. Abdallah Taourati, Tangiers: Dār al-Ḥadīth al-Katānīya, 2017.

Ibn al-Qaṭṭān, *Naẓm al-jumān li-tartīb mā salafa min akhbār al-zamān*, ed. Muḥammad ʿAlī Makkī, Beirut: Dār al-Gharb al-Islāmī, 1990.

al-Qāḍī ʿIyāḍ, Abū l-Faḍl b. Mūsā b. ʿIyāḍ al-Sabtī, *Tartīb al-madārik wa taqrīb al-masālik fī maʿrifat aʿlām madhhab Mālik*, ed. Saʿīd Aḥmad Aʿrāb et al, 8 vols., Rabat: Wizārat al-Awqāf wa-l-Shuʾūn al-Islāmiyya, 1384–1403/1965–1983.

Ibn Taymiyya, *Dār taʿāruḍ al-ʿaql wa-l-naql*, ed. M. Rashād Sālim, 11 vols., Beirut: Dār al-Kunūz al-Adabiyya, 1980.

Ibn Taymiyya, *Majmūʿ fatāwa*, ed. ʿAbd al-Raḥmān ibn Muḥammad ibn Qāsim al-ʿAṣimī, 35 vols., Riyadh: Maṭābiʿ al-Riyāḍ, 1961–1967.

Janssens, Jules, "Al-Ghazzālī's Tahāfut: It it really a rejection of Ibn Sīnā's philosophy?" in *Journal of Islamic Studies* 12 (2001), 1–17.

Krawulsky, Dorothea, *Briefe und Reden des Abū Ḥāmid al-Ghazālī: Übersetzt und Erläutert*, Islamkundliche Untersuchungen, Freiburg im Breisgau: Klaus Schwarz Verlag, 1971.

al-Manūnī, Muḥammad, "Iḥyāʾ ʿulūm al-dīn fī manẓūr al-gharb al-islāmī ayām al-murābiṭīn wa-l-muwaḥḥidīn", in *Abū Ḥāmid al-Ghazālī: dirāsāt fī fikrihi wa-ʿaṣrihi wa-taʾthīrihi*, Rabat: Jāmiʿat Muhammad al-Khāmis, Nasharāt Kuliyyat al-Ādāb wa-l-ʿUlūm al-Insāniyya bi-Rabāṭ, Silsilat Nadawāt wa Munāẓarāt, 1988, 125–137.

Ormsby, Eric, *Theodicy in Islamic thought: The dispute over al-Ghazālī's* "best of all possible worlds", Princeton, Princeton University Press, 1984.

Schreiner, Martin, "Beiträge zur Geschichte der theologischen Bewegungen im Islam", in *Zeitschrift der deutschen morgenländischen Gesellschaft* 52 (1898), 463–512.

Ṭālibī, ʿAmmār, *Ārāʾ Abī Bakr ibn al-ʿArabī al-kalāmiyya*, 2 vols., Algiers: al-Sharika al-Waṭaniyya li-l-nashr wa-l-tawzīʿ, 1974.

Treiger, Alexander, *Inspired knowledge in Islamic thought: al-Ghazālī's theory of mystical cognition and its Avicennian foundation*, New York: Routledge, 2012.

al-Ṭurṭūshī, Abū Bakr, *Kitāb al-Ḥawādit wa-l-bidaʿ* (*el Libro de las novedades y las innovaciones*), trans. and ed. Maribel Fierro, Madrid: Consejo Superior de Investigaciones Cientificas, 1993.

al-Wansharīsī, Aḥmad. *Al-Miʿyār al-muʿrib wa-l-jāmiʿ al-mughrib ʿan fatāwa ʿulamāʾ Ifrīqiya wa-l-Andalus wa-l-Maghrib*, ed. Muḥammad Ḥajjī, 13 vols, Rabat: Wizāra al-Awqāf wa-l-Shuʾūn al-Mamlaka al-Maghribiyya, 1981.

al-Zabīdī, Muḥammad b. Muḥammad, *Itḥāf al-sāda al-muttaqīn bi-sharḥ Iḥyāʾ ʿulūm al-dīn*, 14 vols., Beirut: Dār al-Kutub al-ʿIlmiyya, 2002.

5

القاضي ابن العربي أصوليًا

عبد الغني ادعيكل

[al-Qāḍī Ibn al-ʿArabī as an *uṣūlī*]

[*Abdelghani Daikal*]

إن علم أصول الفقه من العلوم الإسلامية النقدية والمنهجية التي، اشتركت في بناء لبناتها علوم شتى، وصارت بعدها في حاجة إليها، فالفقيه يلجأ إليه في قراءة النص الشرعي وتحديد المقصد منه ثم تنزيله، والمفسر يحتاجه من جهة المنهج والبيان والكشف عن المراد، والمحدث يرجع إليه في التوجيه والتوفيق بين المرويات ونقدها.

ونظرا لأهمية هذا العلم وضرورته فقد أولاه علماؤنا العناية التي تناسب قيمته ومكانته، فبذلوا جهدهم في طلبه وتحصيله، وأفنوا أعمارهم في إدراكه ومعرفته وبيانه، كل بحسب رؤيته والإشكالات التي عنت له، فأصبح للشريعة الإسلامية حاكمية في كل المجالات، واستطاع الفقهاء مسايرة الواقع بإعطاء المستجدات حكمها، واستشعر عموم الناس فضل دينهم وجماليته وصلاحيته وضرورته.

وممن نعتقد أنه أسهم في هذا العلم تقعيدا وتأليفا وتوظيفا وطلبا وعناية: القاضي أبو بكر محمد بن عبد الله ابن العربي المعافري الإشبيلي (543هـ\1148م).

إن هذه الورقة -التي أقدمها- أروم من خلالها الإجابة عن إشكالين: أحدهما متعلق بالقطر الذي ينتمي إليه القاضي أبو بكر، والثاني متعلق بشخصيته الأصولية.

أما بخصوص الإشكال الأول، فإن الدارس لتاريخ أصول الفقه يلحظ أن ثمة تجاوزا للمذهب المالكي، وإغفالا لإبداعات علمائه، واتهامهم بأنهم لم يبدعوا ولم يؤسسوا منهجا أصوليا خاصا بهم، وباستعراض مراجع أصول الفقه في القرون الأولى قلما نجد من يذكر كتب أصول المالكية؛ مثل كتاب حماد بن إسحاق في الرد على الشافعي، وكتاب الأصول والاحتجاج بالقرآن للقاضي إسماعيل بن إسحاق البغدادي، وكتاب الخلاف والحجة لأبي الحسن بن المنتاب، وكتاب اللمع في أصول الفقه للقاضي أبي الفرج عمر بن محمد الليثي، وكتاب الأصول وكتاب إجماع أهل المدينة لأبي بكر

 | DOI:10.1163/9789004706859_007

الأبهري، وكتاب المقدمة في الأصول لأبي الحسن ابن القصار، وغير هؤلاء ممن كان لهم فضل كبير في تقعيد هذا العلم وتطويره.

وإذا كانت هذه التهمة موجهة للمذهب المالكي دون تفريق بين قطر وقطر، فإنها أشد التصاقا بالقطر الأندلسي المغربي، وتزداد عجبا أن تجد هذه التهمة صادرة من أهله، و ممن يحفظ عنه لمزا لعلماء الأندلس والمغرب بهذه النقيصة الإمام ابن رشد؛ قال: "فهذه صناعة أصول الفقه، والفقه نفسه، لم يكمل النظر فيها إلا في زمن طويل، ولو رام إنسان اليوم- من تلقاء نفسه -أن يقف على جميع الحجج التي استنبطها النظار من أهل المذاهب في مسائل الخلاف، التي وقعت المناظرة فيها بينهم في معظم بلاد الإسلام **ما عدا المغرب**، لكان أهلاً أن يضحك منه"([1]).

وأيضا ابن خلدون الذي قال: "وأما المالكية فالأثر أكثر معتمدهم، وليسوا بأهل نظر"[2].

وتبعهما في هذه الدعوى الإمام المقري فقال: "وسمة الفقيه عندهم جليلة، حتى إن الملثمين كانوا يسمون الأمير العظيم منهم الذي يريدون تنويهه بالفقيه، وهي الآن بالمغرب بمنزلة القاضي بالمشرق، وقد يقولون للكاتب والنحوي واللغوي فقيه؛ لأنها عندهم أرفع السّمات. وعلم الأصول عندهم متوسط الحال، والنحو عندهم في نهاية من علو الطبقة"[3].

والناظر في تاريخ هذا العلم عند المالكية في القطر المغربي الأندلسي يرى في هذه الدعوى تجنيا على المذهب المالكي وأهله من وجوه[4]:

أولا: لقد بين الإمام الشافعي في رسالته وكتاب اختلاف الحديث وكتاب الأم أسس مذهبه؛ وهي في الإجمال تتركز في إعطاء أولوية للحديث الصحيح المرفوع، وعدم اعتبار العمل مفسرا له، وعدم حجية عمل أهل المدينة، وعدم اعتبار المصالح التي لم ينص عليها الشرع، وهذه المبادئ الأصولية الشافعية تعتبر طعنا في المبادئ المالكية، فهبّ المالكيون في استخلاص الأسس التي بُني عليها المذهب وتوضيح مناهجه، وتصدى كثير من الأئمة بالغرب الإسلامي لمخالفات الإمام الشافعي كأبي

---

1 فصل المقال: 27–28.
وهذا النقل لا يفهم منه صراحة لمز ابن رشد أهل المغرب بعزوفهم عن علم الأصول، وقد أوردته هنا لاتكاء كثير من العلماء والباحثين عليه في إثارة هذه المسألة.

2 ديوان المبتدأ والخبر: 578/1.

3 نفح الطيب: 221/1.

4 أغلب ما سطرته في دعوى عدم اهتمام المغاربة بعلم أصول استفدته من الدراسات الآتية: مدخل إلى أصول الفقه المالكي لمحمد المختار ولد أباه: 11–16، محاضرات في تاريخ المذهب المالكي لعمر الجيدي: 69–88، المدرسة المالكية الأصولية وإبداع المغاربة فيها لمحمد التمسماني: 90–105، المصادر الأصولية عند المالكية: دراسة في النشأة والمدونات والخصائص لمولاي الحسين بن الحسن الحيان: 174–176.

عبد الله محمد بن سحنون(256هـ\869م)، وأبي العباس عبد الله بن طالب القاضي(275هـ\888م)، وأبي زكريا يحيى بن عمر بن يوسف بن عامر الكناني الأندلسي(289هـ\902م)، وأبي عثمان سعيد بن محمد بن الحداد القيرواني(320هـ\932م)، وأبي بكر محمد بن اللباد القيرواني(333هـ\944م)، وأبي مروان عبد الملك بن أحمد القرطبي(393هـ\1002م)، وأبي عبد الله محمد بن عمر يوسف بن الفخار القرطبي(419هـ\1028م)؛ فهذه الردود والمناقشات والمناظرات تعطي صورة لما كان رائجا في ذلك العصر بالقطر المغربي من شيوع المناظرة والاستدلال والاجتهاد والنقد، وهي كذلك دالة على اهتبال الأئمة المغاربة بهذا العلم إلى حد مناقضتهم لمؤسسه ومنظره؛ وهو الإمام الشافعي.

ثانيا: لا شك أن منهج المتكلمين في الكتابة الأصولية أسسه الإمام الشافعي، وتبعه في ذلك أصوليو الشافعية والمالكية والحنابلة وطوائف الشيعة، وأبدعوا فيه وطوروه من حيث المنهج والبناء، وكثرت تآليفهم ومصنفاتهم فيه، ووسعوا مجال البحث والمناظرة في قضاياه، ولما كانت عناية المتكلمين الشافعية بهذا المنهج أكثر من غيرهم تأليفا ومناظرة وتدريسا تفوق نشاط غيرهم، سلمت لهم الزعامة والقيادة فيه[5]، والحقيقة أن جل المذاهب أسهموا فيه بصورة تقارب عطاءهم أو تفوق، ولا أدل على ذلك من اعتماد أساطين هذا العلم كإمام الحرمين الجويني، وأبي حامد الغزالي، وفخر الدين الرازي، وسيف الدين الآمدي على إمام أئمة هذا العلم؛ وهو القاضي أبي بكر الباقلاني، ومنه نعلم أن الأئمة المالكية عموما والمغاربة خصوصا كانت لهم مشاركة فعالة في تحرير قواعد هذا العلم وإرساء دعائمه، وهي تشهد بطول باعهم فيه ونبوغهم وتفوقهم.

ثالثا: لقد أدرك الأئمة المالكية عموما والمغاربة خصوصا أن قانون التعامل مع النص الشرعي فهما واستنباطا وتنزيلا منوط بالمكنة الأصولية، وهو أمر دعاهم إلى العناية بهذا العلم، ومن مظاهر ذلك افتتاحهم به موسوعاتهم الفقهية مع تحريرهم المسائل وتأصيلها واستخلاص مبادئها، ومن طالع- على سبيل التمثيل -كتب ابن رشد الجد أدرك هذه القضية.

رابعا: إن الباحث في التراث المغربي الأندلسي يجد أمامه ثروة هائلة من المدونات الأصولية، الضاربة بجذورها في أعماق التاريخ، المتفاوتة مضامينها ومشاربها تناولا وأسلوبا ومنهجا؛ منها ما هو موضوع مبتدع ومبتكر من حيث الفكرة والمنهج والمعالجة، ومنها ما هو شروح وحواش وتعليقات ومنظومات، وهذه الحصيلة تدل بلا شك على اهتمام القوم بهذا العلم، وتؤكد أنهم مثل غيرهم من أتباع المذاهب

5 الفكر الأصولي: 446.

الأخرى، يهتمون بهذا العلم ويتنافسون في ارتياد آفاقه، وسبر أغواره، وتقعيد قضاياه. والموضوعية تقتضي الإنصاف والاعتراف بجهودهم، وعدم الانسياق وراء دعاوى التنقيص[6].

خامسا: لا أحد ينكر التلازم الحاصل بين علم الفقه وعلم أصول الفقه، والجميع يقر باستحالة تحصيل مرتبة الفقيه دون المكنة الأصولية، وإذا كان الأمر كذلك، وقد وجد في هذا القطر أئمة فقهاء بلغوا مرتبة الاجتهاد في هذا العلم، فإنه يستبعد جدا أن يكون القطر المغربي الأندلسي خلوا من هذا العلم أو متوسط الحال فيه كما قال المقري.

سادسا: إن المقارنة بين مساحة الجزيرة العربية والبلدان التابعة لها في التمذهب ومساحة المغرب والأندلس تظهر الفرق الكبير بينهما، وأكيد أن هذا كان له دور في التأليف الأصولي والإبداع فيه، ورغم ذلك يبقى القطر المغربي يضاهي الأقطار المشرقية التي كانت أكثر نشاطا وإنتاجا في هذا العلم.

سابعا: يضاف إلى ما قيل سابقا؛ أن أغلب الكتب المصنفة في هذا الفن ما زالت في عداد المفقود، ورغم أن جميع الأقطار تشترك في ذلك، إلا أن وجودها وتداولها يقرّبنا من الصورة الحقيقية التي كان عليها هذا العلم بهذا القطر، ويبصرنا بالشخصية الأصولية في المغرب والأندلس.

فإن قيل[7]: إن كثرة المؤلفات لا تنهض حجة على قوة فهمهم لهذا العلم وتعمقهم فيه، وإتيانهم فيه بالجديد المبتكر، لا سيما إذا كان الأمر يتعلق بالشروح في الحواشي، وهي معظم ما ألف المغاربة.

قلنا: هذا الاعتراض لا يصلح دليلا، وذلك لأسباب:

أولها: أن المؤلفات في هذا الفن ليست كلها من قبيل الشروح والتعليقات، بل فيها مؤلفات كثيرة مبتكرة في أسلوبها ومنهجها وموضوعها.

ثانيها: أن المالكية كان لهم فضل السبق في إبداع هذا العلم وتجديده، والناظر في تاريخه يجد أن أول من أسهم في وضع لبناته وأبدع في بعض قضاياه الإمام مالك؛ قال ابن العربي في شرح الموطأ: "هذا أول كتاب ألف في شرائع الإسلام وهو آخره؛ لأنه لم يؤلف مثله إذ بناه مالك على تمهيد الأصول للفروع، ونبه فيه على معظم أصول الفقه التي ترجع إليها مسائله وفروعه"[8]، ثم بعد الشافعي جدده الإمام القاضي أبو بكر الباقلاني، ثم بعده ظهر منهج المختصرات فبرع فيه الإمام أبو عمرو جمال الدين

6 ينظر ما كتبناه في تقديمنا لأرجوزة أبي عبد الله الفندلاوي ابن الكتاني، فقد أوردت كثيرا من العناوين الأصولية لأئمة أندلسيين ومغاربة، وينظر كذلك ما كتبه أخونا يونس العدلاني في دراسة قيمة بعنوان: "طبقات الأصوليين بعدوتي المغرب والأندلس"، فقد أجاد فيها وأفاد، وحاول استقراء واستقصاء كل الأصوليين في القطر المغربي، وبلغ عدد من ترجم لهم في هذا السفر الحافل بضع وأربعمائة.

7 محاضرات في تاريخ المذهب المالكي: 73.

8 القبس في شرح موطأ مالك بن أنس: 75/1.

عثمان بن عمر بن أبي بكر بن يونس ابن الحاجب الكردي المالكي، ثم بعده جدده الإمام أبو إسحاق الشاطبي الأندلسي المالكي.

ثالثها: أن من يتصدى لشرح أي فن من الفنون أو التعليق عليه، يفترض فيه أن يكون عالما بالفن المشروح، أو المعلق عليه، فالشارح هو الذي يهتم بتوضيح ما غمض من المعاني في الأصل المشروح، والإضافة إليه، ونقد ما قد يكون وقع فيه صاحب الأصل من وهم أو خطأ، وهذا لا يتأتى إلا إذا كان الشارح أو المعلق متمكنا من ذلك العلم، ومالكا لناصيته، وآخذا بزمامه.

أما بخصوص الإشكال الثاني، فقد يتفاجأ كثير من الباحثين عند مطالعتهم للكتب المعتنية بإحصاء أئمة هذا الفن وتأثيرهم فيه عدم ذكر القاضي ابن العربي ضمنهم، فمثلا نجد ابن خلدون يشير عند حديثه عن أصول الفقه والتأليف فيه[9] إلى إسهامات كل من ابن الحاجب (646هـ\1248م)، والقرافي (684هـ\1285م)، ولم يعرج ولو بإشارة خفيفة على إسهامات ابن العربي فيه.

ودراسة هذا الإشكال والجواب عنه يكون من خلال المسائل الآتية:

المسألة الأولى: أصولية ابن العربي من خلال شيوخه الأصوليين.

المسألة الثانية: أصولية ابن العربي من خلال كتب التراجم والفهارس.

المسألة الثالثة: أصولية ابن العربي من خلال المصنفات الأصولية.

المسألة الرابعة: أصولية ابن العربي من خلال توظيفه لعلم الأصول.

المسألة الخامسة: أصولية ابن العربي من خلال ترجيحاته وتفرده.

1 المسألة الأولى: أصولية ابن العربي من خلال شيوخه الأصوليين

إن الجلوس بين يدي الشيوخ ومحاورتهم ومشاهدتهم ومصاحبتهم من أهم العناصر المشكلة لشخصية الطالب خلقا وعلما، فإذا انضاف إلى ذلك شهرة أولئك الشيوخ بالعلوم المطلوبة، وأنهم من الجلة فيه، وأعلام يهتدى بهم في الظلمات، وكان الطالب شعلة تتوقد ذكاء وفطنة وعلما، وله نفس تواقة إلى اكتناه العلوم وتحصيلها، فلا شك أن هذا سيساعد الطالب على ضبط العلم واستيعابه والبروز فيه.

والإمام ابن العربي كان بذرة صادف أرضا خصبة، فترعرع في سماء العلم ونما حتى صار مبرزا في هذا العلم ومن أساطينه، ومن الأئمة الجلة الذين أسهموا في تكوين شخصيته الأصولية:

---

9 ديوان المبتدأ والخبر: 573–577.

– الإمام الحافظ المحدث الفقيه الأصولي أبو بكر محمد بن الوليد الفهري الطرطوشي (520هـ\1126م)، له كتاب "العمد" في أصول الفقه، قال ابن العربي في تتلمذه عليه: "ومشيت إلى شيخنا أبي بكر الفِهْرِي رحمة الله عليه، وكان ملتزماً من المسجد الأقصى" إلى أن قال: "فشاهدت هديه، وسمعت كلامه، فامتلأت عيني وأذني منه، وأعلمه أبي بنيتي فأناب، وطالعه بعزيمتي فأجاب، وانفتح لي به إلى العلم كل باب، ونفعني الله به في العلم والعمل، ويسر لي على يديه أعظمَ أمل"[10].

ومن العلوم التي فتح الله عليه فيها على أبي بكر الطرطوشي علم أصول الفقه، قال ابن العربي: "ولم نزل على تلك السجية، حتى اطلعت بفضل الله على أغراض العلوم الثلاثة: علم الكلام وأصول الفقه ومسائل الخلاف التي هي عمدة الدين والطريق المَهْيَعُ إلى التدرب في معرفة أحكام المكلفين، الحاوية للمسألة والدليل، والجامعة للتفريع والتعليل"[11].

إن اجتهاد القاضي أبي بكر وجده، ودوامه للطلب وحرصه الشديد، ونفسه التواقة إلى المزيد جعله ينبغ في العلوم التي درسها على شيخه أبي بكر الطرطوشي، ومنها أصول الفقه.

وهذا أمر عاينه شيخه فيه ولامسه وأعجب به حتى قال فيه وفي أصحابه: "قَيَّضَتْ[12] فِرَاخُنَا! فقال له عطاء: بل طارت"[13].

بل من الجميل أن تجد الشيخ يثني على طالبه، ويصفه بكل صفات الجلال والجمال، قال الشيخ أبو بكر الطرطوشي: "والفقيه أبو بكر محمد بن عبد الله بن العربي ممن صحبنا أعواماً يدارس العلم ويمارسه، بلوناه وخبرناه، وهو مِمّن جمع العلم ووعاه، ثم تحقق به ورعاه، وناظر فيه وجدّ حتى فاق أقرانه ونظراءه، ثم رحل إلى العراق فناظر العلماء، وصحب الفقهاء، وجمع من مذاهب العلم عيونها، وكتب من حديث رسول الله ﷺ، وروى صحيحه وثابته، والله تعالى يُؤتي الحكمة من يشاء"[14].

وهذا النص ينبئنا بما كان عليه ابن العربي وما حصله من العلوم حفظا وتحققا.

– فخر الإسلام الإمام النظار الفقيه الأصولي أبو بكر محمد بن أحمد الشاشي القفال الفارقيّ الشافعي (507هـ\1113م)، من كتبه الأصولية: "أصول الفقه" و"محاسن الشريعة"، قال ابن العربي في شأن

10 قانون التأويل: 435.

11 قانون التأويل: 438.

12 أي خرجت من البيضة.

13 قانون التأويل: 437.

14 شواهد الجلة والأعيان في مشاهد الإسلام والبلدان؛ اللوحة 18، مخطوط بالخزانة العامة بالرباط د1020.

طلبه العلم عليه: "ثم فاوضت بعد ذلك العلماء، وواظبت المجالس، واختصصت بفخر الإسلام أبي بكر الشاشي، فقيه الوقت وإمامُه، فطلعت لي شموس المعارف، فقلت: اللّٰه أكبر هذا هو المطلوب الذي كنت أصمد، والوقت الذي كنت أرقب وأرصد، فدرست وقَيَّدْتُ وارتويت وسمعت ووعيت"[15].

- الشيخ الإمام البحر، حجة الإسلام، أعجوبة الزمان، النظار المتكلم الفقيه الأصولي أبو حامد محمد بن محمد الغزالي (505هـ\1111م)، رحل إليه القاضي وصاحبه ولازمه حتى ارتوى من علومه، وملأ فؤاده من جميل أحواله، قال متحدثا عن لقائه به: "حتى ورد علينا دَانِشْمَنْد، فنزل برباط أبي سعد بإزاء المدرسة النظامية معرضاً عن الدنيا، مقبلاً على اللّٰه تعالى، فمشينا إليه وعرضنا أمنيتنا عليه، وقلت له: أنت ضالتنا الذي كنا ننشد، وإمامنا الذي به نسترشد، فلقينا لقاء المعرفة، وشاهدنا منه ما كان فوق الصفة، وتحققنا أن الذي نقل إلينا من أن الخبر عن الغائب فوق المشاهدة ليس على العموم"[16]، وقال أيضا: "فقصدت رباطه، ولزمت بساطه، واغتنمت خلوته ونشاطه، وكأنما فَرَغَ لي لأبلُغَ منه أملي، وأباح لي مكانه، فكنت ألقاه في الصباح والمساء والظهيرة والعشاء، كان في بَزَّتِهِ أو بَذْلَته، وأنا مستقل في السؤال، عالمٌ حيث تؤكل كتفُ الاستدلال، وألفيته حَفِيّاً بي في التعليم، وفياً بعهدة التكريم"[17].

وبقي القاضي على هذه الحال، ملازما للشيخ ومقبلا على علمه وحاله حتى طلع له ما كان يبحث عنه من النور، وانجلى عنه ما كان يغشاه من الدَّيْجُور؛ قال: هذا مطلوبي حقاً، هذا بأمانة اللّٰه منتهى السالكين، وغاية الطالبين للعلم المبين، إني تارك لما تطلبون، ونابذ ما كنتم تقولون، وقد علم الإمام أني من السالكين في سبيل المهتدين، فَسَدَّدَنِي إلى سوائها، وأوجد لي معلوم دليلها، وأرشدني إلى لَقَم ظاهرها وتأويلها، وليس التحصيل بطول الصحبة، وإنما هو فضل من اللّٰه وموهبة"[18].

ومما يؤكد ما شعر به القاضي وما حصل؛ تلك الشهادة التي خطها شيخه الغزالي في رسالته المشهورة إلى الأمير يوسف بن تاشفين: "والشيخ الإمام أبو بكر بن العربي قد أحرز من العلم في وقت تردده إليَّ ما لم يحرزه غيره، مع طول الأمد، وذلك لما خص به من صفاء الذهن، وذكاء الحسّ، واتّقاد القريحة، وما يخرج من العراق إلّا وهو مستقل بنفسه، حائز قصب السبق بين أقرانه"[19].

15 قانون التأويل: 449.
16 قانون التأويل: 450.
17 قانون التأويل: 451.
18 قانون التأويل: 452.
19 شواهد الجلة: 14.

- الإمام الفقيه الأصولي أبو القاسم إسماعيل بن عبد الملك الطوسي المعروف بالحاكمي الفقيه الشافعي (529هـ\1135م)، عديل الإمام أبي حامد الغزالي، ويذكر أنه كان أعلم بالأصول منه إلا أنه كان في لسانه ما يمنعه من الكلام[20]، لقيه القاضي وجلس بين يديه ينهل من معينه الفقهي والأصولي، قال القاضي: "ولقينا شيخ الشيوخ وصاحب الباب في العلم والرسوخ: إسماعيل الطوسي"[21].

ومما يدل على المناقشات الأصولية التي كانت بين القاضي وشيخه، ما رواه في مسألة التخصيص بالمصلحة: "ولم يفهم الشريعة من لم يحكم بالمصلحة ولا رأى تخصيص العلة، وقد رام الجويني رد ذلك في كتبه المتأخرة التي هي نخبة عقيدته، ونخيلة فكرته فلم يستطعه، وفاوضت الطوسي الأكبر في ذلك وراجعته حتى وقف، وقد بينت ذلك في المحصول والاستيفاء بما في تحصيله شفاء إن شاء الله تعالى"[22].

- الفقيه النظار الأصولي أبو الوفاء علي بن عقيل البغدادي الحنبلي (513هـ\1119م)، من مصنفاته الأصولية: "الواضح في أصول الفقه"، التقى به القاضي وجلس إليه واستفاد منه كثيرا، وقد أكثر من ذكره في كتبه، ومن المناقشات الأصولية بين القاضي وشيخه؛ ما ذكره في في مسألة التمكن من الفعل؛ هل هو شرط في إلزام الأمر أم لا؟ قال: "فذهب أحمد بن حنبل إلى أن المأمور به يثبت في ذمة المكلف قبل التمكن من الفعل، وقد فاوضتُ في ذلك علماءه؛ فقال لي الشيخ من أهل مذهبه في وقتنا أبو الوفاء بن عقيل وأبو سعد البرداني: إن هذه المسألة صحيحة من مذهبنا، ولكنها غير مذكورة الدليل؛ لأنها لا تجري عندنا في سنن تكليف ما لا يطاق، وإنما هي من باب إلزام المغمى عليه قضاء ما فاته من الصلوات في حال إغمائه فيما لم يدرك شيئا من وقته، فترجع المسألة فقهية؛ وهي في مسائل الخلاف مذكورة"[23].

- الإمام النظار الفقيه الأصولي أبو سعد يحيى بن علي الحُلْوَاني الشافعي (520هـ\1126م)، قرأ المذهب والخلاف والأصول على الشيخ أبي إسحاق، وبرع حتى التحق بالأئمة المناظرين[24]، التقى به القاضي، وناقش في مجلسه أصحابه، ومما حفظه لنا القاضي من ذلك ما ذكره في مسألة إفادة الاسم المشتق العلية؛ قال: "وإذا علق الشارع الحكم على اسم جامد أفاد ما تفيده الإشارة؛

20 تاريخ دمشق: 18/9.
21 قانون التأويل: 452.
22 أحكام القرآن: 279/2.
23 المحصول في الأصول: 57.
24 طبقات الشافعية لأبي بكر تقي الدين بن أحمد بن محمد ابن قاضي شهبة: 296/1.

وهي بيان المحل خاصة، وإذا علق الحكم على اسم مشتق أفاد تعليل الحكم بمعنى الاسم، وهذا بين في الأصول معلوم بالدليل، والثيوبة والبكارة اسمان مشتقان، فعلق الحكم بمعنى البكارة؛ وهو الاستحياء"[25].

فلما سمع شيخه مقالته؛ قال: "وكذلك والله أعربت عن نفسك، وأبنت عن مكانك، وأدنى مجلسي، وبقيت لديه مكرماً حتى فارقته"[26].

هؤلاء أبرز شيوخه الأصوليين، والناظر فيما سقنا يمكنه لحظ أمور كثيرة؛ منها:

أولا: شيوخ ابن العربي من فطاحل علم الأصول، وأكيد أن من درس عليهم سيكون له حظ وافر من ذلك العلم، كيف لا والقاضي كان ذا عقل وذكاء ودهاء وفطنة، ونفس طموحة متطلعة للعلم مشرئبة إلى المعالي، ولعل ما اجتمع لابن العربي في هذا العلم لم يجمع لغيره.

ثانيا: أغلب الشيوخ الذين درس عليهم وجالسهم وفاوضهم أعجبوا به وبفكره، وبالإشكالات والاعتراضات التي كان يوردها مع إحكام الجواب وإتقانه، وهذا ما تنبه له شيوخه ولامسوه، فقدموه وأكرموه وأعطوه المنزلة التي تناسب مقامه وحاله.

ثالثا: تلك الشهادات والتزكيات والتحليات التي صدرت من شيوخ تنبئ عن المكنة الأصولية التي تحلى به القاضي، وهي بإذن الله كافية في الدلالة على المطلوب؛ إذ لعل من أعلم الناس بالطلبة شيوخهم.

2 المسألة الثانية: أصولية ابن العربي من خلال كتب التراجم والفهارس

رأينا سالفا تلك التزكيات التي صدرت من شيوخ ابن العربي في مكنته الأصولية، وبالإضافة إلى ذلك؛ نحاول في هذه المسألة أن نستأنس بإيراد ما ورد في كتب التراجم من الأوصاف المتعلقة بأصوليته تعلما ورواية وتصنيفا وتدريسا، وبما أن كتب التراجم قد احتفت به كثيرا مع كثرتها، فإنني سأقتصر احتفاء البعض خصوصا تلميذيه القاضي عياض وابن بشكوال، وفي ذلك غنية وكفاية.

25 قانون التأويل: 449.

26 قانون التأويل: 449.

قال القاضي عياض (544هـ\1149م): "ودرس الفقه والأصول عند أبي بكر الشاشي وأبي بكر الطرطوشي، وقيد الحديث واتسع في الرواية، وأتقن مسائل الخلاف والأصول والكلام على أئمة هذا الشأن من هؤلاء وغيرهم".[27]

وقال أيضا: "ثم انصرف هو إلى الأندلس سنة خمس وتسعين، فسكن بلده وشوور فيه، وسمّع ودرّس الفقهَ والأصولَ، وجلس للوعظ والتفسير ورحل إليه للسماع".[28]

وقال عنه تلميذه أبو القاسم خلف ابن بشكوال (578هـ\1183م): " كان من أهل التفنن في العلوم والاستبحار فيها والجمع لها، متقدما في المعارف كلها، متكلما في أنواعها، نافذا في جميعها، حريصا على أدائها ونشرها، ثاقب الذهن في تمييز الصواب منها".[29]

قال أبو جعفر الضبي (599هـ\1203م): "فقيه حافظ، عالم متفنن، أصولي محدث مشهور، أديب رائق الشعر رئيس وقته".[30]

كما نجد أيضا في كتب الفهارس اعتناء القاضي بإدخال كتب الأصول إلى الأندلس، ثم التحديث بها ونشرها، قال ابن خير الإشبيلي (575هـ\1179م)- وهو من تلاميذه الذين أكثروا من الرواية عنه-: "كتاب الإرشاد لأبي المعالي عبد الملك بن عبد الله بن يوسف الجويني، وكتاب التلخيص له، وكتاب البرهان له، وكتاب الشامل له؛ حدثني بذلك كله شيخنا القاضي أبو بكر بن العربي"[31]، ثم ساق سند كل كتاب على حدة.

وفيما سبق إيراده تبيين لحال القاضي في علاقته بعلم أصول الفقه؛ فقد كان راويا لكتبه وناشرا لها في الآفاق، ودارسا لمباحثه.ومفاوضا لشيوخه وأقرانه لكشف خباياه، وعاقدا للمجالس ومدرسا لقضاياه.

3 المسألة الثالثة: أصولية ابن العربي من خلال المصنفات الأصولية

لعل من أهم الدلائل الدالة على تمكن المرء من علم ما، ما خلفه من آثار تشهد له من المكنة فيه، وهذه الآثار تتنوع إلى أمرين: أولهما تصنيف الكتب، وثانيهما صناعة الرجال، والأول أهمها ويمكن

27 الغنية في شيوخ القاضي عياض: 67.

28 الغنية في شيوخ القاضي عياض: 68.

29 الصلة في تاريخ أئمة الأندلس: 558.

30 بغية الملتمس في تاريخ رجال أهل الأندلس: 92.

31 فهرسة ابن خير الإشبيلي: 318–319.

حصر الإفادة منه وبيانها في هذه الدراسة، أما الثاني فهو عسِر ويخرجنا عن المطلوب؛ لكثرة طلاب القاضي والآخذين عنه.

وما دام أن الدراسة متعلقة بأصولية ابن العربي، فالمنهج يقتضي إيراد المصنفات التي لها ارتباط بالقضية المدروسة، والكتب التي تأكدنا من اندراجها في هذا المسلك هي:

- كتاب التمحيص: وقد أحال عليه كثيرا، وتلك الإحالات تفيد بمجموعها أنه من المصنفات الأصولية؛ من ذلك مثلا قوله: "وقد اختلف العلماء في المجتهدين في الفروع إذا اختلفوا، هل الحق في قول واحد منهم غير معين، أم جميع أقوالهم حق؟ والذي نراه أن جميعها حق؛ لقوله: ﴿فَفَهَّمۡنَٰهَا سُلَيۡمَٰنَۚ وَكُلًّا ءَاتَيۡنَا حُكۡمٗا وَعِلۡمٗاۚ﴾ [الأنبياء: 79]، وقد مهدنا ذلك في كتاب التمحيص، فلينظر فيه إن شاء الله"[32].

وقال أيضا: "اختلف الناس قديما وحديثا في أقل الجمع، وقد ذكرنا ذلك في التمحيص، وبينا ما وقع من الخلاف بين ابن عباس وعثمان رضي الله عنهما؛ وهما رؤساء الشريعة واللغة، والمحصول من ذلك: أنك إذا نظرت إلى لفظ الجمع فالاشتقاق يعطي أن الإثنين جمع، وإذا نظرت إلى إفراد العرب كل واحد بلفظه علمت أن أقله ثلاثة"[33]، وقال: "وأما اللفظ المشترك إن ورد كما قدمنا- معلقا بحكم يحتمل المتناولات كلها أو بعضها -فهو عندنا مجمل، وقد بينا ذلك كله في التمحيص"[34].

وقال عند حديثه عن مسألة حمل المطلق على المقيد: "ولهذا خالف الخاص العام، فإنهما إذا تعارضا حمل العام على الخاص من غير نظر في دليل سواهما، وقد بسطنا هذا الغرض في التمحيص وهذا محصوله"[35].

وقال في مسألة مراتب الرواة من الصحابة عن النبي ﷺ: "وقد بينا أدلة هذه المراتب في كتاب التمحيص، وخلصنا إلى المقصود منها في المحصول"[36].

وقال في دلالة لفظة "افعل": "ولفظة افعل في لسان العرب تأتي على وجوه كثيرة، ذكرناها في كتاب التمحيص وغيره"[37].

32 أحكام القرآن: 270/3.

33 المحصول: 77.

34 المحصول: 78.

35 المحصول: 108، وجل الإحالات التي أحال فيها على كتابه التمحيص من كتابه النكت: مسائل أصولية.

36 القبس في شرح موطأ مالك بن أنس: 226/1.

37 الناسخ والمنسوخ: 347/2.

هذه النقول وغيرها مما لم نذكره تفيد أمرين:

أحدهما: أن كتاب التمحيص من المصنفات الأصولية.

والثاني: أن الكتاب يعتبر من أوسع الكتب الأصولية للقاضي ابن العربي؛ إذ جل المواضع التي ورد فيها ذكره ينبه على مخلص ما تحدث عنه في التمحيص، ويرشد القارئ إلى طلب التفصيل منه.

- كتاب المحصول في علم الأصول، وقد أحال عليه المؤلف كثيرا في مواضع من كتبه، والذي ترجح لدي من خلال الإحالات أن هذا الكتاب ما زال في عداد المفقود، والذي طبع هو نكت المحصول، وقد تتبعت كثيرا من المسائل التي أحال عليه في كتبه فلم أجدها؛ من ذلك مثلا قوله في القبس: "ههنا أصل من أصول الفقه لم يتفطن له إلا مالك، وهو أن مراتب الرواة من الصحابة عن النبي ﷺ خمس" ثم ذكرها وقال: "وقد بينا أدلة هذه المراتب في كتاب التمحيص وخلصنا إلى المقصود منها في **المحصول**"[38]، فلما رجعت إلى الكتاب المطبوع لم أعثر عليها.

وكذلك لم أعثر على هذه المسألة التي ذكرها في قوله: "يجوز للنبي ﷺ أن يفرض بالاجتهاد على أمته؛ لأنه لو كان وحياً من الله تعالى بنفي أو إثبات لبلغه كان فيه حرج أو لم يكن، وقد مهَّدنا ذلك في كتاب المحصول وغيره"[39].

وأيضا هذه المسألة: "فأطلق المكروه على الحرام، وهو عنده ينقسم إلى ما يحرم فعله وإلى ما تركه أولى من فعله؛ وهو المكروه في إطلاق الأصوليين، إلا أنهم ما عرفوه ولا شرحوه في كتبهم، ولا ضربوا له مثالًا، وإنما يذكرون حده ويتجاوزونه دون بيان له، وهو من المعضلات في الأصول، وقد بيناه في المحصول، وذكرنا حده وأمثلته التي أغفلها العلماء قديمًا"[40].

وتتبع هذه النقول وإيرادها يخرجنا عن المقصود من هذه الدراسة لكثرتها، وفيما أوردناه كفاية في دلالة على صحة ما قلناه.

ومما يقوي هذا القول ويصححه، أن النسختين الخطيتين الموجودتين بدير أسكوريال[41] ومكتبة الملك عبد العزيز بالسعودية ورد في صفحة العنوان: "نكت المحصول في علم الأصول"[42].

38 القبس: 224/1–226.

39 القبس: 212/1.

40 القبس: 829/2–830.

41 رقم المخطوط (1191) في مجموع (60–108).

42 من أراد الاستزادة في المسألة فلينظر في الدراسة التي أعدها الدكتور حاتم باي عند تحقيقه لكتاب نكت المحصول، فقد أجاد وأفاد.

- كتاب نكت المحصول في علم الأصول؛ وهو كتاب مختصر، أورد فيه مهمات القضايا في علم الأصول، مع اعتناء بالحدود، والإعراض عن التفصيل والاستدلال ومناقشة الأقوال والاعتراض عليها، وندرة في النقد والترجيح، وذلك لأن غرض المؤلف منه هو تيسير هذا العلم على المبتدئ، وإعطائه مادة أولية ليترقى بها في سلم المعرفة الأصولية، وينظر بعدها في كتابيه "المحصول" و"التمحيص"، قال: "ويطلع على شيء من أصول الفقه "المحصول" أو "**نكته**" إن استطاله"[43].
- الاستشفا من كتاب الشفا؛ وهو تلخيص لنكت شفاء الغليل في بيان الشَّبه والمخيل ومسالك التعليل للإمام أبي حامد الغزالي، وإذا عرفت مادة الأصل فقد أغنت عن التعريف بمادة الفرع[44].

هذه الكتب الأصولية التي علمنا أنها لابن العربي في هذا الفن، وهي دالة على أصوليته من جهات متعددة؛ أهمها:

أولا: أن ابن العربي كان مثل غيره من الأئمة الذين صنفوا في هذا العلم على طريقة التثليث في التصنيف؛ أعني تأليف بسيط، وتأليف متوسط، وتأليف مختصر، فالكبير هو "التمحيص"، والمتوسط هو "المحصول"، والوجيز هو "النكت"، وهذا النهج في التصنيف كان شائعا إلى عهد ابن العربي، فالغزالي مثلا له ثلاثة كتب: تهذيب الأصول أبسطها، والمستصفى متوسطها، والمنخول أوجزها، وفي ذلك يقول القاضي: "وإن العبد منا ليؤلف كتاباً مُوعَباً في علم، ثم يختصره في طريق، ثم يشير إلى نكته في آخر، فيأتي عمله بسيطاً ووجيزاً وخلاصة، ويدل الأول على الآخر، ويقتضي القليل الكثير".[45]

ثانيا: هذه المصنفات بما ورد فيها من قضايا أصولية أو ما أحال عليه القاضي في مصنفاته الأخرى دالة على علو كعبه في هذا الفن، وناطقة بإمامته فيه، واستقلاليته في اختراع المعاني وابتكار المباني، وأحقيته بلقب "الأصولي"، ولو وقفنا على كتبه البسيطة والمتوسطة لعجبنا من صنيعه، ولتنادى الجميع بجدته وتجديده فيه، كيف لا وهو يقول عن المنهج الذي يقتفى في التصنيف: "ولا ينبغي لحصيف إذ يتصدّى إلى تصنيف أن يعدل عن غرضين: إما أن يخترع معنى، أو يبتدع وصفا ومتنا".[46]

ثالثا: من الإجحاف العلمي أن يكون الإمام ابن العربي بهذه الجلالة في هذا الفن، ثم عند استعراض أصوليي الغرب الإسلامي يقفز عليه أو يشار إليه إشارة دالة دونيته في علم الأصول، ويذكر من هو دونه فيه إبداعا وتأليفا وتجديدا واجتهادا.

---

43 قانون التأويل: 644.

44 الكتاب حققه الباحث المجد الدكتور عبد الله التوراتي، وصدر عن دار الحديث الكتانية.

45 القانون: 464.

46 عارضة الأحوذي بشرح صحيح الترمذي: 4/1.

رابعا: أن ابن العربي ومصنفاته تعطي صورة واضحة عن الدرس الأصولي بالأندلس والمغرب، وأنه كان منافسا لما كان عليه في المشرق، يمتاز بالمناقشة والتأصيل والنقد والإبداع والابتكار، وحتى ولو لم يكن كذلك، فهو ليس بتلك الصورة التي روجت عنه.

4 المسألة الرابعة: أصولية ابن العربي من خلال توظيفه لعلم الأصول

من المقطوع به عند الدارسين في العلوم الشرعية أن الملكة الأصولية لا تتحصل إلا بممارسة العلم وتوظيفه وتنزيله على فروعه، والاجتهاد في رد جزئياته إلى كلياته، والحذق في الاستفادة من مسائله وقضاياه.

والإمام أبو بكر ابن العربي كان له حظ وافر من الملكة الأصولية؛ إذ يمكن القول: إنه تفرد بمنهج في كتبه قلما يحاذيه أحد فيه أو يقاربه أو يجاريه، فهو كان نزاعا إلى اقتناص الأحكام من أصولها، وميالا إلى ربط الجزئيات بكلياتها، وحاذقا في توظيف القواعد والضوابط، وماهرا في التقعيد والتأصيل والاستدلال.

وأيضا كان له اعتناء كبير بالتأصيل، فقد أولاه أهمية كبيرة في مؤلفاته، فلا يكاد يخلو بحث من أبحاث الفقهية من لفت نظر أو توجيه أو فائدة أصولية؛ إذ كان حريصا على ربط تلك الفروع المدروسة بأصولها متى ما وجد إلى ذلك سبيلا.

هذا ويمكن تلمس التوظيف الأصولي عند الإمام من خلال النقط الآتية:[47]

أولا: التأصيل للمسائل الأصولية؛ وهذه من مميزات منهج الإمام؛ إذ تجده مجتهدا وبارعا في رد الفروع إلى أصولها، أو تنبيه على القواعد الكلية التي تتفرع منها الجزئيات حتى يسهل على الباحث استيعاب تلك المباحث، وضبط العلاقات المعرفية بينها، الشيء الذي تنبئ عن التفوق الأصولي في جانبه التوظيفي لهذا الإمام؛ إذ التوظيف والتنزيل غاية هذا العلم وهدفه وثمرته وفائدته.

إن المطالع لتراث الإمام الفقهي وبالخصوص "أحكام القرآن" و"القبس" و"المسالك" و"عارضة الأحوذي" تعترضه كثيرا مصطلحات ذات حمولة تأصيلية تقعيدية؛ كقوله: "تأصيل"، أو "مسألة أصولية" أو "توصيل"، ولعله أراد بهذا أن يوصل مسائل الفقه بحسب الإمكان بالأصول، ويستدل للقضايا الأصولية بالنصوص الشرعية، وينبه إلى مآخذ الأئمة في الأصول والفروع، ويرشد إلى طرق اقتناص الأحكام الكلية والجزئية، وأيضا يريد البرهنة على صحة مقالته حينما قال متحدثا عن الموطأ: "وهو أَوَّلُ كتاب أُلِّفَ في

47 استفدت كثيرا هذه المسألة مما كتبه الدكتور عبد الرحمن الزخنيني في كتابه أصول الفقه المالكي من خلال تطبيقات ابن العربي المعافري: 178، وما بعدها.

الإسلام، لم يُؤَلَّف مثله لا قَبْلَهُ ولا بعدَهُ، إذْ قد بناه مالك على تمهيد الأصول للفروع، وَنَبَّهَ فيه على عِلْمٍ عظيمٍ من مُعْظَم أصول الفقه التي ترجع إليه مسائله وفروعه. وَأَنَا إِنْ شاء الله أنبِّهكمَ على ذلك عِيَانًا، وتُحيطونَ به يقينًا، عند التّنبيه عليه في موضعه إن شاء الله"[48].

ومن الأمثلة التي يمكن الاستئناس بها على تأصيله للمسائل الأصولية والاستدلال لها[49]:

المثال الأول: ما جاء في تعليقه على قوله تعالى: ﴿إلا مَا يُتْلَى عَلَيْكُمْ﴾[المائدة: 1]: "يحتمل قوله: إلا ما يتلى عليكم الآن، أو إلا ما يتلى عليكم فيما بعد من مستقبل الزمان، وفي هذا دليل على جواز تأخير البيان عن وقت لا يفتقر فيه إلى تعجيل الحاجة، وهي مسألة أصولية، وقد بيناها في المحصول، ومعناه أن الله سبحانه أباح لنا شيئا، وحرم علينا شيئا استثناء منه، فأما الذي أباح لنا فسماه وبينه، وأما الذي استثناه فوعد بذكره في حين الإباحة، ثم بينه بعد ذلك في وقت واحد أو في أوقات متفرقة على اختلاف التأويلين المتقدمين، وكل ذلك تأخير للبيان"[50].

المثال الثاني: قوله في العارضة: "أصوله: في قوله: « نهاني ولا أقول نهى الناس »، دليل على نفي نقل الحديث على المعنى واتباع اللفظ"[51].

المثال الثالث: قوله عند حديثه عن الدلوك: "تأصيل: بيَّن مالك في هذا الباب أصلاً من أصول الفقه وهو أن الحكَم إذا تعلَّق باسم له أول وآخر تعلق بأوله، وقد اختلف العلماء في ذلك اختلافاً كثيراً، تتعلَّق به الفروعُ من "كتاب الطهارة" إلى "أُمَّهَات الأولاد"".[52]

المثال الرابع: قوله عند تأسيسه للقواعد العامة التي تندرج تحتها أحكام البيوع: "وإذا انتهينا إلى هذا المقام، فلا بد من تأسيس قواعد عشر نبني عليها معنى الكتاب، ويرجع الناظر إليها في أثناء الأبواب"[53]، ومن تلك القواعد التي أسسها وأصلها: قاعدة القول بالعرف، وقاعدة اعتبار الحاجة في تجويز الممنوع، وقاعدة المصالح والمقاصد.

ثانيا: تصحيح الاستدلال بالقواعد الأصولية على المسائل الفرعية؛ وهذا مما اشتهر به القاضي ابن العربي، فما أكثر المسائل التي صححها بناء على خطأ أصولي بنيت عليه أحكام جزئية؛ وهذا ضرب من

48 المسالك في شرح الموطأ: 330/1.

49 والأمثلة كثيرة جدا، اكتفيت ببعضها تعلقا بالموضوع، ورعيا للمقصود.

50 أحكام القرآن: 15/2، وقد استدل لهذه القاعدة في العارضة: 211/1، عند حديثه عن حديث السائل عن أوقات الصلاة، وأطال النفس في بيان أنواعه واحتمالاته.

51 عارض الأحوذي: 57/2.

52 القبس: 95/1.

53 القبس: 786/2.

التمهير الأصولي الذي اتصف به القاضي، ومن المسائل التي يمكن الاستئناس بها في هذا السياق ما جاء في تعليقه على حديث بيع الخيار: "وقد مهدنا في مسائِل الخلاف تأويلات هذا الحديث، وحققناها من ثمانية أوجه؛ منها: قول مالك فيه: "وليس عندنا في هذا الحديث حد معروف، ولا أمر معمول به" إشارة إلى أن المجلس مجهول المدة، ولو شرط الخيار مدة مجهولة لبطل إجماعًا، فكيف يثبت حكم بالشرع بما لا يجوز شرطًا في الشرع؟ وهذا شيء لا يتفطن إليه إلا مثل مالك، فظن الجهال المتوسمون بالعلم من أصحابنا أن مالكًا إنما تعلق فيه بعمل أهل المدينة، وهذه غباوة"[54].

وفي هذ الذي ذهب إليه رد على من اعتبر مالكا قد رد خبر الواحد بعمل أهل المدينة، وفيه أيضا رد على من اعتمد على هذا الحديث في شروط قبول الاحتجاج بخبر الواحد.

وأيضا تعليقه على قوله تعالى: ﴿لَا يَنْهَاكُمْ اللَّهُ عَنْ الَّذِينَ لَمْ يُقَاتِلُوكُمْ فِي الدِّينِ وَلَمْ يُخْرِجُوكُمْ مِنْ دِيَارِكُمْ أَنْ تَبَرُّوهُمْ وَتُقْسِطُوا إِلَيْهِمْ إِنَّ اللَّهَ يُحِبُّ الْمُقْسِطِينَ﴾ [الممتحنة: 8]: "استدل به بعض من تعقد عليه الخناصر على وجوب نفقة الابن المسلم على أبيه الكافر، وهذه وهلة عظيمة؛ فإن الإذن في الشيء أو ترك النهي عنه لا يدل على وجوبه، إنما يعطيك الإباحة خاصة"[55].

ثالثا: إيراد النكت الأصولية؛ وهذا مما تفرد به القاضي في كتبه الفروعية، وغرضه من ذلك رفع الكلفة والعنة[56]، وإيقاف القارئ على أصول المسائل ومنشأ الخلاف، والتنبيه على أدلة القواعد الكلية، والإشارة إلى العلاقة التلازمية بين علم الأصول وعلم الفقه، وتقديم الفقه بطريقة مبتكرة قائمة على دمج الأصول بفروعها، ونظرا لكثرة الأمثلة التي وردت في كتبه، فإني سأقتصر على بعضها، ومن الأمثلة التي يمكن الاستئناس بها؛ نذكر:

المثال الأول: "نكتة أصولية: إذا اختلف ألفاظ الحديث في الرواية فتأملوا الحديث؛ فإن كان مما يتكرَّر فكل لفظ أصل يمهَّد وتبنى عليه الأحكام، وإن كان مما لا يتكرَّر فيُعلم قطعاً أن النبي ﷺ إنما قال أحدهما، وأن الراوي هو الذي عبَّر عن تلك الحالة الواحدة بألفاظ مترادفة أو متقاربة، فتعرض الألفاظ على الأصول والأدلة، فما استمر منها عليها هو الذي يُبنى عليه الحكم"[57].

54 القبس: 845/2.
55 أحكام القرآن: 288/4.
56 القبس: 1201/3.
57 القبس: 695/2.

المثال الثاني: "نكتة أصولية: وهي تنبني على أصل من أصول الفقه، وهو أنّ الصّاحب إذا أَفتَى بخلاف القياس، هل يكونُ أصلًا يُرجَعُ إليه أم لا؟ والصّحيحُ أنّه لا يرجع إليه، ولضَعفِ المسألةِ استَحَبَّ مالكٌ للرَّاعِفِ إنّ تكلَّمَ ألَّا يبني"[58].

المثال الثالث: "نكتة أصولية: قال علماؤنا: في هذا الحديث دليلٌ عظيمٌ على أصلٍ من أصولِ الفقه، وهو القولُ بالقياس في الدِّين والاجتهاد، ألَّا ترى إلى مشاوَرةِ رسولِ الله ﷺ مع أصحابه في الأذان ولم ينتظر في ذلك وَحْيًا ولا طلبَ منه بيانًا، وإنّما أراد أنّ يأخذ فيه ما عند إصحابه من رَأي يستنبطونَهُ من أصول الشّريعة، ويَنْتَزِعُونَه من أغراضها"[59].

رابعا: النقد والاعتراض؛ وهي سمة جلية في الفكر الأصولي عند القاضي، وقد أكثر من توظيفها في رده على المخالفين، ولا يغادر مسألة له فيها نظر إلا ويبدي فيها وجها من وجوه النقد والاعتراض.

وليس أحد عند ابن العربي يرتفع عن أن يكون محلا للنقد[60]، فهو ينتقد على شيوخه ممن كرع من حياض علمهم، وتخرج عليهم كالغزالي وإمام الحرمين على تقديره لهما وإقراره لهما بالتحقيق والتدقيق والتبحر في هذا العلم، بل تعدى نقده إلى مقالات أبي بكر الباقلاني وأبي إسحاق الإسفرايني، وأحيانا تعدى نقده إلى إمام الأئمة مالك.

وهذه الشخصية العلمية البارزة في ابن العربي تحريرا للمسائل ونقدا للآراء والمذاهب، وتحقيقا لمذهب مالك؛ جعلته من أهل التحقيق، فقد وصفه الزركشي في بحره بذلك، قال: "وقال بعض محققي المالكية..."[61].

والأسلوب الذي يخلعه ابن العربي على اعتراضاته وانتقاداته أسلوب تهويلي، فكثيرا ما يتعجب ويفرط في تعجبه؛ كقوله: "يا لله وللمحققين من رأس التحقيق الجويني"، "وأعجب لرأس المحققين إمام الحرمين يقول..."، "وتعجبوا من رأس التحقيق يعوِّل..."، "ومن يعجب فعجب تعويل رأس المحققين على مثل هذا الأصل المتزعزع"، "وعجبا لهذين الإمامين –يقصد الإسفرايني والجويني -على جلالة قدرهما أصولا ولغة كيف خفي عليهما هذا"، وقال معلقا على قول القاضي الباقلاني: "وهذه هفوة من القاضي".

ومنهجه في النقد والاعتراض أنه يسوق رأي من يريد انتقاده والاعتراض عليه، ثم ينبه على وجه ضعفه، ثم يعقب بعدها بتأصيل القاعدة الأصولية التي يرجح صحتها؛ ومن الأمثلة على ذلك؛ نذكر:

---

58 المسالك: 157/2.

59 المسالك: 313/2.

60 هذه الفقرة واللاتي بعدها استفدتها مما كتبه الدكتور حاتم باي في مقدمة تحقيقه للنكت: 122–124.

61 البحر المحيط: 98/8.

المثال الأول: قال: "**اعتراض:** فإن قيل: هذا الحديث لا حجة فيه، فإن راويه ابن عباس وكان يفتي بأن المرتدة لا تقتل، والراوي إذا أفتى بخلاف ما روى سقطت روايته.

قلنا: هذا سؤال فاسد؛ لأنهم بنوه على مذهبهم، وعندنا أن الراوي في مخالفة روايته كسائر الناس، وهي مسألة أصولية، وقد أوضحناها في "الكتاب "الكبير" وبينا أنهم قد نقضوا هذا الأصل، وأخذوا بمسائل أفتى فيها الراوي بخلاف ما روى"[62].

المثال الثاني: قال: "فإن قيل: إجماع الأمة لا يكون إلا عن حديث سمعوه من النبي عليه السلام، فإذا أجمعوا علمنا ضرورة وجود الأثر.

قلنا: هذا مذهب محمد بن جرير الطبري، وهو ساقط قطعا، فإن الأمة قد تجمع على النظر كما تجمع على الأثر، وقد بينا ذلك في "أصول الفقه""[63].

المثال الثالث: قال: "كان أحمد بن حنبل يروي: أن ضعيف الأثر خير من قوي النظر، وهذه وهلة لا تليق بمنصبه؛ لأن ضعيف الأثر كالعدم لا يوجب حكمًا، والنظر أصل من أصول الشريعة، عليه عوَّل السلف، ومنه قامت الأحكام وبه فصل بين الحلال والحرام"[64].

خامسا: التنبيه على الوهم؛ وهذا طريق آخر من طرق التوظيف الأصولي في الفروع الفقهية، وغرضه من ذلك أن ينبه على أوهام وقعت في الاستدلال لبعض القواعد الأصولية، ومن الأمثلة التي يمكن الاستئناس بها؛ نذكر:

المثال الأول: "تنبيه على وهم: قال بعض علمائنا: هذه الآية في شأن سعد، نسخت ما كان في الجاهلية، وهذا باطلٌ، فإنّ الجاهليّة ليس لها حُكمٌ يثبت حتّى يتطرّق إليه رَفعٌ، ولا يثبت له في الإسلام قدم بيان، فإنّه لو كان مسكوتًا عنه لكان شرعًا، ولو كان شرعًا لما انتزع النَّبيُّ ' من أخي سعد بن الرّبيع ما كان أخذه، فإنَّ حكمَ النّاسخ يثبتُ ساعةَ نزوله ولا يعترض على ما سبقه، فلمّا أمر النّبيُّ ﷺ أخا سعدٍ بأن يردّ ما أخذَ، وهذا يُبيّن أنّها كانت ظُلَامَة"[65].

المثال الثاني: "تنبيهٌ على وَهمٍ: قال بعضُ النَّاس: إنَّ الرَّجْمَ الواردَ في الشَريعةِ ناسخٌ للحبس إلى الموت الّذي كان مشروعًا قبلَهُ، وقد بيَّنّا فسادَ ذلك في "كتب الأصول" من وجوهٍ، أقربُها الآنَ إليكم: أنّ الحبسَ في البيوتِ كان حُكمًا ممدودًا إلى غايةٍ، وكلُّ حُكمٍ مُدَّ إلى غايةٍ فانتهى إلينا، لا يكونُ انتهاؤُهُ

62 المسالك: 6/352–353.

63 المسالك: 6/547.

64 القبس: 1/341.

65 المسالك: 6/548.

نسخًا، وهو أحدُ شروطِ النَّسخِ الأربعةِ الّتي لا يُزادُ عليها، وحُكمُ الغايةِ أنّ يكونَ ما بعدَها مخالفًا لما قبلَها، وإلّاَ فمتَى كانت تكون غايةً"[66].

من خلال ما سبق يمكن القول: إن القاضي أبا بكر كانت له مهارة أصولية فريدة وعجيبة، استطاع من خلالها أن يوظفها في بناء الأحكام الجزئية، وربط الفروع بالأصول، وكشف بعض الأوهام التي وقع فيها من سبقه في التأصيل والاستنباط والتنزيل.

5 المسألة الخامسة: أصولية ابن العربي من خلال ترجيحاته وتفرده

مما امتاز به الفكر الأصولي عند ابن العربي عدم اكتفائه بسرد المسائل الأصولية، بل يتجاوز ذلك إلى النقد والاعتراض والترجيح بين الأقوال، وأحيانا يتفرد فكره بإيراد مسائل لم يسبق إليها، وهذا يبرز لنا جانبا من جوانب أصولية القاضي.

وقضية الترجيح بين الأقوال الأصولية ليست بالأمر الهين يستطيعه كل من اشتغل بهذا العلم، بل الأمر يحتاج إلى مكنة ومهارة وصناعة؛ إذ العملية الترجيحية تتطلب استدعاء معارف متخصصة من مجالات شتى، مع ضبط العلاقات المعرفية بينها، وإلا اختلطت عليه القضايا، وأتى بالغرائب التي تخالف القطعيات وبدائه العقول.

أما قضية التفرد بمسائل أصولية لم يسبق إليها؛ فهي رغم قلتها ناطقة بإمامته في هذا العلم، ولو قدر لكتبه الأصولية الذيوع والنشر لرأينا العجب العجاب، ولبَانَ لنا جلالة قدر هذا العالم وما أنتجته قريحته الأصولية، وإنني أخمن، بل أقطع وأوقن: أن الإمام ابن العربي كان في مصاف المحققين والمبدعين في علم الأصول كالباقلاني والجويني والغزالي.

هذا وبعد التتبع وجدت له ترجيحات كثيرة ومتنوعة يصعب حصرها في هذه الدراسة المختصرة، لذلك سأكتفي بإيراد بعض الأمثلة ثم أردفها ببعض المسائل التي تفرد بها.

فمن ترجيحاته؛ قوله في مسألة اللَّفْظ الْعَام إِذا خص فِي بعض متناوَلاته: "قَالَت الْمُعْتَزلَة: يكون مُجملا لَا يُتَعَلّق بِهِ، وَقَالَ القَاضِي: هُوَ مجَاز، وَقَالَ الشَّافِعِي: هُوَ حَقِيقَة على أَصله. وَالصَّحِيح: جَوَاز التَّعلق بِهِ، فإنه إذا نُسِخ بعضُ الحكم جَازَ التَّعلّق بِهِ، فَكيف فِيمَا خص؟ وَلَو صرح المتلفظ بالتخصيص مَقْرُونا بِاللَّفْظِ مَا كَانَ ذَلِك مجَازا يُضعفهُ، وَلَا إِجْمَالا يُبطلهُ، فَكَذَلِك إِذا جَاءَ بعد حِين"[67].

---

66 المسالك: 107/7.

67 المحصول: 81.

وأيضا قوله في مسألة صيغ العموم: "كما اختلف علماؤنا في صيغة الأمر كذلك اختلفوا في صيغة العموم، وكان الذي مال بعلمائنا الأصوليين رحمهم الله إلى نفي القول بالعموم، وحداهم إلى إنكار صيغه: إلحاح الوعيديه عليهم بكل آية عامة وحديث مطلق يقتضيان معاقبة العصاة وجزاء المذنبين، والذي ندين الله به: أن العموم لها صيغ معلومة وألفاظ معروفة، وما تعلق به الوعيدية ساقط لهم- مع مساعدتهم على ألفاظ العموم -بالأدلة المعلومة في تلك المسألة"(68).

وقوله أيضا في مسألة تخصيص عموم القرآن بالخبر الواحد: "اختلف الناس في جواز تخصيص عموم الكتاب بخبر الواحد؛ فمنعت منه المعتزلة؛ لأن القرآن مقطوع به وخبر الواحد مظنون، وقال القاضي: أنا أتوقف فيه، ومال الفقهاء بأجمعهم إلى جواز تخصيصه به"، ثم قال بعد إيراد حجة الفريقين على سبيل الاختصار: "وهذا لا يصح؛ فإن خبر الواحد مقطوع على وجوب العلم به مقطوع على فحواه، يصح في التخصيص، فيرجح على عموم القرآن، والذي يؤكد ذلك: ما روي من قبول الصحابة رضوان الله عليهم لتخصيص عموم آية المواريث بحديث أبي بكر في حق النبي ﷺ حين قال: سمعته يقول: « لا نورث، ما تركت بعد نفقة عيالي معونة عاملي فهو صدقة »".[69]

ومن أمثلة تفرده في المسائل الأصولية؛ قوله في مسألة القضاء بأمر جديد: "لا خلاف بين المحصلين: أن الأمر بالفعل المؤقت إذا فات ليس فيه تعرض لقضائه بعد ذلك، خلا أن الفقهاء شُغفوا بأن يقولوا: إن القضاء بعد فوات الوقت يجب بالأمر الأول، وهذا لا تشهد له اللغة ولا تقتضيه أدلة العقول، فأما الشريعة فإن جاءت به لزم، وإن سكتت عنه سقط، وكل ما يتعلق به في ذلك من الظواهر فهو حجة لنا؛ لأنه أمر ثان، فأما ثبوت الإثم مع سقوط الفعل فثابت إجماعا الفرقانُ بينهما؛ لأن الأمة أجمعت على وجوب التأثيم وانحتام التعصية"[70].

وأيضا قوله في مسألة ترجيح العموم في خصوص العين على العموم في خصوص الحال: "اختلف علماؤنا في دم الحيض؛ فقال بعضهم: هو كسائر الدماء يعفى عن قليله، ومنهم من قال: قليله وكثيره سواء في التحريم، رواه أبو ثابت عن ابن القاسم...

وجه الأول: عموم قوله تعالى: ﴿أو دما مسفوحا﴾[الأنعام: 145]، وهذا يتناول الكثير دون القليل.

ووجه الثاني: قوله تعالى: ﴿قل هو أذى﴾[البقرة: 222]، وهذا يعم القليل والكثير، ويترجح هذا العموم على الآخر بأنه عموم في خصوص عين.

---

68 المحصول: 73-74.

69 المحصول: 88-89.

70 المحصول: 65-66.

وذلك الأول هو عموم في خصوص حال، وحال المعين أرجح من حال الحال، وهذا من غريب فنون الترجيح، وقد بيناه في أصول الفقه، وهو مما لم نسبق إليه ولم نزاحم عليه"[71].

وأيضا قوله في مسألة الخبر بمعنى الأمر: "هذا منتهى كلامهم، وهو ساقط جداً؛ لأن الخبر لا يجوز أن يكون بمعنى الأمر، كما لا يجوز أن يكون الأمر بمعنى الخبر، كما لا يجوز أن يكون كل واحد منهما بمعنى النهي، ولا يجوز أن يكون النهي بمعناهما؛ لأن الكلام له حقيقة ينفرد بها عن العلم والإرادة، وكذلك أيضاً أقسامه من الأمر والنهي والخبر والاستخبار لها حقائق ينفرد كل واحد منها حقيقة عن صاحبه"[72].

خلاصة الكلام: إن الإشكاليتين اللتين انطلقتْ منهما هذه الدراسة قد تم الجواب عنهما؛ إذ بينا أن علم الأصول عريق في القطر المغربي الأندلسي، وأن رجالاته أسهموا بمصنفات جليلة أعطت نفسا تجديديا لهذا العلم، ومن الذين ضربوا بحظ وافر فيه الإمام ابن العربي، وهو يعتبر بحق أصوليا عقت أرحام النساء أن تلد مثله، فقد أثبتت أنه كذلك من خلال تتبع الشيوخ الذين كرع من حياضهم الأصولية، وأردفت ذلك بنقول عن شيوخه وطلابه وما احتوت عليه كتب التراجم والفهارس تشهد بمكنته الأصولية، ثم عززت أصوليته بما صنفه من كتب فيها، وكيفية استفادته منها واستدعاء قواعدها في الفروع الفقهية، ثم نبهت على بعض ترجيحاته وتفرداته.

لائحة المصادر والمراجع

أحكام القرآن للقاضي أبي بكر محمد بن عبد الله بن العربي المعافري الإشبيلي المالكي، تحقيق: محمد عبد القادر عطا، دار الكتب العلمية-بيروت، الطبعة: الثالثة، 1424هـ/2003م.

أرجوزة في أصول الفقه لأبي عبد الله محمد بن علي بن عبد الكريم الفندلاوي المعروف بابن الكتاني، تحقيق: عبد الغني ادعيكل، دار فارس-الكويت، الطبعة: الأولى، 1443هـ/2021م.

أصول الفقه المالكي من خلال تطبيقات ابن العربي المعافري لعبد الرحمن الزخنيني، مطبعة الجسور-وجدة، الطبعة: الأولى، 2021م.

البحر المحيط في أصول الفقه لأبي عبد الله بدر الدين محمد بن عبد الله بن بهادر الزركشي، دار الكتبي، الطبعة: الأولى، 1414هـ/1994م.

71 أحكام القرآن: 1/224.

72 القبس: 1/397، وهذا الذي ذهب إليه اعتمد عليه كثيرا في بناء الأحكام الجزئية وتوجيه النصوص الشرعية.

بحوث الملتقى الأول: القاضي عبد الوهاب البغدادي شيخ المالكية بالعراق: المصادر الأصولية عند المالكية: دراسة في النشأة والمدونات والخصائص لمولاي الحسين بن الحسن الحيان، المؤتمر العلمي لدار البحوث والدراسات الإسلامية وإحياء التراث بدبي، 2003م.

بغية الملتمس في تاريخ رجال أهل الأندلس لأبي جعفر أحمد بن يحيى بن أحمد بن عميرة الضبي، دار الكاتب العربي-القاهرة، 1967م.

تاريخ دمشق لأبي القاسم علي بن الحسن بن هبة الله المعروف بابن عساكر، المحقق: عمرو بن غرامة العمروي، دار الفكر، 1415هـ/1995م.

ديوان المبتدأ والخبر في تاريخ العرب والبربر ومن عاصرهم من ذوي الشأن الأكبر لأبي زيد ولي الدين عبد الرحمن بن محمد بن محمد ابن خلدون الحضرمي الإشبيلي، تحقيق: خليل شحادة، دار الفكر-بيروت، الطبعة: الثانية، 1408هـ/1988م.

شواهد الجلة والأعيان في مشاهد الإسلام والبلدان منسوب إلى القاضي أبي بكر محمد بن عبد الله بن العربي المعافري الإشبيلي المالكي، مخطوط بالخزانة العامة بالرباط د1020.

الصلة في تاريخ أئمة الأندلس لأبي القاسم خلف بن عبد الملك بن بشكوال، تحقيق ومراجعة: السيد عزت العطار الحسيني، مكتبة الخانجي، الطبعة: الثانية، 1374هـ/1955م.

طبقات الأصوليين بعدوتي المغرب والأندلس ليونس العدلاني، دار الأمان-الرباط ودار الكتب العلمية-بيروت، الطبعة: الأولى، 1441هـ/2020م.

طبقات الشافعية لأبي بكر تقي الدين بن أحمد بن محمد ابن قاضي شهبة، تحقيق: الحافظ عبد العليم خان، عالم الكتب-بيروت، الطبعة: الأولى، 1407هـ.

عارضة الأحوذي بشرح صحيح الترمذي للقاضي أبي بكر محمد بن عبد الله بن العربي المعافري الإشبيلي المالكي، تحقيق: جمال مرعشلي، دار الكتب العلمية-بيروت، الطبعة: الأولى، 1418هـ/1997م.

الغنية فهرست شيوخ القاضي عياض لأبي الفضل عياض بن موسى بن عياض بن عمرون اليحصبي السبتي، تحقيق: ماهر زهير جرار، دار الغرب الإسلامي، الطبعة: الأولى 1402هـ/1982م.

فصل المقال فيما بين الحكمة والشريعة من الاتصال لأبي الوليد محمد بن أحمد بن محمد بن أحمد بن رشد القرطبي الشهير بابن رشد الحفيد، دراسة وتحقيق: محمد عمارة، دار المعارف، الطبعة: الثانية، (ب ت).

الفكر الأصولي: دراسة تحليلية نقدية لعبد الوهاب إبراهيم أبو سليمان، دار الشروق، الطبعة: الأولى، 1403هـ/1982م.

فهرسة ابن خير الإشبيلي، تحقيق: بشار عواد معروف ومحمود بشار عواد، دار الغرب الاسلامي-تونس، الطبعة: الأولى، 2009م.

قانون التأويل للقاضي أبي بكر محمد بن عبد الله بن العربي المعافري الاشبيلي المالكي، دراسة وتحقيق: محمّد السّليماني، دار القبلة للثقافة الإسلاميَّة-جَدّة ومؤسَسَة عُلوم القرآن-بيروت، الطبعة: الأولى، 1406هـ/1986م.

القبس في شرح موطأ مالك بن أنس للقاضي محمد بن عبد الله أبو بكر بن العربي المعافري، تحقيق: محمد عبد الله ولد كريم، الناشر: دار الغرب الإسلامي، الطبعة: الأولى، 1992م.

محاضرات في تاريخ المذهب المالكي في الغرب الإسلامي لعمر الجيدي، منشورات عكاظ، (ب ب).

المحصول في أُول الفقه للقاضي أبي بكر محمد بن عبد الله بن العربي المعافري الإشبيلي المالكي، تحقيق: حسين علي اليدري وسعيد فودة، دار البيارق-عمان، الطبعة: الأولى، 1420هـ/1999م.

مدخل إلى أصول الفقه المالكي لمحمد الختار ولد أباه، دار ابن حزم-بيروت ودار الأمان بالرباط، الطبعة: الأولى، 1432هـ/2011م.

المسالك في شرح الموطأ للقاضي أبي بكر محمد بن عبد الله بن العربي المعافري الإشبيلي المالكي، تحقيق: محمد بن الحسين السُّليماني وعائشة بنت الحسين السُّليماني، دَار الغَرب الإسلامي، الطبعة: الأولى، 1428هـ/2007م.

المصادر الأندلسية لعلم الكلام: المسار والتحولات والخصائص لخالد زهري، منشورات الرابطة المحمدية للعلماء، الطبعة: الأولى، 1441هـ/2020م.

الناسخ والمنسوخ في القرآن الكريم للقاضي أبي بكر محمد بن عبد الله بن العربي المعافري الإشبيلي المالكي، تحقيق: عبد الكبير العلوي المدغري، مكتبة الثقافة الدينية، الطبعة: الأولى، 1413هـ/1992م.

ندوة التراث المالكي في الغرب الإسلامي: المدرسة المالكية الأصولية وإبداع المغاربة فيها لمحمد التمسماني، منشورات كلية الآداب والعلوم الإنسانية-عين الشق، الدار البيضاء، مطبعة المعارف الجديدة-الرباط، 1998م.

نفائس الأصول في شرح المحصول لشهاب الدين أحمد بن إدريس القرافي، تحقيق: عادل أحمد عبد الموجود وعلي محمد معوض، مكتبة نزار مصطفى الباز، الطبعة: الأولى، 1416هـ/1995م.

نفح الطيب من غصن الأندلس الرطيب، وذكر وزيرها لسان الدين بن الخطيب لشهاب الدين أحمد بن محمد المقري التلمساني، تحقيق: أحسان عباس، دار صادر-بيروت، الطبعة 1997م.

نكت المحصول في علم الأصول للقاضي أبي بكر محمد بن عبد الله بن العربي المعافري الإشبيلي المالكي، تحقيق: حاتم باي، دار ابن حزم، الطبعة الأولى، 1438هـ/2017م.

6

الاستشفا من كتاب الشِّفا: بيانٌ للنظر المعافريّ في مَشْرَع أبي حامد الأُصُولي

عبد اللّٰه التَّوْرَاتِي

[*al-Istishfā min Kitāb al-Shifā*]

[***Ibn al-ʿArabī's Analysis of al-Ghazālī's* uṣūlī *Project***]

[*Abdallah Taourati*]

تقدمة:

تقصد هاته الورقة إلى الكشف عن حلقات الوصل في الدرس الأُصولي بين أبي بكر ابن العربي (543هـ\1148م) وأبي حامد الطُّوسِي (505هـ\1111م)، من خلال جملة من الأنظار الأُصولية المعافرية، والتي بثّها في تفاريق كُتُبه، ومنثور مسائله، محدّدة أساس هذا الاتصال، وجوانب التمايز بين مشارع الرجلين.

وتعتمد هاته الأنظار -أساساً- على الكتاب الجديد الذي كان في حُكم المفقود، وهو الاستشفا من كتاب الشفا، لتلحظ جملة من الأنقاد المعرفية لمَشْرَع أبي حامد الغزالي، والتي تمتاح من أصول الفقه المالكي، وتعوّل على استثماره في مسائل الخلاف والجدل الفقهي.

1 الحاجةُ إلى أبي حامد في الأندلس

ورث أبو حامد تراث المدرسة الجُوَينية، وسار بسيرها، وعوَّل على قواعدها، وقد كانت بواكير عمله الأُصولي على نَظَرٍ من شيخه أبي المعالي، في كتاب المنخول، والذي وسمه في آخره باسم: المنخول من تعليق الأُصول، وكتاب شيخه اسمُه: التعليقُ في الأُصول، وشُهر بكتاب البرهان، وذَكر أبو حامد ثمة أن عمله لا يزيد عن تفصيل الفصول، وتبويب الأبواب[1].

1 المنخول: 504.

 | DOI:10.1163/9789004706859_008

وعلى مَشرع المنخول كان كتابه المنتخل، وهو خاصٌّ بمسائل الجدَل، ورأى ابن العربي رأياً خاصّاً في كتابه هذا، فجعله فصلاً في كتاب المحصول، وهو كتابه الأوسط في الأصول.

وفي كتابه نُكت المحصول كان مسايراً لأنظار شيخه أبي حامد، ولم يُخْلِ تلك الأنظار من نقد، ومراجعة، وإعلانٍ عن نفسه ورأيه في مسائل خاصة، يجدُها المطالعُ للكتاب منثورة فيه، من ابتدائه إلى انتهائه.

وتزوّد ابنُ العربي في رحلته من كُتُبه ومصنفاته، فأدخل جملة منها، لم يسبقه أحدٌ إليها، وذَكر تفصيل ذلك في سراج المريدين[2].

والذي نلحظه في خصوص الصلة بأبي حامد وجنسها، أن قصد ابن العربي كان متّجهاً إلى أمور ثلاثة:

أولها: حَصْرُ الامتداد الظاهري في مجتمع إشبيلية والتضييقُ عليه، وأشار أبو بكر إلى فُشُوّ القول بالظاهر بحضرته في جملة من كتبه: المتوسط في الاعتقاد[3]، وترتيب المسالك[4]، والعواصم[5]، والعارضة[6]، والاستشفا[7].

ثانيها: نفخُ الدرس الأصولي نفحةً خراسانية، ولم يكن مرتضياً سيرة الأصول بالأندلس، وذلك أن هاته البلاد كانت على مذهب واحد، وأنظار متوحدة، فلم تنشط بها مجالس النظر والجدل، عكس البلاد المشرقية، وقد لَقِن ابن العربي عن مشيخته الخراسانية سُبُل النظر، وطرائق الجدل، وناظر وحاقق، وجادَل وباحَث، وهاته الطرائق لم تكن مهيّأة في بلاده، فلما رجع أراد أن يشدّ من عضد الأصول بها، وينفحها نفحة تجعلها مطيقة للتصدي للظاهرية ومقالاتها.

ثالثها: الاعتراف بمكانته العلمية في الأصول، وهي التي كان يجادله فيها متفقهة إشبيلية ومُفتوها، وكان ينكرون عليه أموراً، لأنه لم يكن لهم ارتياض في ساحاتها، وقد حاولها قبله أبو الوليد الباجي (474هـ\1081م)، إلا أنه لم يتهيّأ له ذلك، وكان أكثر ما يُعيق حركة ابن العربي طعن علماء بلده فيه، لما غلب عليهم من الحسد والمعاندة.

---

2 سراج المريدين: 407/4.

3 المتوسط في الاعتقاد: 408.

4 ترتيب المسالك: 330/1.

5 العواصم: 249.

6 العارضة: 388/9.

7 الاستشفا: 88.

وأسوق هنا نصّاً يدلّنا على ما قدَّمنا من الأسباب الداعية له إلى الانفتاح على النظر الغزالي في الأصول، قال القاضي: «وقد علمتَ -أيها المسترشد- أنَّ أهلَ بلادنا -تداركهم اللّٰه- ما بين جامد على رَسْمِ الظاهِر، أو سائرٍ في المعاني من غير مُظاهِر فيها ولا مؤازِر، فذاك يسفُل إلى الهاوية، وهذا يقتحم صعودًا عالية، فهو ذو شِدَّة لا تُعينها مِرَّةُ عَيْسٍ ليس دونه وليّ.

يقول: الشروطُ الثلاثة عندي حاصلة؛ فأين الرابعةُ التي لا تكون هذه الطائفة إليها واصلة؟ متى باحثَ وناظر؟ متى حاقَق وجادل؟ أيَّ إمامٍ لَقِي؟ وفي أيّ درَج من الحقائق رَقِي؟ ما طالَع مِن الأصول؟ مع مَن تدرّب في التعليل والتمثيل؟ من أَرشده إلى لَقَمِ الدَّليل؟ »[8].

2 الدلالة على موارد أبي المعالي وأبي حامد في الأصول

ذَكر أبو بكر في طالعة «الاستشفا» أن أبا المعالي (478هـ\1085م) أفاد من أبي زيد الدبُّوسي (430هـ\1039م)، ونهج بنهجه في كتبه الخلافية والأصولية، فنظر في الخلاف إلى كتاب أسرار المسائل، وخلّص منه كتاب الأساليب[9]، وفي الأصول كان نظره إلى تقويم الهداية[10]، ومنه استخرج كتاب البرهان.

وهذه من ابن العربي دعوى، يُحتاج في الركون إليها إلى سبر لها، وتفتيش فيها، وذلك لا يكون إلّا في زمان ممتدّ، فكتاب أسرار المسائل لم يُنشر بعد، وكتاب الأساليب لا نعلمه موجودًا.[11]

وأما **البرهان والتقويم** فمشهوران، ولا أعرف دراسةً وازنتْ بينهما، وإن كان كلامُ ابن العربي هذا داعٍ إلى اختبارهما، ومعارضتهما، حتى يتحقّق قولُه من عدمه.

ولن يعجز من رام استكناه جملةٍ من مسائله أن يقف على دلائل تُرشده إلى أخذ أبي المعالي من أبي زيد واحتوائه على نخيلة فِكره، من ذلك قولُ أبي زيد: «إن الحوادث ممدودة، والنصوص معدودة »[12]، وقولُ أبي المعالي فيها: «وهي على الجملة متناهية، ونحن نعلم -قطعاً- أن الوقائع التي يتوقّع وقوعُها لا نهاية لها »[13].

---

8 الاستشفا: 89.

9 كتاب الأساليب من مرويات ابن العربي، ينظر: المسالك: 181/6.

10 نشر باسم تقويم أصول الفقه وتحديد أدلة الشرع، ويُعرف بتقويم الأدلة.

11 بعد كتابة المقال، تبين أن كتاب الأسرار طبع وأن كتاب الأساليب وجد قسم منه.

12 تقويم الأدلة: 127/1.

13 كتاب البرهان: 743/2.

وقد يشكّك في هذا الذي تلوتُه تعويلُ أبي المعالي على كُتب أبي بكر الباقلّاني (403هـ\1012م) ومقالاته، إذ قد يتابعه في مسائل مخصوصة تخالف مذهب إمامه أبي عبد الله الشافعي (204هـ\819م)، وربما أحنق ذلك بعض الشافعية، كما تراه في صنيع أبي المظفر السمعاني (489هـ\1096م) في مسألة الحديث المرسَل، إذ قال: « وقد وجدتُ بعض من شُغف بكلام أبي بكر الباقلّاني، ويجعله الإمام والقدوة في عامة ما ذكره في أصول الفقه، حتى كأنه رضي لنفسه أن يقلّده وينصبه إمامًا في العقائد .. »[14].

أقول هذا، ولا أذهب إلى مخالفة أبي بكر ابن العربي كل المخالفة، لعلمي أن الرجل مطّلع على طرائق النظّار من الأصوليّين، وعارف بمسالك التأليف عندهم، وموارد المشيخة في كتبهم، وفي تفاريق المسائل ما يؤيد نظره ويؤكده، ولكنه تأكيد جُملي، لم أَجْرِ فيه على قانون الاستقراء، وإنما هي مسائلُ منتخبة، وصُوَرٌ من المباحث معدودة.

وقد يكون هذا الذي ذكره القاضي آنفاً ممّا فاوض فيه أصحابَ أبي المعالي، كأبي سعد الزنجاني (492هـ\1099م)، والطُّوسي الأكبر (529هـ\1135م)، وقد نص ابنُ العربي على ذلك في بعض كتبه، قال: « وفاوضتُ الطوسي الأكبر في ذلك، وراجعتُه حتى وقف »[15].

وقد يكون هذا الذي بدأ به أبو بكر بن العربي كتابَه ممّا شافهه به أبو حامد، في تلك المسائل التي كان يسأله عنها، وقد عُلم تنقيره في سبيل ذلك، وأخصّ ما كان منه في كتاب البرهان.

ولكن الذي يُغبّر في وجه ذلك الإمكان المبتدأ به -من مشافهة الطُّوسي لابن العربي في خصوص مسالك البرهان- أن أبا بكر جعل أبا حامد على مَهْيَع شيخه أبي المعالي في الإفادة، فكان كتابه شفاء الغليل معارضًا لكتاب القياس من تقويم الهداية، وعلى ترتيبه وتبويبه، غير أنه أربى على أبي زيد، وجاء فيه بكل بديعة[16]، كما قال أبو بكر.

وبموازنة هذين الكتابين؛ يرى الأثر الكبير لأبي زيد؛ في المسائل، والمثُل، وفي الأنقاد، غير أن أبا حامد توسع كثيرًا في فصول كتابه، وكثّر من الأمثلة، وإن كان قد انطوى كلامه على جملةٍ صالحةٍ من النقد لأبي زيد.

وأبو حامد ناظرٌ في كل ذلك إلى كتاب البرهان، جالبٌ للمقالات منه، وإن كانت الحوالة عليه لا تكاد توجد، ولكن تلك كانت عادتهم؛ لاشتهار الكتاب، ومعرفة الخاص والعام به.

14 قواطع الأدلة: 2/440.
15 أحكام القرآن: 2/755.
16 الاستشفا: 93.

فإن كان من جديد في كتاب الاستشفا فهو تعيينه لموارد أبي المعالي وأبي حامد، إذا سلّمنا دعوى ابن العربي فيها، وتبيينه لفحولة أبي زيد فيما ارتسمه من مناهج العلم وطرائقه.

وقد قال مُبينًا عن ذلك: «اعلَم -علَّمك اللّٰه- أن رَبَّ النعمة على أهل خراسان، بل على أهل العلم في غابر الأزمان، في الكَشف عن أسرار العلومِ مِن فَنِّ الفقه وأُصُولِه؛ القاضي أبو زَيْد عُبَيد اللّٰه بن عُمَر الدَّبُوسِي؛ فإنه انتدب بذِهن ثاقب، وعقل وافر، وحِفظ ظاهر، وجِدّ صادق، وفراغ متواصل، وتوفيق حاصل؛ لكَشْفِ غوامِض العِلْمَيْن، وإبراز دقائِق الفَنَّيْن، وأقام على هذه الحال خمسة أعوام، متّصلة الليالي بالأيام، لا يقطعه عنها قاطعٌ من أهل ولا ولَد، ولا اطَّلع على مكانه أحَد، كما بيّناه في كتاب عِيان الأعيان، ثم أسفر اجتنانُه عن نُورَين ساطعيْن؛ كتاب تقويم الهداية في أصول الفقه، وكتاب أسرار المسائل في الخلاف، فما بَقِي أحدٌ من المحقِّقين إلَّا نَسج على مِنواله، وتَسَرْبَلَ بسِرباله في نِظامِه، واقتفى آثارَ مراسِمه في إحكامِه»[17].

ثم قال: «وجمع الإمام أبو المعالي كتاب البرهان في أصول الفقه على أغراض تقويم الهداية، وخالف ترتيبه، وصنف الإمام أبو حامد هذا -أي: شفاء الغليل- معارضاً لأبي زيد في فَنّ القياس من أصول الفقه، على ترتيبه، وفي طريقه وتبويبه».

وهذا التأصيل سابقةٌ من سوابق أبي بكر ابن العربي، وفائدةٌ من فوائد رحلته التي قصد بها إلى الرِّفعة لعلوم الإسلام ببلاده.

3 النقدُ لمَشْرَع أبي حامد في الأصول

قصَد الإمام أبو بكر في كتابه الاستشفا إلى التعريف بمسالك العلة وطرقها، وبثِّها في بلدٍ كانت معاقد علم الأصول فيها ضعيفة المدارك، إنما قصارى أمرهم التعلق بمباحث عامة من ذلك العلم، من غير غوص على مشكلاته، وخوض في مزمناته.

وقد رأى أن غالب المؤلفين فيه من أهل بلده لم يكن لهم ارتياض في تلك المساحات، ولا قطعوا في سبيل ذلك أقفر المسافات، وإنما كان النفوذ يقع لمن كانت له رحلة سابقة؛ كأبي محمد الأصيلي (392هـ\1001م)، وأبي الوليد الباجي (474هـ\1081م)، وآخرون إنما كان تعلقهم في سبيل معرفتهم بالرواية والحديث، أما هاته المسالك التي شرع في تقريبها فهي بعيدة عنهم؛ حسًّا ومعنى.

17 الاستشفا: 92.

وإنما وقع بعض النفح في هذا العلم بعد زمان متطاول؛ إبان العناية بكتاب « البرهان » و« المستصفى »، وكان بعضٌ من هؤلاء المتصدّرين لتقرير مباحثهما من مدارس القيروان، وما صاقبها من البلدان، وبعضهم من الأندلس، وإن كان ذلك بتقتير واحتشام.

وقد رأى أبو بكر أن بعضًا من أهل بلاده ربما يتشوّفون إلى ما يكشف لهم عن تلك المسالك، ورأى أن بعضًا ممن رام كشفها لم يتناولها بحقها، فقصّر في سبيل ذلك، فتحتّم عليه أن ينظر فيما قصّروا فيه، وأن ينبّه المعتني بذلك إلى حقّ الأمر وشَرطه.

قال: « فإني رأيتُ النَّاس قد هجمُوا على كتاب شفاء الغليل، في بيان مسالك التعليل، من المناسِب والمؤثر والشَّبَه والطرد والمُخِيل، وتسوَّروا على لسانه من غير هادٍ ولا دليل، ونقضوا عُهَده في ابتدائه، كأنهم ما قرؤوه من أوَّلِه؛ لأنهم أَتَوه من غير بابِه ».

وقد ذكر أبو حامد جملة من الشرائط التي يتعيّن على الناظر في كتابه الاحتواء عليها، والاستقلال بها؛ من اتقاد القريحة، واستكداد الفكر، ومجانبة العناد، وسابق المعرفة بأغراض الأصوليّين وطريقتهم في المناظرة والمباحثة.

وبرز نقدُ ابن العربي لأبي حامد في ثلاثة أمور:

3.1 الأول: في ابتناء الكتاب مالكيًّا

وقصدُه بذلك المسائل الخلافية؛ إذ كان نظرُ أبي حامد إلى الاستدلال لمذهبه ونصرته، وردَّه أبو بكر إلى أصول المالكية في الخلاف والاستدلال، ولم يقصد أبوابَ الكتاب، فقد أبقاها على أصلها في شفاء الغليل.

والمؤلفات الخلافية عند ابن العربي لها مذاق خاص، فهي محتوية على تلك المسائل، ومعها أدلتها، وجوابات أسولة المخالفين، مع تحرير النزاع، والاستطراد في تأصيل الأصول، وبناء النظر على تحرير الأصل المتنازع عليه.

وهذا ميل إلى طريقة العراقيّين، وهي طريقة المناظرين، وعليها كان يعوّل أصحاب الحِجاج عن المذهب؛ كالأبهري (375هـ\985م)، وابن القصّار (398هـ\1007م)، والقاضي عبد الوهاب[18] (422هـ\1031م).

وقد رام الإمام أبو بكر إعلاءَ درسِ الخلاف ومسائله بالأندلس، لكنه لم يجد له رجالًا يقومون به، وذلك لجملة أسباب، أهمها: بُعد هذا الصقع الغربي عن مواطن المذاهب والفرق، وانحسار التمذهب

18 المستوعب لتاريخ الخلاف العالي: 725/2، 728.

في مذهب واحد، وهو المالكي؛ إذ لم يكن ينازعه أحدٌ في سلطانه، إلّا ما كان من الحزمية، وهم قوم لا معرفة لهم بالجدل، ولا رأس لهم في فنّ المناظرة في ذلك الزمان.

وتأكّد هذا في تآليف أبي بكر الأصولية، إذ لم نر عالمًا من علماء الأندلس قاصدًا إلى ما قصد إليه أبو بكر في مؤلّفاته الثلاثة في الأصول: التمحيص، والمحصول، ونُكَته.

وقد شاع عند العارفين بالدرس الأصولي أن التأليف في هذا الفنّ مقصور على النظّار، أصحاب الاختيار، أما غيرهم من المتفقهة؛ فقصارى أمرهم أن يؤلفوا مختصرًا في الأصول، يكون على سبيل التذكرة لا التحقيق، وعلى طريق التقريب لا التحرير.

وهذه منازع قلّما يُتفطّن إليها، وذلك أن المؤرخ لحال الأصول ببلاد الغرب يرى كثرة المؤلفات فيها، ولكن يقصر نظره عن أن يحيط بأسباب إنشائها، ومعاقد ابتنائها، فيطلق تلك الأحكام التي نسمعها بين الفينة والأخرى، وهي أن المغرب ريّان بالمؤلفات الأصولية، وخزائن كتبه ملأى بالمصنّفات في ذلك، وهو أمر قد تستوجبه الحَمِيّة البلدية، وحُمَيّا القُطرية، ولعل ما في كتاب فهرس الكُتب المخطوطة في علم أصول الفقه المحفوظة بالخزانة الحسنية ما يؤكّد هذه الدعوى.

ولعل هذا الذي قدّمنا من الاعتلال لطريقة المالكيّين بالأندلس في منحاهم الأصولي والخلافي ما يفسّر لنا سبب انصراف أبي الوليد الباجي عن التأليف في مسائل الخلاف[19]، لأنه رأى أن ذا الغرس لا يثمر، بل ولا يورق، وعلى طريقته كان من قبل أبو محمد الأصيلي.

ولكن ابن العربي أراد أن يؤسّس لهذا السبيل، وهو بعدُ محاصر في إشبيلية، مع ما مُني به من حسدة لا يفتؤون عن أذيته، والتحريض عليه، فلم يفرغ لما فرغ له أبو الوليد الباجي، وأبو محمد الأصيلي.

ولا يعدم الناظر في كُتب أبي بكر نقده لطريقة الأندلسيّين في التفقه، وفي بناء المسائل على النظر الأصولي، في كتبه: أحكام القرآن، وترتيب المسالك، وتلخيص الطريقتين، وغيرها من كتبه الكِبار.

وهو بهذا على طريقة شيخه أبي بكر الطُّرطوشي (520هـ\1126م)، فلم يكن هو -أيضًا- مرتضيًا لمسلك الأندلسيّين في الأصول والمسائل، وقد عايَن في بيت المقدس حلقات المناظرة، ومجالس المجادلة، بين المذاهب والطوائف كلها، فلَقِن عنه ابن العربي نقداته لتلك المسالك، فعزم عليها في نفسه، ثم أشرع لها في مؤلفاته، وعَقد لها في مصنّفاته.

وهذا النهج كان قائمًا عند أبي بكر على ابتناء المسائل على الأصول والاعتلال لها، وهو منحى قد يفيدنا بأسبقية التأليف -عنده- في مسائل الخلاف على التأليف في الأصول.

19 ذكر القاضي عياض لأبي الوليد كتابًا في مسائل الخلاف سمّاه: السراج في عُمَد الحِجاج، لكنه لم يتم، ترتيب المدارك: 124/8.

ولعله في هذا البناء المالكي، لذاك الأصل الشافعي، كان ناقدًا لمن حاول مطاولة الكتاب، لكنه لمّا كان قاصرًا عن مراده، نهج نهجه، وسار سيره، ولم يُمكنه أن ينفكّ من أثر أبي حامد، إذ مطاولة ذانَشْمَند مفتقرة إلى مَدد عال من النظر الخلافيّ والأصولي، وهو الذي أخلّ به الناظر في كتاب الشفا.

3.2 الثاني: في الدَّرْكِ على أبي حامد

ولاح ذلك في مسائل الخلاف بين الشافعية والمالكية، فاعتلّ أبو بكر لمذهبه، وبيّن عِلل الأحكام، ودلائلها، وأحال في أغلبها إلى كُتبه الخلافية؛ الإنصاف، وتلخيص الطريقتين.

وأعانه على النقد نظره في كُتب الأحناف الخلافية، فهي التي أمدّته بمداخل الدرْك والنقض على الشفعوية.

ومن المسائل المنتقدة على أبي حامد ما اعتلّ به في مسألة الربا، قال أبو بكر: « ذَكر الإمامُ أبو حامد في هذا المثال مسألة الربا، ونقل عن الشافعي وابن سُرَيج كلمات في المسألة، قال: أردنا بها أن العلماء تعلّقوا قديمًا بأوصاف لا تناسب، وليس كما زعم الإمام، بل كُلُّ واحد من العلماء تعلّق بإيماء أو مناسب في زعمه، وقد أطنب الإمام أبو حامد وشيخه أبو المعالي في ذلك في مسائل الخلاف، وبه رجّحوا، وعليه عوّلوا، حسب ما قرّرناه في تلك المسألة في الإنصاف، والتلخيص، وغيرهما، فانظره هنالك، تجده بيّنًا موضّح المسالك، ما بين إيماء أو مناسبة ».

وقال -أيضًا-: « ذَكَر الإمام أبو حامد رضي الله عنه الانفصال على مذهب الشافعي، وعندنا أن الجنس وصف علة كما قدَّمناه، وقد قال الإمام أبو حامد في فصل بعد هذا ما يُمكن أن يُجعل جوابًا عنه، وهو أن السبب إذا لم يكن مناسبًا تعلَّق الحُكم بالجميع، وكذلك القُوت والجنس؛ فإنهما: إمَّا أن يكونَا مناسبيْن، فهما وصفَا العلة، أو غير مناسبيْن، فكذلك، وتحقيقُ المناسبة في أحدهما دون الآخر ممكن، ولئن ظهرت مصلحة خفية في الطُّعم أو القوت فإنها في الجِنس لا الطُّعم، حسب ما قرّرنا في مسائل الخلاف ».

وفي موضع آخر: « قال أبو حامد: فإن قيل: فهلَّا جمعتم بين عِلل الربا؟

قُلنا: قال ابن سُرَيج: ذلك يُخرجنا عن قول العلماء، والتعليل في مِثل هذا المقام بعِلّتين غير جائز، إلا أن يجعل عِلّة، كما قال الشافعي: إن العلة هو الطُّعم، والتقدير في قول، وقال مالك: القُوت، وفيه الجمعُ بين الكُلّ.

قال ابن العربي: لم يمتنع جمعُ العلل لأنه يُخرج عن قول العلماء، إنما امتنع لقيام الدليل على أن الكيل ليس بعِلّة، وكذلك الوزن، وبقي الترجيح بين الطُّعم والقوت، فكان القوت أوقع؛ لما بيّناه في مسائله ».

وهكذا في سائر المسائل المنتقدة، إذ أغلبها من صنف الخلافيات، وكان اقتداره على نقدها راجعًا إلى تصنيفه في مسائل الخلاف، وتحريره لها على النظر المالكي.

ومن المسائل المنتقدة -ويُليح منها إلى نقد ولاة الأندلس- توظيف الأموال على الأغنياء زيادة على المقدّر الشرعي إذا صَفِر بيت مال المسلمين[20]، وقد ذهب أبو حامد إلى اعتبار هاته المسألة من جملة المصالح الملائمة القطعية، إذ لا يتمارى منصف في وجوب اتباعها[21].

وهذا سياق المسألة ونقدها عند ابن العربي، قال:

«فإن قيل: فإذا احتاج الجيشُ الذي لا بدّ منه، وصَفِر بيتُ المال الذي المرجع إليه، كيف وجهُ العمل؟

قُلنا: يستسلف الإمام على القرائن الشرعية، كما فعل رسولُ اللّٰه صلى اللّٰه عليه وسلم.

فإن قيل: فإن لم يَفِ بحاجة الجُند؟

قلنا: هذا محالٌ شرعًا؛ لأن اللّٰه وظّف الوظائف، ورتّب الواجبات، وشرّع الشرائع، وأوجب الحماية وقطع المعذرة، فلو لم يطّرد نظامُ ذلك، ولا اتّضح منهاجه؛ لكان تقصيرًا في الحكمة، ونقضًا لموضوع الشريعة، ولكن الولاة والجند أسرفوا في الإنفاق، واستبدّوا بالأرزاق، فعظمت لذلك حاجاتهم، حتى اتّسعتْ نفقاتهم، وصار ذلك بمثابة أرباب الأموال إذا اتسعت شهواتهم، وقصرت عنها غلاتهم، لم يكن بُدٌّ من أن تخرج الأصول عن أيديهم، أو يحتبسون الأموال على وجه يُضلهم ويُرديهم، فالتقصير من سَفَه الوالي، لا من حكمة الباري.

فإن قيل: فإذا وقعت الواقعة، وصار الخلقُ يتعدّى إلى هذه الهافية، ولم يكن بُدٌّ من الرجوع إلى الأغنياء، فماذا تقولون؟

قلنا: ليس هذا حُكْمًا شرعيًّا فيُفتى فيه، إنما هو بلاء نصبر عليه، ونسأل اللّٰه التوبة الكاشفة له.

وإنما هذا بمنزلة من يقول: إن الحدود الموضوعة للزجر من قِبل الشرع لا تَفِي بكَفّ الطغاة، فلا بدّ من تعدّيها إلى قَطع من لم تثبت سرقته، وقَتل من لم يصحّ قتلُه، ونعوذ باللّٰه من هذه المقالة، فما وضع الحكيمُ هذه الزواجر إلا نهايةً في الحكمة، وغايةً في الزجر والمصلحة، ولا يُمكن حكيمًا عالمًا بالعواقب، مطّلعًا على الموارد والمصادر، ومحيطًا بالجليّات والخفيّات؛ أن يضع فوق هذا النظام، فقائلُ ذلك خارجٌ عن دين الإسلام.

فإن قيل: فما بال الحدود لا تفي، ولا تقوم بالردع؟

---

20 شفاء الغليل: 236.

21 شفاء الغليل: 240.

قُلنا: لأنها لا تُستوفى، وإنما تباع وتُشترى ».

وهذا نقدٌ شديدٌ لما كان عليه حال الولاة والقضاة بالأندلس؛ من التسلّط على أموال الناس، وجمعها بدعوى الجهاد، وإنفاقها في غير سبيل المؤمنين، وشيوع الرُّشَى، وإقامة الحدود على الضعفاء، والميل بها عن الأعيان.

وقد حاول البُرْزُلي أن يتعلّق بكلام أبي حامد، فقال بجواز توظيف الأموال زجرًا للمقترفين، فنقض كلامه ابنُ الشمّاع[22] (833هـ\1429م)، وكذلك فعل أبو محمد الونشريسي (950هـ\1543م)، وغيرهما.[23]

3.3 الثالث: في تكميل النقص وتتميم الفوت

وأوفى ما أشرف عليه القاضي أبو بكر في مسلك التكميل والتتميم عقدُه فصلًا مختصًّا بالمصلحة عند مالك بن أنس -رضوان الله عليه-.

قال القاضي في ختامه: « وهذا الفَصْلُ المالكيُّ في نهاية الإبداع، وحسن الاختراع، وجودة البيان، وهو لكم في هذا المختصر -معشر المالكية- خيرٌ من كتاب شفاء الغليل للشفعوية ».

وقد يفيد كلامه هذا أنه أول متكلّم على هذا الأصل من المالكية، وألمح إلى ذلك في بعض كتبه الأخرى، فقال -متحدثًا عن قول مالك بن أنس بالمصلحة وسدّ الذرائع-: « ولم يساعده على هذين الأصلين أحدٌ من العلماء، وهو في القول بهما أقوم قيلًا، وأهدى سبيلًا، وقد بيّنا وجوب القول بهما والعمل بمقتضاهما في أصول الفقه، ومسائل الخلاف »[24].

وقال في موضع آخر: « وأما المقاصد والمصالح فهي -أيضًا- ممّا انفرد بها مالك دون سائر العلماء، ولا بدّ منها؛ لما يعود من الضرر في مخالفتها، ويُدخل من الجهالة في العدول عنها، وقد مهّدنا ذلك في أصول الفقه »[25].

ويؤيّد هاته الأوَّلية قولُه -مبيّنًا للمصلحة عند مالك-: « وهذا فَنٌّ لم يتفطّن له مالكيّ، وقد حقّقناه في أصول الفقه »[26].

يزيده تأكيدًا: ما ذكره في الاستشفا من إنكار بعض الفقهاء لهذا الأصل، والنزوع إلى نفي صحته عن مالك -رحمه الله-، قال أبو بكر: « هذا بابٌ غريب، جرتْ فيه نادرة غريبة مثله، وذلك أنّي ذكرتُه

22 مطالع التمام: 137.

23 جواب عن مسألة العقوبة بالمال لمحمد العربي الفاسي: 33.

24 كتاب القبس: 779/2.

25 كتاب القبس: 786/2، وترتيب المسالك: 22/6.

26 أحكام القرآن: 206/1.

أوّل قدومي في سنة خمس مائة، وفي البلد قاض يَدَّعي أنه من الأصوليّين، وله صديق بلديّ، تقدَّم في الفهم بكل علم، وما له فيه علم، ولما سمِعوا ذلك منّي نقضوا رؤوسهم، واستغشوا ثيابهم، وقالوا: ما سمعنا في كُتب المغرب والعراقيْن بهذا الاختلاف، فقلتُ: قد ذكره أبو المعالي في البرهان، قالوا: ما صنَّفه قطّ، ولا عرّفه ».

وحُجّة أبي بكر هنا أن أبا المعالي ذَكره، فلعلّه اطّلع عليه في كُتب المالكية بالعراق، أو رآه مسلكًا مطّردًا في مسائل مالك، وكذلك ذكره عنه أبو حامد الطوسي في شفاء الغليل[27].

ونسبه إلى مالك -من المالكية- أبو الحسن الإبياري (616هـ\1219م) في « شرحه على البرهان »[28]، والشهاب القرافي (684هـ\1285م) في « شرحه للمحصول »[29]، وأبو إسحاق الشاطبي (790هـ\1388م) في « الاعتصام »، قال: « وذهب مالكٌ إلى اعتبار ذلك وبناء الأحكام عليه على الإطلاق »[30].

ونسبه إليه -أيضًا- ابن جُزَي في « تقريب الوصول إلى علم الأصول »، وقال: « وهو حجّة عند مالك؛ خلافًا لغيره »[31].

وكذلك صنع أبو بكر ابن عاصم (827هـ\1424م)، في نظميْه في الأصول؛ « مرتقى الوصول، إلى الضروري من الأصول »، إذ قال[32]:

وذا يسمّى عندهم بالمرسَلِ وكم له كمالكٍ من مُعمِلِ

وفي « مَهْيَع الوصول، لمن يريد الأخذ في الأصول »، إذ قال[33]:

فذاك حجةٌ رآه مالكْ وغيرُه خالفه في ذلكْ

ولكنّ ابن الشمّاع (833هـ\1429م) جادَل في نسبة هذا الأصل إلى مالك، وقال: « أمّا نسبة القول بالمصالح المرسلة إلى مذهب مالك فقد علمتَ ما فيه من الخلاف، والظاهر أن ابن العربي إنما اعتمد في تقويل مالك بها على ما في البرهان لإمام الحرمين فقط؛ فإنه ذَكر في كتاب الاستشفاء -الذي له في

27 شفاء الغليل: 207.
28 شرح البرهان: 144/4.
29 شرح المحصول: 4095/9.
30 كتاب الاعتصام: 6/3، 54، والموافقات: 690/3.
31 تقريب الوصول: 124.
32 مرتقى الوصول: 154.
33 مهيع الوصول -بخط ابن عاصم-: ق48/ب، ونسبه إليه من العصريّين الفقيه الحافظ المدني ابن الحُسْنِي الرباطي، في مقدمة كتابه: منار السبيل إلى مختصر خليل بالحجة والدليل: 18.

القياس- أن أهل المذهب أنكروا ذلك من مالك، وأنكروا أن تدلّ مسائله عليه، حتى أغرب هو عليهم بما في البرهان »[34].

واعتلّ لذلك بجملة أمور:

الأول: أن أبا المعالي لم يقطع به في نسبته إلى مالك، وإنما قال: « نُقل عن مالك الأخذُ بها ».

الثاني: خلوّ هذا الأصل من كُتب المالكية في الأصول.

الثالث: أكثر المالكية على نفي نسبته إلى مالك.

أما الأوّل منها فصحيح عن أبي المعالي، فما زال يورد من تلك الصيغ الضعيفة، ثم يستدركها بقوله: « إن صح ما روي عنه »[35]، وقال في موضع آخر: « فرئي يثبت مصالح بعيدة عن المصالح المألوفة »[36].

وقوله: رُئي، هو على غير سبيلٍ من قوله: نُقل، على ما أسنده إليه ابن الشمّاع، بل ويُقربّه من القَطع -دون تلك الصيغ التمريضية- قولُه: « وأفرط مالكُ بن أنس في القول بالاستدلال »[37].

فلا يعوّل إذن على هذا التعليل، ولا ينهض مانعًا من المصير إلى قول مالكٍ بالاستدلال المرسَل.

والثاني قد تبيّن ضعفه، فقد أوردنا عددًا من المالكية القائلين بذلك، وقد نسبوه إلى إمام المذهب، ونصّوا على من خالفه من الأئمة الكبار.

وأما الثالث فقولٌ صحيح معتبر، ولكنه لا يُسعف في ردّ هذا الأصل، ولا في الطعن في هذا الانتساب.

وبقي وجهٌ آخَر، ينفي عن أبي بكر بن العربي مسايرته التامّة لأبي المعالي في تقرير ذلك الأصل، وهو ردّه لما زعمه في حق مالك بن أنس؛ من القول بقتل ثُلث الأمة استصلاحًا، وأخذ المال بمصالح يقتضيها في غالب الظن، وغيرها من تلك المسائل.

فلا أظنّ -والأمر على هذا- أن يصحّ ما جادل فيه ابن الشمّاع، في نقضه على البُرْزُلي (841هـ\1437م)، لأن ما ساقه لا ينهض دليلًا لنفي قول مالك بالاستدلال المرسَل.

يزيده تأكيدًا: أن بعض من اطّلع على كتاب الاستشفا كابن عرفة الورغمّي (803هـ\1400م) وأبي العبّاس الونشريسي (914هـ\1508م)، لم يتعقّبا أبا بكر في هذه النسبة، ولم ينقُداها.

34 مطالع التمام: 155.

35 البرهان: 1119/2.

36 البرهان: 1113/2.

37 البرهان: 1113/2.

فتعيّن أن هذا الأصل من أصول مالك في مذهبه، معروف عنه، ملحوظ في مسائله ونوازله، حتى شاع وذاع وانتشر، فتلقّفه أبو المعالي ودوّنه عنه في **كتاب البرهان**، ومِن بعده صاحبه أبو حامد الطُّوسِي في شفاء الغليل، إلى أن انتهى إلى أبي بكر في استشفائه.

تخليص:

أمكن من خلال هاته الورقة لَحْظُ جملة أمور:

الأول: توسل أبي بكر بن العربي بأنظار أبي حامد ومشارعه في مفاوضاته ومناظراته للظاهرية بالأندلس.

الثاني: الاستمداد من أبي حامد في التأسيس لمكانته الاجتماعية وسلطته المعرفية.

الثالث: قراءة مشارع أبي حامد ونقدها، ومدخل هذا النقد هو مسائل الخلاف بين المالكية والشافعية.

الرابع: التأسيس لمدرسة أصولية مالكية قائمة على النظر والاعتبار، والاستمداد من مدرسة الخراسانيين والبغداديين.

الخامس: دوران جملة من مصنفات ابن العربي على كتب أبي حامد؛ اختصاراً، ومعارضة، وانتقاداً، واستمداداً.

السادس: الاستنباط لأصول مالك، والتنبيه عليها، والاحتفاء بها، وهو ما صنعه في كتاب القبس وترتيب المسالك.

السابع: أنظار ابن العربي الأصولية إنما تؤخذ من كتبه الخلافية، وهو الشيء الذي أكدَّه في كتاب الاستشفا، وكان دائم التَّرداد له.

فِهْرِسْت المصادر والمراجع

أحكام القرآن، لأبي بكر محمد بن عبد الله ابن العربي المعافري، تحقيق محمد علي البجاوي، دار الفكر.

الاستشفا من كتاب الشفا، لأبي بكر محمد ابن عبد الله بن العربي المعافري، تحقيق عبد الله التوراتي، دار أسفار (قيد النشر).

البرهان في أصول الفقه، لأبي المعالي عبد الملك بن عبد الله بن يوسف الجُوَيني، تحقيق عبد العظيم الديب، قطر، الطبعة الأولى، 1399هـ.

التحقيق والبيان في شرح البرهان، لأبي الحسن علي بن إسماعيل الإبياري، تحقيق علي بن عبد الرحمن بسام، وزارة الأوقاف، قطر، الطبعة الأولى، 1434هـ/2013م.

ترتيب المدارك وتقريب المسالك لمعرفة أعلام مذهب مالك، لأبي الفضل عياض ابن موسى اليحصبي السبتي، الثامن، تحقيق سعيد أعراب، وزارة الأوقاف، الرباط، 1403هـ/1983م.

ترتيب المسالك في شرح موطأ مالك، لأبي بكر محمد بن عبد الله ابن العربي المعافري، تحقيق محمد السليماني وعائشة السليماني، دار الغرب الإسلامي، الطبعة الأولى، 1427هـ/2007م.

تقريب الوصول إلى علم الأصول، لأبي القاسم محمد بن أحمد بن جُزَي الغرناطي، تحقيق نزار حمادي، دار الإمام ابن عرفة، تونس.

تقويم الأدلة، لأبي زيد عُبَيد الله بن عمر الدَّبُّوسي، تحقيق عبد الرحيم يعقوب، مكتبة الرشد ناشرون، الطبعة الأولى، 1430هـ/2009م.

جامع مسائل الأحكام، لأبي القاسم بن أحمد البلوي البُرْزُلي، تحقيق محمد الحبيب الهيلة، دار الغرب الإسلامي، 2002م.

جواب عن مسألة العقوبة بالمال، لمحمد العربي بن يوسف الفاسي، تحقيق سلمان الصمدي، جمعية البحث التاريخي والاجتماعي، القصر الكبير، 2021م.

سراج المريدين في سبيل الدين، لأبي بكر محمد بن عبد الله ابن العربي المعافري، تحقيق عبد الله التوراتي، دار الحديث الكتانية، الطبعة الأولى، 1439هـ/2017م.

شفاء الغليل في مسالك التعليل، لأبي حامد محمد بن محمد الغزالي الطوسي، تحقيق حمد الكبيسي، وزارة الأوقاف، العراق، 1390هـ/1971م.

عارضة الأحوذي في شرح كتاب الترمذي، لأبي بكر محمد بن عبد الله ابن العربي المعافري، تحقيق طارق الشيباني وجماعة، نسخة مرقونة.

العواصم من القواصم، لأبي بكر محمد بن عبد الله ابن العربي المعافري، تحقيق عمار طالبي، مكتبة دار التراث، الطبعة الأولى، 1417هـ/1997م.

قانون التأويل لعلوم التنزيل، لأبي بكر محمد بن عبد الله ابن العربي المعافري، تحقيق محمد السليماني، دار الغرب الإسلامي، 2010م.

القبس في شرح موطأ مالك بن أنس، لأبي بكر محمد بن عبد الله ابن العربي المعافري، تحقيق محمد عبد الله ولد كريم، دار الغرب الإسلامي، 1992م.

قواطع الأدلة في أصول الفقه، لأبي المظفر منصور بن محمد السمعاني، تحقيق عبد الله بن حافظ الحكمي، مكتبة التوبة، الطبعة الأولى، 1419هـ/1998م.

كتاب الاعتصام، لأبي إسحاق إبراهيم بن موسى اللخمي الشاطبي، تحقيق مشهور آل سلمان، مكتبة التوحيد.

الكتاب المتوسط في الاعتقاد، لأبي بكر محمد بن عبد الله ابن العربي المعافري، تحقيق عبد الله التوراتي، دار الحديث الكتانية، الطبعة الأولى، 1436هـ/2015م.

كتاب الموافقات، لأبي إسحاق إبراهيم بن موسى اللخمي الشاطبي، تحقيق الحسين آيت سعيد، منشورات البشير بن عطية، فاس، 1438هـ/2017م.

المختصر الفقهي، لأبي عبد الله محمد بن عرفة الوَرْغَمِّي، تحقيق حافظ عبد الرحمن محمد خير، مركز الفاروق، 1435هـ/2014م.

مُرتقى الوصول إلى الضروري من الأصول، لأبي بكر محمد بن محمد بن عاصم الغرناطي، تحقيق أحمد مزيد الجكني البوني الشنقيطي، الدار المالكية، 1439هـ/2017م.

المستوعب لتاريخ الخلاف العالي ومناهجه عند المالكية، لمحمد العلمي، مركز الدراسات والأبحاث وإحياء التراث، الطبعة الأولى، 1431هـ/2010م.

مطالع التمام ونصائح الأنام، لأبي العباس أحمد بن محمد الشمّاع الهنتاتي، تحقيق عبد الخالق أحمدون، وزارة الأوقاف، الرباط، 1424هـ/2003م.

المعيار المعرب والجامع المغرب، لأبي العباس أحمد بن يحيى الونشريسي، تحقيق جماعة من الباحثين، وزارة الأوقاف والشؤون الإسلامية، الرباط، 1401هـ/1981م.

المنخول من تعليقات الأصول، لأبي حامد محمد بن محمد الغزالي الطوسي، تحقيق محمد حسن هَيتو، دار الفكر، بدون تاريخ.

7

De Bagdad à Séville : Étude d'une généalogie du *kalām*, d'al-Ashʿarī à Ibn al-ʿArabī

Ilyass Amharar

Le but de cette contribution est non pas d'analyser le contenu de la pensée ashʿarite du Qāḍī mais plutôt sa filiation à l'école éponyme telle que le Sévillan la revendique lui-même. Reconstituer l'arborescence de ces maîtres jusqu'à al-Ashʿarī offrira certainement un aperçu des principaux canaux de transmission de l'ashʿarisme d'Orient vers l'Occident et des différentes connexions qu'ils ont pu tisser. Qu'elle soit ou non l'objet d'une construction postérieure, analyser cette filiation permet d'apporter de nouveaux éclairages sur certaines zones d'ombre de la vie du Qāḍī, en particulier ses premiers rapports à l'ashʿarisme, mais aussi au ẓāhirisme, sujets que nous aborderons après quelques rappels sur la postérité d'al-Ashʿarī et le transfert de la doctrine en Occident musulman.

1 Rappels sur la postérité d'al-Ashʿarī

Le *Tabyīn* d'Ibn ʿAsākir (571/1171) est sans doute l'une des sources les plus complètes sur les disciples d'al-Ashʿarī[1]. Son aspect hagiographique va de pair avec sa conception, puisqu'il s'agit d'un panégyrique visant à restituer l'éclat que devrait mériter, selon Ibn ʿAsākir, la figure d'al-Ashʿarī[2]. En dépit de cet aspect partisan, il recèle néanmoins d'éléments factuels précisément datés qui nous permettent d'établir un tracé continu joignant les premiers disciples les plus célèbres du théologien, classés en générations (*ṭabaqāt*).

Comme en témoignent les nombreuses questions qui lui sont adressées de régions diverses (Syrie, Egypte ou même Oman)[3], la célébrité d'al-Ashʿarī s'est exportée au-delà de Bagdad avant même son décès, que Ibn ʿAsākir fixe

1 Il a récemment réédité à partir d'un manuscrit comportant des passages de la main de l'auteur. Ibn ʿAsākir, *Tabyīn kadhib al-Muftarī fīmā nasaba ilā l-imām al-Ashʿarī*, éd. Sharqāwī, Damas: Dār Taqwā, 2018.

2 Ibn ʿAsākir réagit par ce livre aux propos jugés mensongers proférés par un certain al-Ahwāzī à l'encontre d'al-Ashʿarī.

3 Ibn ʿAsākir, *Tabyīn* 707.

 | DOI:10.1163/9789004706859_009

à l'an 324/936 dans ladite ville[4]. Pourtant, si l'on en croit Ibn ʿAsākir, les seuls dix-neuf disciples que composent la « première génération » (*ṭabaqa ūlā*) ne forment pas encore ce que l'on pourrait qualifier d'école. Comme le souligne Thiele, aucun savant de cette génération est connu pour avoir contribué à une large diffusion de l'ashʿarisme[5]. Nous allons le voir, c'est avec une figure comme al-Bāqillānī et ses disciples que l'appellation « ashʿarite » va massivement se diffuser[6].

Dans cette première génération donc, les deux premiers noms que cite Ibn ʿAsākir sont ceux d'Ibn Mujāhid et Abū l-Ḥasan al-Bāhilī, tous deux morts dans les environs de l'an (370/980). Si les données sont très minces à leur sujet, deux éléments sont à retenir sur ces deux savants[7].

La figure d'Ibn Mujāhid, juriste malékite, est importante dans la transmission de la doctrine ashʿarite au Maghreb. Elle symbolise aussi les premières manifestations de la rencontre entre mālikisme et ashʿarisme[8].

Quant à Abū l-Ḥasan al-Bāhilī, on ne sait pas non plus grand chose de ce savant, si ce n'est qu'il fut anciennement chiite, puis converti au sunnisme des suites d'un débat avec al-Ashʿarī. L'élément remarquable se trouve dans l'identité de certains de ses disciples, qui représentent les trois figures[9] les plus célèbres de ce qu'Ibn ʿAsākir appelle "la deuxième génération" (*ṭabaqa thāniya*) : Abū Isḥāq al-Isfarāyīnī (417/1027) auteur d'ouvrages volumineux dont la plupart ont disparu[10], Ibn Fūrak[11] (406/1015) connu pour son *Mujarrad maqālāt al-Shaykh Abī l-Ḥasan al- Ashʿarī*[12] ou son *Kitāb Mushkil al-ḥadīth*[13],

---

4 Voir : Ibn ʿAsākir, *Tabyīn* 162.

5 Thiele, Jan, « Between Cordoba and Nīsābūr: The emergence and consolidation of Ashʿarism (fourth-fifth/tenth-eleventh century) », dans S. Schmidtke (éd.), *Oxford handbook of Islamic theology*, Oxford: Oxford University Press, 2016, 225-241, à p. 230.

6 Ibn ʿAsākir, *Tabyīn* 705.

7 Pour quelques éléments biographiques sur les deux savants, voir : Ibn ʿAsākir, *Tabyīn* 345-348.

8 Ceci si l'on n'admet pas l'hypothèse, défendue par certains membres de cette école, selon laquelle al-Ashʿarī était malékite. Rien ne permet réellement de savoir de quel *madhhab* se réclamait al-Ashʿarī. *Wa-kull yaddaʿī waṣl bi-Laylā !*

9 Sur la contribution de ces trois figures clés, voir : Thiele, « Between Cordoba and Nīsābūr » 229-234.

10 Tels : *al-Jāmiʿ bayna al-jalī wa-l-khafī*. Un crédo assez bref qui lui est attribué a été édité par R. Frank. Voir : Frank, Richard, « Al-Ustādh Abū Isḥāḳ: An *ʿaḳīda* together with selected fragments », dans *Mélanges de l'Institut dominicain d'études orientales* 19 (1989), 129-202.

11 Voir : Watt, « Ibn Fūrak » dans *EI2*.

12 Edité par Gimaret en 1987 pour le compte de Dār al-Machreq (Beyrouth). Voir : Gimaret, Daniel, « Un document majeur pour l'histoire du kalām: le *Muǧarrad maqālāt al-Ašʿarī* d'Ibn Fūrak », dans *Arabica* 32 (1985), 185-218.

13 Ibn Fūrak, *Kitāb Mushkil al-ḥadīth*, éd. Gimaret, Damas: Publications de l'Ifpo, 2003.

et enfin al-Bāqillānī (403/1012), connu tant pour ses traités de *kalām*, tels le *Tamhīd* ou la *Hidāyat al-mustarshidīn*, que pour ses livres d'*uṣūl al-fiqh*, tels le *Taqrīb wa-l-irshād*[14]. Notons que les deux premiers étaient shāfiʿites et le dernier malékite.

Ce IVe/Xe siècle donc, dans lequel ont vécu la plupart des savants de cette deuxième génération, marque en effet un tournant dans l'exportation de la doctrine au-delà de Bagdad, via le retour à leur domicile des étudiants étrangers. Abū Isḥāq al-Isfarāyīnī et Ibn Fūrak vont ainsi devenir les "représentants" de l'ashʿarisme à Nishapur, capitale intellectuelle et économique du Khorassan[15]. Ce développement est aussi favorisé par un soutien politique. Le gouverneur Samanide Nāṣir al-Dawla (357-8/968-9) fait construire une *madrasa* dans ladite ville pour Ibn Fūrak. Quant à al-Bāqillānī, qui était comme al-Ashʿarī originaire de Bassorah, il fut convié à joindre la cour des Bouyides à Bagdad et fut nommé juge par ʿAḍud al-Dawla. Plus encore, il l'inclut dans une délégation diplomatique sensée le représenter à Byzance[16]. Ibn ʿAsākir rapporte que ses qualités de théologien mais également de théoricien du droit lui ont fait acquérir une impressionnante popularité par rapport aux deux autres, au point d'être nommé par un Ḥanbalite "la Voix de la Communauté" (*lisān al-umma*)[17]. D'ailleurs, il est celui des trois dont l'œuvre a le plus été conservée. Nous l'avons dit, Ibn ʿAsākir reconnait que ce n'est qu'avec al-Bāqillānī que le vocable « ashʿarite » renvoyant à l'école caractérisée par ce système de doctrines devient véritablement populaire. Al-Bāqillānī a contribué à l'évolution des enseignements de l'école « *by broadening its conceptual framework and by further developing ideas of the school's founder* »[18]. La réputation d'al-Bāqillānī attire à Bagdad des étudiants locaux et d'autres venus de contrées éloignées profiter de l'enseignement de celui qui était considéré alors par certains comme « l'*imām* de l'époque ».

Les noms les plus célèbres de la génération suivante sont les disciples des trois figures citées. Parmi les disciples d'Abū Isḥāq al-Isfarāyīnī, on pourra citer

---

14 Pour une bibliographie de Bāqillānī : Gimaret, Daniel, « Un extrait de la *Hidāya* d'Abū Bakr al-Bāqillānī: le *Kitāb at-Tawallud*, réfutation de la thèse muʿtazilite de la génération des actes », *Bulletin d'études orientales* 58 (2009), 259-313. (La bibliographie se trouve à la troisième partie). Pour une description du contenu de certains livres : Thiele, « Between Cordoba and Nīsābūr » 231-232.

15 Voir : Broderson, « Abū Isḥāq al-Isfarāyīnī » dans *EI3*. Nous maintenons la *fatḥa* sur la *hamza* en l'absence d'indication explicite d'une autre orthographe.

16 Allard, Michel, *Le problème des attributs divins dans la doctrine d'al-Ašʿarī et de ses premiers grands disciples*, Beyrouth: Imprimerie Catholique, 1965, 290-295.

17 Voir : Ibn ʿAsākir, *Tabyīn* 417.

18 Thiele, « Between Cordoba and Nīsābūr » 231-232.

deux hérésiographes, Abū Manṣūr al-Baghdādī[19] (429/1037) et Abū l-Qāsim al-Isfarāyīnī (452/1060), célèbres entre autres pour leur traité en la matière, à savoir respectivement *al-Farq bayna al-firaq* et *al-Tabṣīr fī l-dīn*. Parmi ceux d'Ibn Fūrak, nous pouvons citer le savant du *ḥadīth* al-Bayhaqī (458/1065), célèbre compilateur de traditions (*sunan*). Parmi ceux d'al-Bāqillānī, on peut citer Abū Jaʿfar al-Samnānī (444/1052) ou encore Abū Isḥāq al-Shīrāzī[20] (476/1083). Enfin, parmi les Ashʿarites de cette génération n'ayant pas étudié chez l'un des trois, on peut compter Abū Muḥammad al-Juwaynī (438/1046), qui a appris auprès d'Abū l-Ṭayyib al-Ṣuʿlūkī (369/980), lui-même disciple d'un élève direct d'al-Ashʿarī, Abū Sahl al-Ṣuʿlūkī (396/1006), et qui est surtout connu pour être le père d'une des plus importantes figures de la quatrième génération : Abū l-Maʿālī al-Juwaynī[21] (478/1085). Théologien, théoricien du droit et juriste shāfiʿite, al-Juwaynī fils est l'auteur de nombreux ouvrages dans les fondements du *fiqh*[22] et la théologie[23]. Il prit la place de son père et maître en tant qu'enseignant à Nishapur. Après une vague de diabolisation de l'ashʿarisme orchestrée par le vizir de la dynastie seldjoukide[24] Tughril Bek (455/1063), al-Juwaynī fait le choix, avec d'autres Ashʿarites tel Abū l-Qāsim al-Qushayrī, de quitter Nishapur[25]. En 450/1058, il a le privilège d'enseigner aux deux lieux saints (la Mecque et Médine), ce qui lui vaut le surnom honorifique d'*Imām al-Ḥaramayn.*

Cet épisode difficile, ou *miḥna*, qui a contraint plusieurs savants ashʿarites à l'exil, va être succédé par un autre plus heureux et décisif dans le développement de la doctrine. Le nouveau vizir seljūqide, Niẓām al-Mulk (485/1092) mène une véritable politique de mécénat à l'égard de l'ashʿarisme : il ordonne la construction d'écoles enseignant ladite doctrine dans tout l'Irak, le Khorassan

---

19 Abū Manṣūr était le maître du célèbre soufi Abū l-Qāsim al-Qushayrī.

20 Voir : Chaumont, « al-Shīrāzī » dans *EI2*.

21 Voir : Thiele, Jan, « al-Juwaynī », dans H. Lagerlund (éd.), *Encyclopedia of medieval philosophy*, Dordrecht: Springer, 2018, 1028-1032.

22 Citons : *al-Waraqāt* (Traduit en 1930 par Léon Bercher pour la Revue Tunisienne et récemment réédité par les Editions Iqra), *al-Talkhīṣ fī uṣūl al-fiqh* (qui semble être l'abrégé du *Taqrīb* de Bāqillānī) ou encore le *Burhān*.

23 Citons : *al-Kāmil, al-Shāmil, Lumaʿ al-adilla*, ou encore *al-ʿAqīda al-Niẓāmiyya* (édité au Caire en 1948 par M.Z. al-Kawtharī. Toujours au Caire, elle a été traduite vers l'allemand dix ans plus tard par Klopfer, H., *Das Dogma des Imām al-Haramain al-Djuwainī und sein Werk al-ʿAqīda an-niẓāmiya*). Son traité de théologie *al-Irshād* gagna une grande popularité au Maghreb. Sur l'apport d'al-Juwaynī à l'école ashʿarite, voir l'introduction d'Arfa Mensia, « Regards d'Ibn Rushd sur al-Juwaynī: questions de méthode », dans *Arabic sciences and philosophy* 22 (2012), 199-216.

24 Ceux-ci ont conquis Nishapur en 428/1037.

25 Nguyen, Martin, *Sufi master and Qur'an scholar: Abū 'l-Qāsim al-Qushayrī and the Laṭā'if al-Ishārāt,* Oxford: Oxford University Press, 2012.

et la péninsule arabique. Il fait revenir al-Juwaynī à Nishapur avec les honneurs, en lui faisant construite sa propre école.

La figure de proue de la cinquième et dernière génération citée par Ibn ʿAsākir est bien sûr al-Ghazālī[26] (504/1111) qui fut l'élève d'al-Juwaynī. Cette génération, correspondant approximativement au 6e/12e siècle, marque, d'après Ibn Khaldūn, un tournant décisif dans les sciences de *kalām* qui sonne la fin de la « voie des anciens » et la naissance de la « voie des modernes », qu'inaugure donc d'après lui al-Ghazālī[27]. L'élément déterminant de celle-ci est l'introduction du syllogisme aristotélicien, et plus généralement l'emprunt assumé de certaines notions à la *falsafa*[28].

Au fil du temps, la doctrine ashʿarite s'est donc décloisonnée, dépassant Bagdad pour ensuite s'implanter dans le Khorassan, avant de gagner le Moyen-Orient, puis progressivement tout l'Occident musulman.

2 L'ashʿarisme en Occident musulman : essai de chronologie

Plusieurs chercheurs avant nous se sont interrogés sur les modalités de l'ashʿarisation de cette zone géographique. Idris, sans doute l'un des premiers, s'est intéressé à la propagation de cette doctrine en Ifrīqiya, soit un territoire qui inclut la Tunisie actuelle[29].

Dans son étude sur la transmission de quelques traités ashʿarites en al-Andalus, Fórneas prouve que cette doctrine était déjà présente dans la péninsule dès le début du Ve/XIe siècle[30]. Lagardère proposa de se concentrer sur l'étape inaugurale de l'introduction de cette doctrine, pour arriver à la

---

26 La figure de Ghazālī et son œuvre ont été abondamment étudiées. Pour une synthèse sur sa vie et son œuvre, voir : Griffel, Frank, *Al-Ghazālī's philosophical theology*, 2009, New York, Oxford University Press.

27 Sur cette distinction, voir : Ibn Khaldūn, *al-Muqaddima*, 3e éd. [Beyrouth, 1967], pp. 834-5. Voir aussi : Gardet, L., et M.M. Anawati, *Introduction à la théologie musulmane*, Paris: Vrin, 1948, 72-76.

28 Influence dont al-Juwaynī a été le précurseur. Al-Juwaynī semblait représenter « non le chef de file des 'modernes' mais plutôt se situait dans l'entre-deux ». Voir : Arfa Mensia, « Regards » 1999n9.

29 Voir : Idris, H.R., « Essai sur la diffusion de l'ašʿarisme en Ifrîqiya », dans *Les cahiers de Tunisie* 1 (1953), 128-30, ainsi que Idris, H.R., « Deux juristes kairouanais de l'époque zīrīde: Ibn Abī Zaid et al-Qābisī (Xe-XIe siècle) », dans *Annales de l'Institut d'études orientales* 12 (1954), 122-198.

30 Fórneas Besteiro, J.M., « De la transmisión de algunas obras de tendencia ašʿarī en al-Andalus », dans *Awrāq* 1 (1978), 4-11.

conclusion qu'Abū l-Walīd al-Bājī ne pouvait en être l'unique acteur[31]. Serrano s'est intéressée à l'adhésion progressive de l'ashʿarisme en al-Andalus[32], mais aussi au rôle de la dynastie almoravide dans la diffusion de cette doctrine[33]. Thiele s'est intéressé à la transmission de l'ashʿarisme en al-Andalus en marge de son travail sur l'ashʿarisme d'al-Bājī. Il a également, avec Hassan Ansari, étudié la réception maghrébo-andalouse d'un traité d'al-Bāqillānī, le *Tamhīd*[34].

Enfin, parmi les études arabes, citons Aḥnāna et son livre sur le développement de la doctrine ashʿarite en Occident Musulman[35] et l'article de Muḥammad Naṣīr qui met en évidence le rôle prééminent d'al-Bāqillānī dans la constitution des chaînes de transmissions (*asānīd*) des Ashʿarites du Maghreb[36].

Il est très difficile de dater la pénétration de l'ashʿarisme au Maghreb. Le processus d'ashʿarisation s'est sans doute effectué par plusieurs vagues. L'un des éléments déterminants ayant facilité cette propagation est sans doute le fait qu'al-Bāqillānī, et avant lui Ibn Mujāhid, incarnaient l'alliance du mālikisme et de l'ashʿarisme, alliance qui marqua pour des siècles le Maghreb et al-Andalus et continue d'être d'actualité[37].

Pour des raisons de proximité géographique évidente, la zone ouest de l'Afrique du Nord, recouvrant l'espace nommé jadis Ifrīqiya a sans doute été

31 Lagardère, Vincent, « Une théologie dogmatique de la frontière en al-Andalus aux XIe et XIIe siècles: L'ašʿarisme », dans *Anaquel de estudios árabes* 5 (1994), 71-98.

32 Voir : Serrano Ruano, Delfina, « Later Ashʿarism in the Islamic West », dans S. Schmidtke (éd.), *The Oxford handbook of Islamic theology*, Oxford: Oxford University Press, 2016, 514-539.

33 Voir : Serrano Ruano, Delfina, « La diffusion de l'ashʿarisme et la réforme du crédo malikite à l'époque almoravide: Ibn Rushd al-Jadd, Abū Bakr Ibn al-ʿArabī et le Qāḍī ʿIyāḍ », dans C. Aillet et B. Tuil Leonetti (éds.), *Dynamiques religieuses et territoires du sacré au Maghreb médiéval*, Madrid: CSIC, 2015, 79-102. Voir également : Serrano Ruano, Delfina, « Los almorávides y la teología ašʿarí: ¿Contestación o legitimación de una disciplina marginal?", dans de la Puente (éd.), *Identitades marginales*, Madrid: CSIC, 2003, pp. 461-516.

34 Voir : Thiele, « Between Cordoba and Nīsābūr » 233-234. Voir également : « Notes sur l'ashʿarisme d'Abū l-Walīd al-Bājī », et Ansari, Hassan et Jan Thiele, « Discussing al-Bāqillānī's theology in the Maghrib: ʿAbd al-Jalīl b. Abī Bakr al-Dībājī al-Rabaʿī's *al-Tasdīd fī sharḥ al-Tamhīd* », dans *al-Qanṭara* 39 (2018), 127-168.

35 Voir : Aḥnāna, Yūsuf, *Taṭawwur al-madhhab al-ashʿarī fī l-gharb al-islāmī,* Rabat: Publications du Ministère Marocain des Affaires Islamiques, 2007.

36 Naṣīr, Muḥammad, « Dawr al-Bāqillānī fī talqīn al-asānīd al-ashʿariyya bi-bilād al-maghrib », dans Jamāl ʿAllāl al-Bakhtī (éd.), *Al-fikr al-ashʿarī bi-l-maghrib: Khilāl marḥalatay al-taʾsīs wa-l-tarsīm*, Rabat: Publications de la Rābīṭa Moḥammedienne des Savants, 2016, 65-94.

37 Les ministères des affaires islamiques de tous les pays du Maghreb proclament l'ashʿarisme et le mālikisme comme composante fondamentales de l'orientation religieuse de leur pays.

l'un des premiers lieux de contact entre Maghrébins et Ashʿarites. Du fait de sa position de trait d'union entre l'Orient et l'Occident, Kairouan a eu un rôle important comme métropole culturelle accueillant des Maghrébins et Andalous, qui, au départ sur le chemin de l'aller ou du retour du pèlerinage, ont pu y faire halte, parfois des années, afin de profiter de l'enseignement dispensé sur place. C'est ainsi qu'al-Darrās b. Ismāʿīl (357/968), savant de Fès, y séjourna une période suffisamment longue pour y enseigner avant de regagner sa ville d'origine et y introduire la *Mudawwana*[38]. Contemporain d'al-Ashʿarī, rien ne nous indique s'il l'a rencontré durant ses voyages en Orient, mais l'épitre qui lui est attribuée faisant l'apologie d'al-Ashʿarī porte à croire qu'il avait adhéré à ses thèses[39] et qu'il les transmettait à ses disciples.

Parmi eux, deux savants qui vont progressivement faire figure d'autorité dans le mālikisme : Ibn Abī Zayd al-Qayrawānī (386/997) et Abū l-Ḥasan al-Qābisī (403/1012)[40]. Avec eux deux, c'est l'inverse du phénomène cité précédemment qui se produit : leur réputation dans le *fiqh* dépasse l'Afrique du Nord et parvient aux oreilles de leur contemporain ashʿarite, le bagdadien Ibn Mujāhid. Dès 368/979, Ibn Mujāhid et Ibn Abī Zayd entretiennent une correspondance, dans laquelle l'Irakien, qui déplore de ne pouvoir se rendre jusqu'à lui en personne, demanda au Kairouanais de lui offrir deux de ses travaux sur la *Mudawwana* devenus célèbres en Irak[41]. Ibn Abī Zayd chargea deux

38 La *Mudawwana* est une des sources incontournables du mālikisme : c'est la compilation faite par Saḥnūn b. Saʿīd (m. 240/854) rassemblant des *fatāwā* de Mālik b. Anas (m. 179/795) entendues par son disciple Ibn al-Qāsim (m. 190/806) ainsi que les avis propres à ce dernier. Souvent décrit comme le premier à avoir introduit ce livre à Fès, Darrās est tenu pour être à l'origine de la malékisation de la capitale idrisside. Voir : al-Qāḍī ʿIyāḍ, *Tartīb al-madārik wa-taqrīb al-masālik li-maʿrifat aʿlām madhhab Mālik*, Rabat: Publications du Ministère des Habous et des Affaires Islamiques, 1965-1983, vi, 81-82. Il est, d'après Sanseverino, le premier saint marocain connu de la période post-idrisside. Voir : Sanseverino, R.V., *Fès et sainteté, de la fondation à l'avènement du proctectorat (808-1912)*, Rabat: Publications du Centre Jacques Berque, 2014, 145-160.

39 Ibn Ḥazm a d'ailleurs répondu à cette épître. Idris, « Deux juristes » 134.

40 Idris, « Deux juristes ». Leur contribution au mālikisme n'est pas sans rappeler celle (bien plus tardive) du duo al-Rāfiʿī /al-Nawawī au shāfiʿisme. Notons que dans les deux écoles, ce duo de savants est désigné par *al-shaykhān* (« les deux maîtres »).

41 Lagardère, « Une théologie dogmatique » 72. Les deux livres sont un abrégé de la *Mudawwana* et un commentaire encyclopédique de celle-ci en plusieurs volume intitulé : *al-Nawādir wa-l-Ziyādāt*. Notons que cette lettre et sa réponse ont été éditées, avec le fac-similé de la *Risāla* d'Ibn Abī Zayd, par Tawrātī. Voir : Ibn Abī Zayd, *al-Risāla*, éd. Tawrātī, Tanger : Dār al-Ḥadīth al Kattāniyya, 2015, 20-25. Sur le mālikisme irakien et sa connexion au Maghreb, voir : Benhima, Yassir, « Le mālikisme irakien et le Maghreb: Réseaux, circulations et transferts entre Orient et Occident musulmans », dans M. Meouak et C. de la Puente (éds.), *Connected stories: Contacts, traditions and transmissions in premodern Mediterranean Islam*, Berlin: De Gruyter, 2022, 111-130.

de ses élèves, de faire le déplacement de la Mecque à Bagdad pour lui offrir les volumes ainsi qu'une *ijāza* (autorisation de transmettre) écrite de sa main. L'un des deux, Ibn ʿAdhrā, devint disciple d'Ibn Mujāhid, qui lui enseigna sa fameuse *Risāla fī ʿuqūd ahl al-sunna*. Ibn ʿAdhrā va la transmettre à son retour au Maghreb[42]. Réciproquement, un élève d'Ibn Mujāhid, Ibn ʿAbd al-Mu'min s'installa à Kairouan étudier auprès d'Ibn Abī Zayd. Plus qu'une simple relation cordiale avec un ashʿarite, les productions d'Ibn Abī Zayd témoignent de son affiliation à la doctrine dont il était l'un des ardents défenseurs. Lui est entre autres attribuée une épître dans laquelle il réfute un Muʿtazilite et défend al-Ashʿarī, qu'il déclare être « attaché à la *Sunna*[43] » tout en lui reconnaissant un statut d'*imām* en la matière[44]. Lorsqu'Ibn Abī Zayd est pris dans la tourmente d'une controverse[45], c'est al-Bāqillānī qu'il consulte ; le fait de lui demander d'arbitrer sur une question de théologie l'opposant à ses pairs montre bien qu'à Kairouan, la parole du *mutakallim* irakien était déjà considérée comme argument d'autorité. Tout ceci rend donc crédible l'hypothèse selon laquelle Ibn Abī Zayd était un ashʿarite convaincu.

Il en est de même concernant Abū l-Ḥasan al-Qābisī, décrit comme un *mutakallim*[46]. Interrogé par des Tunisois sur la doctrine d'al-Ashʿarī, celui-ci répond par une épître dans laquelle il fait l'apologie de celle-ci et se targue d'en être un connaisseur, invalidant tout propos dévalorisant cette croyance et insistant sur le bien-fondé de celle-ci[47]. Pour les biographes orientaux, al-Qābisī n'est donc pas un simple sympathisant de la doctrine mais bien un membre à part entière de l'école ; certains le font d'ailleurs figurer parmi les Ashʿarites de la troisième génération, au même titre qu'Abū Manṣūr al-Baghdādī[48].

---

42 Lagardère, « Une théologie dogmatique » 72.

43 Naṣīr, « Dawr al-Bāqillānī » 69.

44 Ibn Abī Zayd, *al-Risāla* 14.

45 Pour cette controverse, voir : Fierro, M., « The polemic about the *karāmāt al-awliyāʾ* and the development of Ṣūfism in al-Andalus (fourth/tenth-fifth/eleventh centuries) », dans *Bulletin of the School of Oriental and African Studies* 55 (1992), 236-249. Plus récemment, voir l'étude introductive de Tawrātī dans son édition facsimilée de la *Risāla* (p. 13). Ibn Abī Zayd aurait réprimandé un homme prétendant avoir vu Dieu à l'état d'éveil, ce qui suscita le courroux de plusieurs juristes de Kairouan jugeant qu'Ibn Abī Zayd niait l'existence même des prodiges des saints (*karāmāt al-awliyāʾ*), ce qui est une des thèses caractéristiques des Muʿtazilites. Al-Bāqillānī répondra par un ouvrage en deux tomes intitulé *al-Farq bayna muʿjizāt al-anbiyāʾ wa-karāmāt al-awliyāʾ* (la distinction entre miracles des prophètes et prodiges des saints) en introduction duquel il innocentera Ibn Abī Zayd de ce qui lui a été attribué. Sur ce livre, voir : Thiele : « Notes sur l'ashʿarisme » 2.

46 Naṣīr, « Dawr al-Bāqillānī » 71.

47 Naṣīr, « Dawr al-Bāqillānī » 71.

48 Naṣīr, « Dawr al-Bāqillānī » 73.

Ainsi, à la fin du IVème/Xème siècle, les textes ashʿarites sont déjà enseignés sur le sol maghrébin.

Durant le Vème/XIème siècle, les contacts avec les Ashʿarites Orientaux de la « deuxième génération » s'amplifient. Un certain ʿAbd al-Malik al-Ṣaqallī étudia le livre *Mushkil al-ḥadīth* directement auprès de son auteur, Ibn Fūrak, et l'enseigna à son tour au Maghreb. Une copie maghrébine de ce texte date de 459/1066[49]. Dans le même temps, la doctrine se développe dans le Maghreb. Certains andalous ayant séjourné à Kairouan transmettent à leur retour chez eux des traités ashʿarites, à l'image d'al-Ghassānī (498/1105), qui reçut la fameuse *Risāla* d'Ibn Mujāhid de la part d'un élève d'Ibn ʿAdhrā, un certain Abū Marwān ʿAbd al-Malik al-Ṭubunnī (455/1063)[50]. Parmi les élèves d'al-Qābisī[51], Abū ʿImrān al-Fāsī (430/1038), savant marocain né à Fès, décide de prendre la route pour Bagdad, qu'il atteint en l'an 399/1008. Il y assiste aux séances d'*uṣūl* données par al-Bāqillānī, qui le gratifie d'une *ijāza* l'autorisant à transmettre tous ses livres[52]. Revenu à Kairouan, il se met à son tour à transmettre l'enseignement d'al-Bāqillānī, mais également ceux d'al-Ashʿarī ou Ibn Mujāhid à un public constitué de maghrébins et d'andalous.

Pendant ce temps, à Bagdad, des pressions que font subir des Ḥanbalites à al-Bāqillānī vont pousser ce dernier à orienter sa prédication vers de nouvelles directions. Il décida, à cet effet, d'envoyer au Maghreb deux de ses disciples : Ibn Ḥātim al-Adharī[53] (423/1023) et Ibn Muḥammad al-Shāfiʿī al-Baghdādī (448/1056)[54]. Avec Abū ʿImrān, tous les trois ont un rôle majeur dans la diffusion et l'enseignement des livres d'al-Bāqillānī, et, plus généralement la pensée ashʿarite à Kairouan.

En 449/1057 les Hilaliens[55] s'emparent de Kairouan, ce qui va contraindre à l'exil une partie des savants locaux et un retour au pays des savants andalous sur place[56]. La pierre angulaire de l'ashʿarisme occidental se déplace alors vers le centre et l'Ouest du Maghreb, où les dynasties almoravides et ḥammādides soutiennent les savants de l'école[57].

49 Thiele, « Notes sur l'ashʿarisme » 2.

50 Lagardère, « Une théologie dogmatique » 72.

51 Idris, « Deux juristes » 122-123.

52 Naṣīr, « Dawr al-Bāqillānī » 77.

53 Il existe une confusion sur l'orthographe de son nom entre al-Azdī (de la tribu arabe Azd) et al-Adharī (azéri). Voir : Maḥfūẓ, M., « al-Azdī am al-Adharī? », dans *Majallat al-Fikr* 17 (1971), 56-60.

54 Naṣīr, « Dawr al-Bāqillānī » 77.

55 Il s'agit des tribus bédouines de Banū Hilāl et Banū Sulaym. Voir : Idris, H.R., « L'invasion hilālienne et ses conséquences », dans *Cahiers de civilisation médiévale* 11 (1968), 353-369.

56 Ansari et Thiele, « Discussing » 129.

57 Serrano Ruano, « Los almorávides » 461-516.

Notons qu'Abū ʿImrān joue un rôle important dans la naissance du mouvement almoravide. Alors que Yaḥyā b. Ibrāhīm, chef de la tribu berbère des Lamtūna, revenait de pèlerinage et était de passage à Kairouan pour suivre l'enseignement d'Abū ʿImrān, il adressa une requête à ce dernier : désigner parmi ses élèves qui pourrait l'aider à répandre l'enseignement islamique en territoire berbère. Abū ʿImrān chargea son disciple Waggāg b. Zalū, originaire de Sous, de s'occuper de l'enseignement de ʿAbd Allāh Yāsīn, et devint l'enseignant attitré de celui qui fonda la dynastie almoravide[58].

Parmi les Ashʿarites de Kairouan ayant un rôle dans la diffusion de l'ashʿarisme sous les Almoravides, citons bien sûr Abū Bakr Murādī al-Ḥaḍramī[59] (489/1095), qui enseigna d'abord à Cordoue en 487/1094, avant de rejoindre l'émir almoravide Abū Bakr b. ʿUmar al-Lamtūnī dans son entreprise de prédication au Sahara et vers l'Afrique de l'Ouest. Il se fixe à Azkī, aux portes du désert saharien. Son crédo « scolaire », découvert par al-Bakhtī et édité en 2012 sous le titre de *ʿAqīdat Abī Bakr al-Murādī al-Ḥaḍramī*, est l'une des premières productions pédagogiques ashʿarites maghrébines[60]. Al-Murādī a très certainement un rôle indéniable dans le processus d'ashʿarisation des contrées aux rives du Sahara (comme le Mali) et du Maghreb de l'Est de manière plus générale. Cependant, l'orientation essentiellement pédagogique de son œuvre ne doit pas nous induire en erreur : il est un *mutakallim*, plus encore, d'après certains biographes, le dernier véritable *mutakallim* maghrébin[61].

Parmi les disciples d'al-Adharī ayant quitté Kairouan après son invasion hilalienne, nous pouvons citer ʿAbd al-Jalīl Rabaʿī al-Dībājī (vivant en 478/1085), qui, par ailleurs, avait également appris auprès d'Abū ʿImrān al-Fāsī, mais aussi d'al-Juwaynī. Il s'installa à Fès et contribua à la diffusion des œuvres ashʿarites, parmi lesquelles, le *Tamhīd* d'al-Bāqillānī, dont il produit un commentaire[62].

C'est donc durant ce vème/xième siècle qu'al-Andalus connait une affluence de nombreux savants ashʿarites, composés d'exilés kairouanais[63], tels Makkī b. Abī Ṭālib (437/1045), élève d'al-Qābisī, ou d'élèves directs d'al-Bāqillānī

58 Lagardère, « Une théologie dogmatique » 80.

59 Voir : Ould Cheikh, Abdel Wedoud et Bernard Saison, « Vie(s) et mort(s) de al-Imām al-Ḥaḍramī », dans *Arabica* 34, (1987), 48-79.

60 Il a été édité à la Rābīta Muḥammadiyya (Rabat).

61 Lagardère, « Une théologie dogmatique » 80.

62 Sur le commentaire du *Tamhīd* par Dībājī, voir : Ansari et Thiele, « Discussing » 127-168. Ce commentaire a été récemment édité par Rashīd ʿAmmūr et publié en 2022 par la Rābiṭa Muḥammadiyya (Rabat). Sur la diffusion du *Tamhīd* en Andalousie, voir : Fórneas Besteiro, J.M., « *Al-Tamhīd* de al-Bāqillānī y su transmisión en al-Andalus », dans *Miscelánea de estudios árabes y hebraicos* 26-28 (1977-1979), 433-40.

63 Parmi lesquels les discipes d'Abū ʿImrān. Lagardère a pu en répertorier 19. Lagardère, « Une théologie dogmatique » 75-76.

venus de Bagdad, comme ʿAlī al-Tibrīzī, ou enfin Abū ʿAbd Allāh Shāhrazūrī (427/1035), qui compte parmi les natifs andalous à s'être expatrié vers les Ashʿarites Orientaux[64].

Parmi ces derniers, citons Abū l-Walīd al-Bājī[65] (474/1081). Formé d'abord à Cordoue auprès d'ashʿarites de « l'école de Kairouan » pour la plupart disciples dudit Makkī b. Abī Ṭālib, il avait quitté la péninsule en 426/1034 à la rencontre de trois disciples d'al-Bāqillānī, parmi lesquels Abū Dharr al-Harawī[66] (434/1043), qu'il rencontra à la Mecque. Ensuite, lors de son passage de trois ans à Bagdad, al-Bājī profita de l'enseignement d'Abū Isḥāq al-Shīrāzī (476/1083)[67]. Enfin, il apprit à Mossoul auprès du juge ḥanafite Abū Jaʿfar al-Simnānī[68] (444/1052). De retour en al-Andalus, il ajouta aux traités déjà enseignés d'autres textes, parmi lesquels ceux d'al-Simnānī[69]. Signalons aussi que c'est lui qui introduisit le traité d'*uṣūl al-fiqh al-Taqrīb wa-l-Irshād*[70] d'al-Bāqillānī.

Parmi les disciples andalous d'Abū ʿImrān, Abū ʿAlī al-Ghassānī se mit à son tour à transmettre la doctrine, à travers la *Risāla* d'Ibn Mujāhid qu'il reçut d'Ibn ʿAdhrā, mais également le *Tamhīd* et l'*Inṣāf fī-mā yajibu ʿtiqāduh wa-lā yajūz al-jahl bih* de al-Bāqillānī, introduit à Cordoue via Abū Ṭāhir al-Wāʿiẓ[71]. Le public d'al-Ghassānī était constitué d'élèves qui occuperont dans les grandes métropoles de la péninsule des fonctions de premier plan au sein de l'administration almoravide, entre Cadi ou Grand Cadi. Parmi ces disciples, notons la présence du grand-père d'Averroès, Ibn Rushd al-Jadd (520/1126), et d'Ibn Ḥamdīn (521/1127) qui furent l'un après l'autre Grands Cadis de Cordoue, ou encore celle d'Abū l-Ḥasan Shurayḥ al-Ruʿaynī (539/1144), Grand Cadi de Séville. Ce soutien d'une partie de la judicature n'est pas négligeable : lorsqu'une remise en cause de l'orthodoxie de l'ashʿarisme ainsi que de ses figures

64 Thiele, « Notes sur l'ashʿarisme » 3.

65 Sur ce savant, voir : Fierro, M., « al-Bājī, Abū l-Walīd », dans Delgado et al (éds.), *Enciclopedia de al-Andalus: Diccionario de autores y obras andalusíes*, Grenade: Fundacion El Legado Andalusi, 2002, 118-123.

66 Il est l'un des premiers *Mutakallimūn* à avoir transmis ces doctrines aux pèlerins à la Mecque. Il était par ailleurs versé dans le *ḥadīth* et compte parmi les transmetteurs du *Ṣaḥīḥ* de Bukhārī. Naṣīr, « Dawr al-Bāqillānī » 74.

67 Sur la vie et l'œuvre de ce savant, voir : Chaumont, « al-Shīrāzī », dans *EI2*, et son édition traduite et commentée du *Lumaʿ*, édité en 1999 à Berkeley pour Robins Collection.

68 Voir : Gimaret, « al-Simnānī, Abū Djaʿfar Muḥammad », dans *EI2*.

69 Thiele, « Notes sur l'ashʿarisme » 5.

70 Ce traité est décliné en un *Taqrīb kabīr, saghīr* et *awsaṭ*. Seul le *saghīr* a été partiellement édité.

71 Fórneas Besteiro, « *Al-Tamhīd* de al-Bāqillānī » p. 439.

commence à grandir, un haut fonctionnaire comme Ibn Rushd al-Jadd n'hésite pas à émettre une *fatwā* innocentant l'école[72].

Entre temps, la mort d'al-Bāqillānī n'a pas mis fin aux déplacements massifs de savants andalous vers les ashʿarites orientaux. Le fait qu'al-Juwaynī ait été alors enseignant à la Mecque a grandement facilité l'affluence de savants andalous et maghrébins, qui font sa connaissance lors du pèlerinage et deviennent ses élèves, à l'image d'al-Dībājī, qui obtient une *ijāza* de sa part et retourne en Andalus avec une partie de son œuvre[73].

Ainsi, à la fin du vème/xième siècle, l'ashʿarisme est déjà enraciné en al-Andalus. Il n'est pas une grande métropole qui n'abrite en son sein un *mutakallim*, fusse-t-il un étudiant andalou ayant fini sa formation en Orient, ou le disciple d'un élève d'al-Adhari ou d'Abū ʿImrān. La haute fonction qu'occupent plusieurs de ces Ashʿarites dans l'administration almoravide permet de conforter « officiellement » la légitimité de cette doctrine lorsque celle-ci est remise en cause. Les œuvres d'al-Bāqillānī, Ibn Fūrak, ou plus tard al-Juwaynī y sont massivement enseignées. Les déplacements vers l'Orient vont continuer après la mort d'al-Juwaynī, d'autant qu'avec la réputation grandissante de son disciple et successeur al-Ghazālī, beaucoup d'étudiants ambitionnent de le rejoindre à Bagdad et profiter de son enseignement. Parmi eux, Abū Bakr Ibn al-ʿArabī.

3 L'ashʿarisme d'Ibn al-ʿArabī avant la *riḥla* (avant 485/1093) : un angle mort ?

Remonter cette « généalogie ashʿarite » nous permet de faire un certain nombre d'observations, concernant des zones d'ombres autour de la figure d'Ibn al-ʿArabī ou même de son père. Dans l'introduction de notre ouvrage sur les *Nukat al-Maḥṣūl*, nous proposions un découpage de la vie du Qāḍī[74] qui n'était finalement pas éloigné de celui que le savant lui-même propose dans sa *riḥla* ainsi que dans d'autres passages autobiographiques qu'il égrène dans son *Qānūn al-taʾwīl*: 1°) sa vie à Séville avant son voyage (469-485) ; 2°) son séjour en Orient (485-496) 3°) sa vie en al-Andalus postérieure à ce séjour (496-513).

72 Voir l'article déjà cité de Serrano Ruano, « Los almorávides », à completer, plus récemment, avec : Gómez-Rivas, Camilo, *Law and the Islamization of Morocco under the Almoravids: The fatwās of Ibn Rushd al-Jadd to the Far Maghrib*, Leyde: Brill, 2015, 86-93.

73 Naṣīr, « Dawr al-Bāqillānī » 85.

74 Voir : Amharar, Ilyass, *Langage et théologie chez Ibn al-ʿArabī: Les informations subtiles de la somme de théorie légale* (*Nukat al-Maḥṣūl fī ʿilm al-uṣūl*), Piscataway NJ: Gorgias Press, 2023, p. xvi.

Si Ibn al-ʿArabī nous offre d'abondants passages nous renseignant sur les maîtres ashʿarites qu'il a pu avoir en Orient, parfois même avec le détail des traités étudiés auprès d'eux, il est en revanche beaucoup moins éloquent sur les théologiens qui l'ont influencé dans la première partie de sa vie. Plus encore, dans les quelques passages décrivant la formation soutenue qu'a pu lui offrir son père et les nombreux cours dispensés à domicile, notre auteur évoque la langue arabe, les mathématiques, les lectures coraniques … mais jamais la théologie. Le rapport du Qāḍī à l'ashʿarisme dans sa jeunesse semble donc faire office d'un angle mort.

Relever le défi d'apporter de nouveaux éclairages sur l'évolution du rapport d'Ibn al-ʿArabī à l'ashʿarisme est extrêmement compliqué, compte tenu des larges zones d'ombre enveloppant une phase pourtant décisive de la vie d'Ibn al-ʿArabī : celle de sa vie avant la *riḥla*, qui correspond aux seize premières années de la vie du Sévillan, celles durant lesquelles il a acquis ses premiers enseignements.

Pour procéder à telle analyse, la consultation des catalogues de transmission (*barāmīj, fahāris*) s'avère particulièrement utile. Malheureusement, une telle investigation est malaisée par la rareté, voire l'absence de littérature de première main : le *barnāmaj* d'Ibn al-ʿArabī ainsi que son autobiographie, le *Tartīb al-riḥla*, sont toujours disparus. Plus encore, il semble que le texte dans sa version intégrale ait été perdu dès le vivant de son auteur[75]. Fort heureusement, nous savons qu'Ibn al-ʿArabī a réécrit une partie de ses textes, quelques fois en les modifiant, ou en y ajoutant des extraits supplémentaires d'autres ouvrages aujourd'hui indisponibles. Ainsi, si l'on en croit son introduction du *Qānūn al-taʾwīl*, qu'il a écrit – comme l'indique littéralement le titre – en vue d'établir un *canon* permettant d'accéder aux sciences de la Révélation, Ibn al-ʿArabī a réintégré dans ce dernier les éléments du *Tartīb* qu'il avait encore à l'esprit[76].

Le contenu de son enseignement pré-*riḥla* est décrit dans le premier chapitre, intitulé *Dhikr ibtidāʾ ṭalab al-ʿilm*[77] :

> Un des signes du destin favorable que Dieu m'a accordé est que mon père, que Dieu lui fasse miséricorde, a engagé un enseignant de Coran pour moi, alors que je n'étais encore qu'aux prémices de ma jeunesse. Ainsi,

---

75 « *Raʾaynā an nujaddida mā salima fī l-ruqāʿ al-mawjūda …* » (« Nous avons choisi de réécrire [sur ce sujet] à partir des fragments ayant subsisté … »). Ibn al-ʿArabī, Abū Bakr, *Qānūn al-taʾwīl*, éd. M. Slīmānī, Tunis: Dār al-Gharb al-Islāmī, 1990, 414.

76 « *… maʿ mā ḥaḍar fī l-dhikr* » (« … avec ce qui restait [en mémoire] »). Ibn al-ʿArabī, *Qānūn al-taʾwīl* 414.

77 Ibn al-ʿArabī, *Qānūn al-taʾwīl* 414-420.

> à neuf ans, j'étais très doué en ce qui concerne le Coran. Mon père m'a ensuite entouré de trois enseignants : un dans la maîtrise de la typographie du Coran (*ḍabṭ al-Qur'ān*) selon ses sept lectures ; un deuxième en langue arabe ; et un troisième en mathématiques. Ainsi, dès le début de mon adolescence, à seize ans, je récitais déjà près de dix lectures coraniques [...]. J'ai étudié l'ensemble des disciplines de la langue arabe, et je m'y suis exercé. J'ai entre autres étudié le livre *al-Īḍāḥ* d'al-Fārisī, *al-Jumal*, un livre d'al-Naḥḥās, *al-Uṣūl* d'Ibn al-Sarrāj, le livre d'al-Duraywad, d'al-Thumālī, le livre *al-Ṣināʿa* qu'al-Khalīl a entamé puis que Sībawayh a réordonné et achevé. J'ai lu de nombreux poèmes parmi lesquels les six [*Muʿallaqāt*], la poésie d'al-Ṭā'ī, d'al-Juʿfī et de nombreux textes de poètes arabes classiques ou plus récents. J'ai étudié dans la lexicographie le livre de Thaʿlab, *Iṣlāḥ al-manṭiq, al-Amālī* et d'autres textes encore. J'ai étudié de nombreux *ḥadhīth-s* auprès de maîtres. J'ai également étudié les mathématiques : les opérations, l'algèbre, les mathématiques appliquées à l'héritage, puis le livre d'Euclide [...]. J'ai également [appris à] examiner l'astrolabe, la projection orthogonale, etc. Les enseignants se succédaient chez moi, à partir du temps de la prière de l'aube jusqu'à l'appel à la prière du *ʿaṣr*. Puis je prenais congé d'eux jusqu'au *ṣubḥ* du jour d'après. Je ne me laissais aucun temps libre sans l'occuper à lire, réviser, ou encore consigner toute information utile ; jeune et insouciant, je rassemblais à travers ces notes ce qui était bon comme ce qui ne l'était pas, mais mon destin a été tel que j'ai malgré tout pu en tirer en profit par la suite dans ma réfutation des athées et dans mon initiation aux fondements de la foi (*uṣūl al-dīn*). Puis cette situation exceptionnelle s'est transformée en un bouleversement général, lorsque les Almoravides sont entrés dans notre pays, en l'an 484 [...] et que nous ne pouvions alors plus y rester.

Ce long passage appelle à plusieurs observations. La plus immédiate est bien sûr l'absence de quelconque précision sur une étude potentielle de *kalām* ou d'ashʿarisme. Ensuite, compte tenu du fait qu'Ibn al-ʿArabī dresse ici un long et foisonnant inventaire des disciplines et traités étudiés, le fait qu'il aille jusqu'à citer les sciences profanes et détailler les sujets abordés (comme la projection orthogonale) semble bien traduire une intention de dresser une liste la plus précise possible, ce qui rend peu pertinent de justifier l'absence d'évocation du *kalām* dans cet inventaire par un simple oubli. Il est également peu probable qu'il s'agisse d'une auto-censure : l'apologie de l'ashʿarisme que l'on peut trouver dans ses textes, à commencer par ce même *Qānūn*, laisse peu de place au doute quant au fait qu'il ne s'est jamais caché d'avoir été un théologien. Enfin, la fin de cet extrait est presque un aveu explicite que le Qāḍī n'a alors pas étudié, ou du moins n'a pas approfondi l'étude de l'ashʿarisme : en déclarant que son appétence à consigner un certain nombre d'informations plus ou moins immédiatement pertinente s'est avérée par la suite utile dans son étude des fondements de la foi et l'apologie anti-athées, comme si « le destin lui avait caché ces informations » (*al-qadar yukhabbiʾuhā lī*) et lui en avait révélé l'intérêt que plus tard, Ibn al-ʿArabī laisse entendre que cette étude du *kalām* a été postérieure à cet enseignement premier.

Saʿīd Aʿrāb s'est intéressé avant nous aux premiers enseignants d'Ibn al-ʿArabī[78], et est parvenu à en identifier deux : son oncle maternel Abū l-Qāsim al-Ḥasan al-Hawzanī (512/1134) et Abū ʿAbd Allāh al-Saraqusṭī (500/1122). Pour le premier, Aʿrāb renvoie à une notice du *Bughyat al-multamis* d'al-Ḍabbī (599/1203)[79] :

> 639. al-Ḥasan b. ʿUmar al-Hawzanī al-Ishbīlī : juriste savant issu d'une famille noble. Il est mort en 512, à l'âge de 80 ans environ. Le Ḥāfiẓ Abū Bakr Ibn al-ʿArabī, dont il était l'oncle maternel, à transmis par son truchement l'abrégé des lectures coraniques écrit par Abū Ḥafṣ ʿUmar, son père. Il a transmis d'après son père ʿUmar.

Aʿrāb ne cite pas sa référence pour al-Saraqusṭī mais on retrouve sa notice dans le même ouvrage[80] :

> 191. Muḥammad b. ʿAbd al-Raḥmān b. ʿAbd Allāh al-Saraqusṭī : juriste et spécialiste de la récitation du Coran. Le Ḥāfiẓ Abū Bakr Ibn al-ʿArabī et d'autres ont transmis son savoir. Il a transmis le Coran d'après Muḥammad Muhallab et d'autres.

Cette dernière information (qu'il était enseignant de Coran) est confirmée par Ibn al-Abbār (658/1280), qui ajoute dans sa *Takmila* qu'Ibn al-ʿArabī l'a lui-même signalée dans son propre catalogue de transmission (*barnāmaj*)[81].

Il est donc fort possible qu'al-Saraqusṭī et al-Hawzanī figurent parmi les trois enseignants[82] décrits par Ibn al-ʿArabī[83]. Les données ne sont pas assez

78 Aʿrāb, Saʿīd, *Maʿa al-Qāḍī Ibn al-ʿArabī*, Tunis: Dār al-Gharb al-Islāmī, 1987, 12.

79 Al-Ḍabbī, *Bughyat al-multamis fī tārīkh rijāl ahl al-Andalus*, le Caire: Dār al-Kitāb al-ʿArabī, 1968, p. 265.

80 Al-Ḍabbī, *Bughyat al-multamis* 101.

81 Ibn al-Abbār, *al-Takmila li-kitāb al-Ṣila*, éd. al-Harrās, Beyrouth: Dār al-Fikr, 1995, i, 316.

82 On peut s'interroger sur le chiffre donné par Ibn al-ʿArabī et, surtout, à quoi il renvoie. Il pourrait, par exemple, s'agir de trois enseignants quotidiens pour ces trois matières, sans pour autant que

83 Concernant l'enseignant d'arabe, il pourrait s'agir d'Ibn al-Akhḍar al-Tanūkhī (514/1119), qu'Ibn al-ʿArabī désigne par « *shaykh al-ʿarabiyya ʿindanā* » et dont il laisse entendre avoir été l'étudiant. Voir : Ibn al-ʿArabī, *Qānūn al-taʾwīl* 425.

abondantes à leur sujet, mais, compte tenu de leur domaine de prédilection, il est probable que ces deux savants aient surtout enseigné au Qāḍī les lectures coraniques. Al-Ḍabbī l'a affirmé explicitement concernant al-Hawzanī. Pour al-Saraqusṭī, c'est un disciple d'Ibn al-ʿArabī, Ibn Bashkuwāl (578/1200) qui le confirme[84] :

> Notre maître l'imam le Qāḍī Abū Bakr Ibn al-ʿArabī a étudié les lectures coraniques auprès de lui. Il a mentionné à son sujet qu'il était un maître vertueux et qu'il enseignait les lectures coraniques (*yuqriʾ*) à Séville. Il est mort en 500.

Il est donc assez peu probable qu'Ibn al-ʿArabī ait étudié l'ashʿarisme auprès d'al-Saraqusṭī. Concernant al-Ḥasan al-Hawzanī, un curieux extrait de la *Fahrasa* d'Ibn Khayr, qui semble avoir échappé à Aʿrāb, a retenu notre attention :

> 1158. [...] un poème sur les bonnes manières selon la Loi (*al-ādāb al-sharʿiyya*) ainsi que les articles de la foi des Gens de la Tradition (*ʿaqāʾid Ahl al-Sunna*), du vizir, le juriste, pèlerin et martyr Abū Ḥafṣ ʿUmar b. al-Ḥasan al-Hawzanī. Le Qāḍī Abū Bakr Ibn al-ʿArabī – que Dieu lui fasse miséricorde – me l'a rapporté d'al-Ḥasan, qui le rapporte de son père (Abū Ḥafṣ)[85].

Ibn al-ʿArabī a donc étudié un poème écrit par son grand-père, Abū Ḥafṣ al-Hawzanī, incluant de la *ʿaqīda*. Ceci est, dans l'état actuel des choses, la seule donnée factuelle associant le jeune Abū Bakr à une quelconque étude de la *ʿaqīda*. Cette donnée établit un lien scientifique avec son grand-père maternel Abū Ḥafṣ, par l'intermédiaire d'Abū l-Qāsim, le fils de ce dernier. Ce qui nous amène à une tout autre question : le rapport des deux Hawzanī à l'ashʿarisme et aux ashʿarites. Là encore, les sources[86] ne nous aident pas beaucoup : les quelques lignes qui leur sont consacrées ne les évoquent jamais comme des *mutakallimūn* ni même n'indiquent une orientation ashʿarite. Elles sont néanmoins plus précises concernant Abū Ḥafṣ. Un autre disciple d'Ibn al-ʿArabī, le Qāḍī ʿIyāḍ (544/1149), considère leur place légitime dans sa prosopographie consacrée aux Mālikites, et propose de faire un « tir groupé » en consacrant une seule notice aux deux. ʿIyāḍ n'écrit malgré tout que deux lignes sur l'oncle

84 Ibn Bashkuwāl, *al-Ṣila fī tārīkh aʾimmat al-Andalus*, éd. al-ʿAṭṭār al-Ḥusaynī, le Caire: Maktabat al-Khanjī, 1955, i, 533.

85 Ibn Khayr, *al-Fahrasa*, éd. Fuʾād Manṣūr, Tunis: Dār al-Gharb al-Islāmī, 2009, 508.

86 Voir, parmi elles : Ibn Bashkuwāl, *al-Ṣila* 137-138.

maternel. Pour le reste, outre le rôle politique d'Abū Ḥafṣ et sa relation houleuse avec al-Muʿtaḍid, que nous évoquions déjà dans notre travail sur les *Nukat al-Maḥṣūl*[87], la notice du Qāḍī ʿIyāḍ nous offre surtout quelques passages d'importance, que nous traduisons ici[88] :

> Abū Ḥafṣ ʿUmar b. Ḥusayn al-Hawzanī comptait parmi les habitants de Séville, et était son plus grand juriste. Il était doué dans de nombreuses disciplines. En plus de son savoir en droit et de sa transmission du *ḥadīth*, il s'est également intéressé aux sciences anciennes (*ʿulūm qadīma*). Il était également un bon littéraire, auteur d'une belle poésie, d'une brillante prose et d'aphorismes rassemblés. En al-Andalus, il a étudié auprès de l'autorité savante du pays : le Qāḍī Abū ʿAbd Allāh al-Bājī. Puis il effectua le pèlerinage et rencontra les savants de Sicile et d'Egypte. Il étudia le *ḥādīth* à la Mecque et ailleurs auprès d'Abū Muḥammad b. al-Walīd, ainsi qu'Ibn Manṣūr al-Shāhrazūrī. Il habita en al-Andalus orientale, sous la protection des Banū Ṭāhir, qui contrôlaient la région, et eût, durant son séjour là-bas, quelques litiges avec Abū l-Walīd al-Bājī. Il retourna à Séville où il officiait comme jurisconsulte et auprès duquel les gens étudiaient le *ḥadīth* [...]. Son fils, Abū l-Qāsim al-Ḥasan était, en son temps, le savant en chef de la ville. Il a étudié auprès de son père, d'Ibn Manẓūr et d'autres de la même ville. Il a voyagé et étudié auprès de plusieurs savants. Il a reçu une licence de transmission (*ijāza*) de Muḥammad b. al-Walīd, d'Abū Manṣūr al-Shāhrazūrī, de qui il a entendu le *ḥadīth*. Il est décédé en 512.

Là où l'oncle maternel était exclusivement décrit comme spécialiste des lectures coraniques, et, accessoirement, comme un juriste ayant étudié le *ḥadīth*, Abū Ḥafṣ, lui, est considéré comme un savant polyvalent. Même s'il reste surtout connu pour sa contribution dans la transmission du *ḥadīth* en al-Andalus – il est le premier à y avoir introduit le *Jāmiʿ* d'al-Tirmidhī et figure parmi les importants rapporteurs du *Ṣaḥīḥ* de Muslim en Occident, ʿIyāḍ, insiste sur le fait qu'Abū Ḥafṣ était doué dans différentes sciences. Ibn Bashkuwāl[89], lui, précise qu'Abū Ḥafṣ avait acquis un savoir abondant et une compréhension fine dans chaque discipline du savoir, dont la théologie fait bien sûr partie. Plus encore, et c'est là l'un des intérêts de notre extrait, ʿIyāḍ cite qu'il était

87 Amharar, *Langage et théologie* xvii-xviii.

88 Al-Qāḍī ʿIyāḍ, *Tartīb al-madārik* viii, 156-157.

89 Ibn Bashkuwāl, *al-Ṣila* 381.

versé dans les *ʿulūm qadīma*. On ne peut savoir précisément ce que visait ʿIyāḍ par ces termes, même si on peut tout à fait penser qu'il fait ici référence aux savoirs hérités des Grecs, incluant la médecine, mais aussi, peut-être, la *falsafa*. Compte tenu de ces hypothèses, il n'est donc pas absurde de supposer qu'Abū Ḥafṣ ait pu avoir certaines compétences comme *mutakallim*, et a pu rencontrer des maîtres ashʿarites lors de son séjour en Orient, qu'Ibn Bashkuwāl situe à l'an 444/1052. Nos recherches sur la liste des maîtres d'Abū Ḥafṣ citées par ʿIyāḍ ont été infructueuses, mais ce même Ibn Bashkuwāl ajoute à celle-ci un certain Abū Muḥammad al-Shantajyālī. Ce dernier, au même titre qu'Ibn Manẓūr, le maître d'al-Ḥasan, ont un point commun : tous deux[90] comptaient parmi les disciples d'Abū Dharr al-Harawī (434/1042).

Nous devons dire ici quelques mots d'Abū Dharr, qui avait, si l'on puit dire, cette « double casquette » de *mutakallim-muḥaddith*. Natif d'Herat, il étudie à Bagdad puis passa majoritairement sa vie à la Mecque, où il sera reconnu comme une autorité dans le *ḥādīth* : il est nommé « *muḥaddith al-Ḥaramayn* » (le spécialiste du *ḥadīth* des deux lieux saints). Dans la postérité, les savants musulmans le désignent souvent comme détenteur de la plus authentique transmission du *Ṣaḥīḥ* d'al-Bukhārī[91]. C'est sans doute ce qui a attiré à lui une génération de savants des quatre coins du monde musulman qui profiteront de leur pèlerinage pour étudier auprès de lui. L'Occident musulman n'est pas une exception à la règle[92]. Ibn Manẓūr et al-Shantajyālī font donc partie de cette génération andalouse d'étudiants d'al-Harawī, dans laquelle on compte un ashʿarite andalou célèbre : Abū l-Walīd al-Bājī, sur lequel nous reviendrons ultérieurement. Nous l'avons dit, al-Harawī était également un théologien ashʿarite. Il était le disciple d'al-Bāqillānī et d'Ibn Fūrak[93]. On comprend de quelques sources postérieures qu'al-Harawī n'était pas simplement ashʿa-rite d'obédience mais faisait l'apologie de la doctrine : il est connu pour avoir introduit l'ashʿarisme à la Mecque[94]. Plus encore[95], il est tenu comme l'une

90 Notons qu'Ibn Manẓūr était également le disciple d'al-Shantajyālī.

91 Brown, Jonathan, *The canonization of al-Bukhārī and Muslim: The formation and function of the Sunnī Ḥadīth canon*, Leyde: Brill, 2007, 143.

92 Citons parmi les savants siciliens partis à la rencontre d'al-Harawī ʿAbd al-Ḥaqq al-Saqallī. Kaddouri, Samir, « Refutations of Ibn Ḥazm by Mālikī authors from al-Andalus and North-Africa », dans C. Adang, M. Fierro et S. Schmidtke (éds.), *Ibn Ḥazm of Cordoba: The life and works of a controversial thinker*, Leyde: Brill, 2013, 539-599, à p. 553.

93 Al-Qāḍī ʿIyāḍ, *Tartīb al-madārik* vii, 231.

94 Ibn al-Jawzī déclara: "Huwa awwal man adkhala al-Ḥaram al-madhhab al-ashʿarī". Ibn al-Jawzī, *al-Muntaẓam fī l-tārīkh*, Beyrouth: Dār Ṣādir, 1934, vii, 194.

95 Ibn Ṭāhir, qui a écrit un article sur sujet, cite Ibn Taymiyya et al-Dhahabī affirmant le rôle d'al-Harawī dans la transmission de l'ashʿarisme en Occident musulman. Ibn Ṭāhir, « Abū

des principales causes de la diffusion de l'ashʿarisme en al-Andalus à cette époque, par le biais de cette génération d'étudiants andalous, que l'on présume nombreux, parmi lesquels al-Bājī donc, mais aussi, on peut le supposer, par le truchement d'Ibn Manẓūr et al-Shantajyālī, Abū Ḥafṣ et son fils Abū l-Qāsim, ce qui pourrait faire de ce dernier le premier maître ashʿarite d'Ibn al-ʿArabī. Malheureusement, en l'absence de données supplémentaires, il est difficile d'affirmer ceci de façon certaine, d'autant qu'un élément peut troubler cette hypothèse.

Revenons à la notice du Qāḍī ʿIyāḍ. On comprend qu'Abū Ḥafṣ et Abū l-Walīd al-Bājī se sont côtoyés à Murcie, le fief en question des Banū Ṭāhir, mais qu'il y avait entre eux un certain nombre de litiges (*munāzaʿāt*). La nature de ces litiges pourrait nous aider à en savoir plus sur l'orientation théologique d'Abū Ḥafṣ. S'ils ont pu simplement être d'ordre humain ou politique, ils ont tout à fait pu reposer sur de vraies divergences doctrinales. Selon cette hypothèse, l'opposition entre les deux savants a deux configurations possibles : il s'agissait d'une divergence entre deux partisans soit d'une même doctrine (un débat entre deux mālikites, ou deux ashʿarites) ou alors de deux doctrines opposées, comme ce serait le cas pour un *disputatio* ashʿarite/ẓāhirite.

Malheureusement, là encore, il est impossible d'en savoir plus. Un deuxième élément doit être pris en compte : le ẓāhirisme du gendre d'Abū Ḥafṣ, ʿAbd Allāh Ibn al-ʿArabī.

4 Le père d'Abū Bakr et le ẓahirisme

Tout ce qui précède doit être mis en relation avec une information pour le moins surprenante que l'on trouve dans une notice d'al-Dhahabī. Ce dernier attribue à ʿAbd Allāh, le père d'Abū Bakr, d'avoir dit : « J'ai été le compagnon d'Ibn Ḥazm durant sept ans. J'ai entendu de lui la totalité de son œuvre, excepté le dernier tome du livre *Fiṣal*. Nous avons aussi lu quatre tomes du *Kitāb al-Īṣāl*. Hormis cela, je n'ai manqué aucun de ses livres[96] ».

Dharr al-Harawī, al-ḥāfiẓ al-muḥaddith al-mālikī, awwal man adkhala al-ʿaqīda al-ashʿariyya li-l-ḥaram al-Makkī wa-awwal man baththahā fī l-Maghāriba", dans *Majallat al-fiqh al-mālikī* 11 (2011), 167.

96 Al-Dhahabī, Muḥammad, *Siyar aʿlām al-nubalā'*, Beyrouth: Mu'assasat al-Risāla, 2001, xx, 201.

L'idée que le père ait pu être ẓāhirite interpelle[97], quand on sait que le fils est justement célèbre pour son animosité virulente vis-à-vis de ce groupe, qu'il n'a pas hésité, dans un vers célèbre, à comparer à du fourrage pour bovins[98] !

Cette notice est pourtant corroborée par un document signalé par Tawrātī[99], retrouvé en marge d'un manuscrit d'un livre d'Ibn al-ʿArabī conservé à la Bibliothèque Ibn Yūsuf de Marrakech contenant la description du père d'Ibn al-ʿArabī comme étant un des compagnons Ibn Ḥazm. Dans son étude sur l'autodafé des livres d'Ibn Ḥazm[100], Kaddouri signale une notice très intéressante tirée du *Muʿjam al-udabāʾ* de Yāqūt al-Ḥamawī (619/1222). Al-Ḥamawī y dit avoir eu accès à un autographe de Muḥammad b. Ṭarkhān (446/1054 ; 513/1119), disciple d'Abū Isḥāq al-Shīrāzī et par ailleurs maître d'Abū Bakr Ibn al-ʿArabī[101]. En fait, cette notice est corroborée par celle que rapporte Ibn ʿAsākir dans son *Tārīkh Dimashq*[102]. Ibn Ṭarkhān fait dire à ʿAbd Allāh plusieurs informations sur Ibn Ḥazm : 1 ° la date et le lieu exacts de son décès ; 2 ° sa date de naissance et d'investiture au poste de vizir ; 3 ° l'anecdote qui le poussa à apprendre le *fiqh* ; 4 ° ce que ʿAbd Allāh a appris auprès d'Ibn Ḥazm, et l'*ijāza* qu'il obtînt de celui-ci.

Cette quatrième information, recoupée à celle rapportée par al-Dhahabī, offre des ajouts précieux. L'une des lectures de ʿAbd Allāh auprès d'Ibn Ḥazm

97 Camilla Adang commence son article sur la diffusion post-caliphale du ẓāhirisme en al-Andalus par une très intéressante citation d'Abū Bakr Ibn al-ʿArabī laissant croire que celui-ci a "découvert" cette doctrine une fois revenu de sa *riḥla* en Orient, ce qui est plus que surprenant compte tenu du fait, comme nous allons le voir plus bas et comme le souligne Adang, que le père d'Ibn al-ʿArabī comptait parmi les principaux disciples d'Ibn Ḥazm. Voir : Adang, Camilla, « The spread of Ẓāhirism in post-caliphal al-Andalus: The evidence from the biographical dictionaries », dans Sebastian Günther (éd.), *Ideas, images, and methods of portrayal: Insights into classical Arabic literature and Islam*, Leyde: Brill, 2005, 297-346, à p. 303.

98 Ibn al-ʿArabī a écrit treize vers célèbres contre les Ẓāhirites, qu'il conclut par : *bīnū ʿan al-khalq lastum minhumū abadan*** mā li-l-anām wa-maʿlūfin min al-baqarī !* (Eloignez-vous des gens, vous n'en faites pas partie ; qu'ont à faire les hommes d'un fourrage pour bovins ?) Voir : Ibn al-ʿArabī, Abū Bakr, *ʿĀriḍat al-aḥwadhī bi-sharḥ Ṣaḥīḥ al-Tirmidhī*, Beyrouth: Dār al-Kutub al-ʿIlmiyya, 1997, x, 113.

99 Ibn al-ʿArabī, *al-Amad al-aqṣā fī sharḥ asmāʾ Allāh al-ḥusnā*, éd. Tawrātī, Tanger: Dār al-Ḥadīth al-Kattāniyya, 2015, 19.

100 Kaddouri, Samir, « Dissimulation des opinions politiques sous contrôle: Le cas d'Ibn Ḥazm à Séville », dans *al-Qanṭara* 35 (2014), 135-150.

101 Al-Subkī le classe dans la cinquième génération de shāfiʿites. Voir : al-Subkī, Tāj al-dīn, *Ṭabaqāt al-Shāfiʿiyya al-kubrā*, éd. Ṭanāḥī et Ḥalū, le Caire: Maktabat al-Bābī al-Ḥalabī, 1963-1976, n ° 637.

102 Ibn ʿAsākir, *Tārīkh madīnat Dimashq*, éd. ʿAmrāwī, Beyrouth: Dār al-Fikr, 1995-2000, xxxii, 232.

y est datée : elle remonte à 456/1064. Ensuite, ʿAbd Allāh signale avoir potentiellement manqué certains livres qu'Ibn Ḥazm aurait écrit hors de chez lui, lors des voyages qu'entreprit ce dernier vers l'Ouest de la péninsule. On peut en déduire que ʿAbd Allāh n'a pas voyagé avec Ibn Ḥazm, mais qu'il a étudié auprès de lui lorsqu'ils se trouvaient dans la même ville. D'après Kaddouri, Ibn Ḥazm aurait gagné Séville à partir de 450/1058[103].

Lorsqu'al-Muʿtaḍid, qui lui aurait d'abord fait bon accueil, lui commande une copie de son *Fiṣal* pour sa bibliothèque royale, Ibn Ḥazm, bien au fait des pratiques sanguinaires du souverain[104], sait pertinemment que pour rester en vie, il doit nécessairement procéder à une réécriture et effacer tous les passages incriminant les rois des Taifas en général et le régime ʿAbbādite en particulier. Kaddouri part d'une analyse des quinze manuscrits du *Fiṣal* pour arriver à la conclusion que cette deuxième version a bien été écrite à Séville de 452/1060 à 454/1062[105], période dans laquelle Ibn Ḥazm était en principe en bons termes avec le souverain. Un hypothétique débat à Séville contre al-Bājī[106] pointant l'hostilité à l'ashʿarisme d'Ibn Ḥazm aurait précipité son exil forcé ainsi que la mise au feu de ses livres, deux évènements que Kaddouri situe approximativement entre 455/1063 et 456/1064.

Ce ẓāhirisme paternel est également à mettre en relation avec ce qu'Abū Bakr dépeint comme étant l'élément déclencheur de cette *riḥla*[107] :

> Alors que j'étais un jour avec quelques enseignants, mon père s'assit avec nous et pris de son temps, malgré ses lourdes occupations, pour contrôler le niveau de science que j'avais atteint. D'autres personnes qui voulaient le voir se sont également assises ; parmi elles, un commerçant avec l'un des deux bras chargés d'un sac plein de livres à vendre. Parmi ces livres, il y avait un ouvrage d'al-Simnānī et un autre d'al-Bājī. Je les ai alors tous entendu dire : « Voilà de grands ouvrages et des sciences éminentes !

103 Kaddouri, « Dissimulation » 140.

104 Il aurait eu un jardin où étaient exposés les têtes empalées de chaque personne qu'il faisait tuer. Kaddouri, « Dissimulation » 142.

105 Pour des exemples de cette auto-correction, Kaddouri, « Dissimulation » 141-142 et 142n36 pour la datation de la deuxième version.

106 Nous ne parlons pas ici du célèbre débat auquel s'est intéressé Turkī et qui, de façon globalement admise, a pris fin à Mayorque en 440/1048. Voir : Turki, *Polémiques entre Ibn Ḥazm et Bājī sur les principes de la Loi musulmane. Essai sur le littéralisme ẓāhirite et la finalité malikite,* Alger: Etudes et documents, s.d. Kaddouri souligne qu'Ibn al-Abbār situe un autre débat, postérieur à l'an 452/1060, ce qui va dans le sens de la notice de al-Subkī, qui déclare que al-Bājī et d'autres savants ont été à la manœuvre de l'expulsion et l'autodafé des livres d'Ibn Ḥazm à Séville.

107 Ibn al-ʿArabī, *Qānūn al-taʾwīl* 76-77.

> Al-Bājī les a apportés d'Orient. » Ces phrases retentirent dans mon cœur et ont secoué mon esprit. Puis, ils se sont mis à en discuter, à dire que les juristes de notre pays ne comprennent pas ces livres et n'en saisissent rien. Je fis alors intérieurement le vœu d'atteindre ce niveau, de rencontrer ces hommes et d'étudier auprès d'eux les différentes doctrines et crédo. Mon intention est restée constante dessus.

Ce passage met en lumière l'intérêt particulier que vouait ʿAbd Allāh à ce fameux livre non-identifié de Samnānī, introduit en Andalus par son disciple, al-Bājī. Or, ce livre était très important aux yeux d'Ibn Ḥazm. Dans son étude sur les sources d'Ibn Ḥazm, Schmidtke[108] souligne qu'il est très probable que celui-ci, qui n'a pas entrepris de voyage en Orient, n'ait pu prendre connaissance de textes ashʿarites que via al-Bājī, parmi lesquels un fameux livre de Simnānī, qui ne semble pas être le *Bayān ʿan uṣūl al-īmān wa-l-kashf ʿan tamwīhāt ahl al-ṭughyān*, récemment édité[109].

Nous pouvons à présent, à la lumière de ces données conjecturer sur la doctrine de ʿAbd Allāh. Il est très peu probable que le père ẓāhirite ait laissé son fils, dont il contrôlait l'enseignement, embrasser une doctrine considérée hostile, à moins que celui-ci ait également grandi dans le ẓāhirisme, puis l'aurait abandonné durant sa *riḥla*, ce qui n'est indiqué dans aucune source disponible, et qui n'aurait pas manqué d'être cité si c'était le cas, en raison de ce que ce fait aurait apporté de « romanesque » à la vie d'Abū Bakr, dont le parcours aurait ainsi été comparé à celui d'al-Ashʿarī avec le muʿtazilisme. On n'imagine pas non plus qu'un homme adepte d'une doctrine conspuée par al-Muʿtamid ait pu devenir ainsi vizir auprès de son successeur, dont la politique ne confirme en rien une hypothétique réhabilitation du ẓāhirisme. On sait en outre que ʿAbd Allāh a appris auprès de savants de Cordoue, ville d'Ibn Ḥazm, et de Séville[110]. Nous suggérons donc que ʿAbd Allāh a connu Ibn Ḥazm très jeune[111] ; soit à Cordoue, lors de son séjour pour étude, soit l'année où le ẓāhirite entra à Séville, vers 450/1058 : ʿAbd Allāh avait approximativement quinze ans. Il l'aurait suivi durant sept ans, et au moins jusqu' à son exil : on sait, via Ibn Khayr,

108 Schmidtke, S., « Ibn Ḥazm's sources on Ashʿarism and Muʿtazilism », dans C. Adang, M. Fierro et S. Schmidtke (éds.), *Ibn Ḥazm of Cordoba: The life and works of a controversial thinker*, Leyde: Brill, 2003, 375-401, à p. 384.

109 Il a été édité au Koweit en 2014 par al-Ayyūb pour le compte de Dār al-Ḍiyāʾ.

110 Cano Ávila, P., A. García Sanjuán, et A. Tawfiq, « Ibn al-ʿArabī », dans *Biblioteca de al-Andalus*, Almería: Fundación Ibn Ṭufayl de Estudios Árabes, 2009, ii, 129-158.

111 Ibn Ḥazm incitait ses disciples à inviter à ses cours les jeunes étudiants. Kaddouri, « Dissimulation » 139n29.

qu'Ibn Ḥazm aurait, lors de son départ de Séville, déclamé à ʿAbd Allāḥ, sans doute en guise d'adieu, ce vers devenu célèbre :

> Je suis le soleil éclairant dans les sciences
> Mon défaut est que mon lever est l'Occident[112]

On peut imaginer que ʿAbd Allāh ait abandonné la doctrine ẓāhirite, après le départ d'Ibn Ḥazm, convaincu par la doctrine ashʿarite (sans doute lors du débat présumé avec al-Bājī), conviction qui a pu être facilitée par la peur de souffrir de la même persécution de son ancien maître, d'autant que la fin de cette cinquième décennie se clôt par un incident tragique pour ʿAbd Allāh : al-Muʿtaḍid fait assassiner son beau-père Abū Ḥafṣ en 460/1068[113]. Tout ceci l'aurait donc amené à davantage « montrer patte blanche » au régime ʿabbādite, ouvrant ainsi une nouvelle page avec l'avènement du nouveau souverain, al-Muʿtamid. Selon cette hypothèse, ʿAbd Allāh aurait donc abandonné le ẓāhirisme huit ans environ avant la naissance de son fils, ce qui a priori lui aurait laissé le temps de changer de bord complètement, ce qui explique aussi l'intérêt porté aux livres ashʿarites importés par al-Bājī, y compris celui d'al-Simnānī. Cette hypothèse est cohérente avec l'idée qu'Abū Bakr ait « grandi avec la haine du ẓāhirisme »[114].

Dans le cas où les litiges entre Abū Ḥafṣ et al-Bājī n'étaient pas d'ordre religieux, il est possible que ce soit Abū Ḥafṣ lui-même qui ait converti ʿAbd Allāh à l'ashʿarisme avant de lui marier sa fille, et que son fils, Abū l-Qāsim, ait pris la relève de l'enseignement du jeune Abū Bakr en lui inculquant, à travers des textes parmi lesquels la fameuse *qaṣīda* d'Abū Ḥafṣ, les bases de la *ʿaqīda* avant la *riḥla*.

---

112 Ibn Khayr, *Fihrasat Ibn Khayr*, Tunis, Dār al-Gharb al-Islāmī, 2009, 544.

113 Abū Ḥafṣ était parmi les notables dotés d'autorité à Séville avant la dynastie ʿabbādite et était vu comme un concurrent au trône. Abū Ḥafṣ se sentant en danger de mort, il a d'abord choisi l'exil, prétextant le pèlerinage. Al-Muʿtaḍid le fit ensuite revenir à Séville en l'amadouant (il le nomma membre de la *shūrā*). C'est une harangue très mal perçue écrite par Abū Ḥafṣ incitant al-Muʿtaḍid à faire le *jihād* qui précipita sa mort. Voir : Farrūkh, ʿUmar, *Tārīkh al-adab al-ʿarabī*, iv, 570-572.

114 Elle semble cependant peu compatible avec le fait qu'al-Rāḍī, un des fils d'al-Muʿtamid, ait été vraisemblablement ẓāhirite, si l'on en croit des chroniqueurs postérieurs comme Ibn al-Abbār. Adang va même jusqu'à considérer la probabilité que le père d'Ibn al-ʿArabī ait été l'enseignant d'al-Rāḍī ! Voir : Adang, « The spread of Ẓāhirism » 312-313.

5 Les maîtres ashʿarites d'Ibn al-ʿArabī à partir de l'an 485/1093

Fort heureusement, les sources sont beaucoup plus abondantes en ce qui regarde la phase de la *riḥla* qui recouvre tant son voyage depuis al-Andalus que ses différents séjours dans des grandes villes du monde musulman comme Alexandrie, Jérusalem et Bagdad.

Ainsi, on sait d'Ibn al-Abbār qu'Ibn al-ʿArabī était en lien de transmission avec Ibn Sābiq al-Ṣaqallī, disciple sicilien de ʿAbd al-Jalīl al-Rabaʿī al-Dībājī[115] qui a commenté le *Tamhīd* d'al-Bāqillānī. On sait également grâce à la *Fahrasa* d'Ibn Khayr qu'Ibn al-ʿArabī a étudié auprès d'Abū l-Faḍā'il Ibn Ṭawq, disciple du célèbre Abū l-Qāsim al-Qushayrī[116].

Encore une fois, le *Qānūn al-taʾwīl* est particulièrement utile[117]. D'après le Qāḍī lui-même, c'est à Mahdiyya qu'il a vraiment commencer à étudier le *kalām* de façon approfondie. Il faut dire que c'est là-bas qu'il rencontre et étudie auprès de plusieurs *mutakallimīn* originaires de Kaïrouan, en particulier les disciples d'Ibn al-Qadīm, enseignant du *Tamhīd* d'al-Bāqillānī, qu'il transmettait d'après al-Adharī, disciple d'al-Bāqillānī lui-même[118]. C'est donc à Mahdiyya qu'Ibn al-ʿArabī se met *vraiment* à étudier la théologie et prendre part aux débats avec ses camarades (*fa-akhadhtu fī qirāʾat shayʾ min uṣūl al-dīn wa-l-munāẓara fīhā maʿa al-ṭālibīn*). Ensuite, c'est à Jérusalem, et auprès d'Abū Bakr al-Ṭurṭūshī (520/1127), disciple d'al-Bājī, qu'Ibn al-ʿArabī consolide ses acquis en *kalām* et *uṣūl al-fiqh*, lui permettant dorénavant de prendre part aux débats de haut niveau face aux Muʿtazilites. Il y étudie également auprès d'Abū Saʿīd al-Zanjānī, disciple de l'hérésiographe Abū l-Muẓaffar al-Isfarāyīnī[119]. Enfin, à Bagdad, il a l'opportunité d'étudier à la Niẓāmiyya auprès des disciples d'al-Juwaynī, comme al-Ṭūsī, qui lui enseigne *Kitāb al-Irshād* ou encore son maître le plus célèbre, qu'il aimait à appeler son *Dānashmand* (=maître à penser en persan), Abū Ḥāmid al-Ghazālī.

Cette tentative de reconstitution est, nous l'avons vu, très compliquée, surtout en ce qui regarde la première phase d'apprentissage d'Ibn al-ʿArabī. La pauvreté des données nous force certes à spéculer là où nous aurions aimé pouvoir affirmer des faits. Cependant, les résultats de notre étude nous permettent de proposer une nouvelle esquisse offrant un aperçu de quelques

115 Ibn al-Abbār, *al-Takmila*, ii, 256: il semble qu'Ibn al-ʿArabī ait bien reçu une *ijāza*. Notons qu'Ibn Sābiq était aussi connu sous le nom de Abū Bakr al-Jazīrī.

116 Ibn Khayr, *Fahrasa* 264.

117 Ibn al-ʿArabī, *Qānūn al-taʾwīl* 427-428.

118 Voir : Ibn ʿAṭiyya, *Fahras Ibn ʿAṭiyya*, Tunis : Dār al-Gharb al-islamī, 1983, 76.

119 Voir : Ibn al-ʿArabī, Abū Bakr, *al-ʿAwāṣim min al-qawāṣim*, éd. Ṭālbī, Alger: ENAG Editions, 2013, 27.

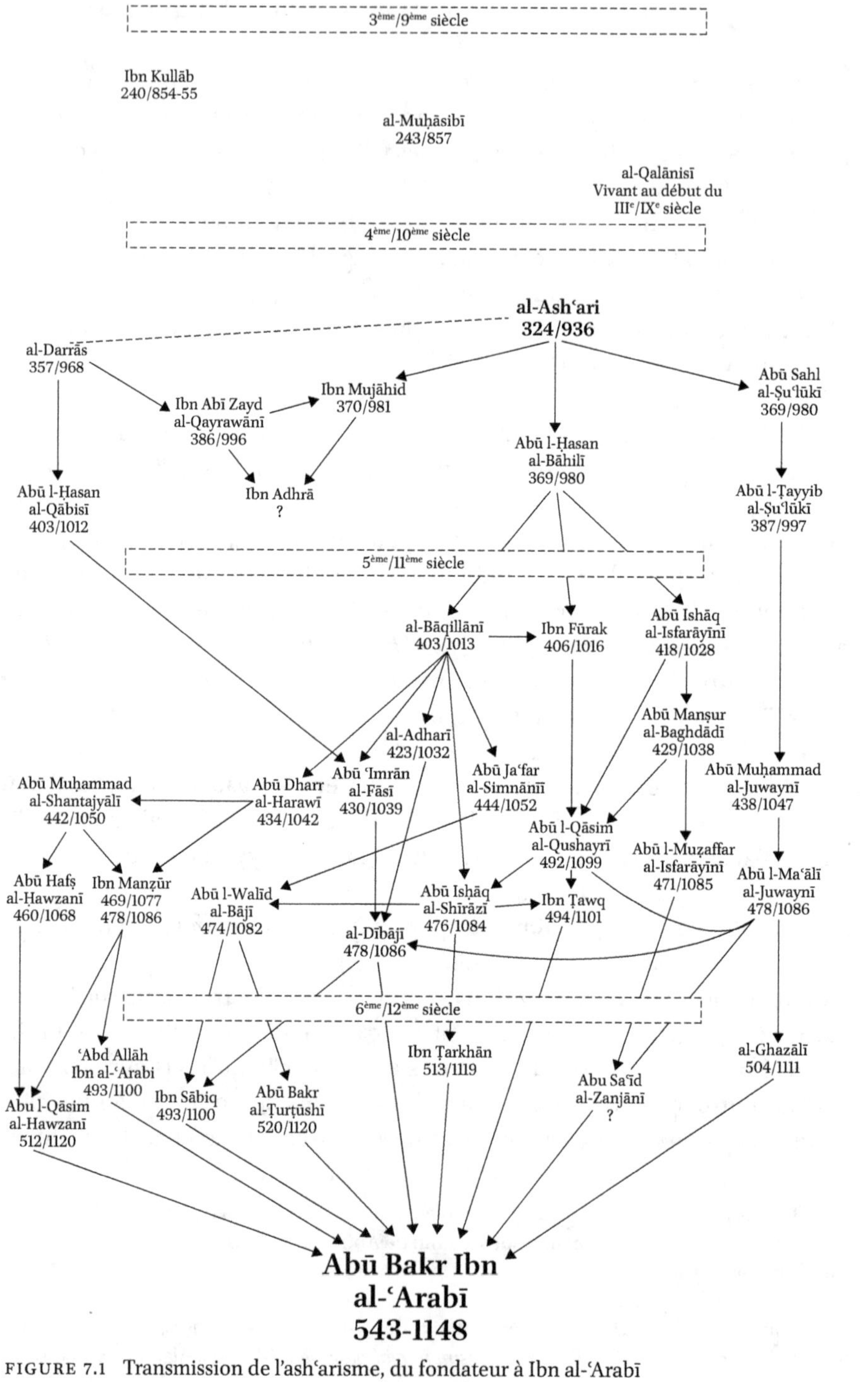

FIGURE 7.1 Transmission de l'ash'arisme, du fondateur à Ibn al-'Arabī

voies de transmission de l'ashʿarisme d'al-Ashʿarī à Ibn al-ʿArabī, de Bagdad à Séville. Nous les récapitulons dans le schéma ci-contre. Chaque flèche entre deux savants indique une relation maître-étudiant, la pointe de chaque flèche indiquant l'étudiant. La ligne en pointillés indique qu'il n'a pas été possible de prouver une telle relation entre al-Ashʿarī et al-Darrās, malgré le lien d'influence déjà évoqué du second par le premier.

Bibliographie

ʿAbd al-Ḥalīm, M.A.S., «Early Islamic theological and juristic terminology: *Kitāb al-Ḥudūd fī 'l-uṣūl, by Ibn Fūrak», dans Bulletin of the School of Oriental and African Studies* 54 (1991), 5-41.

Adang, Camilla, «The spread of Ẓāhirism in post-caliphal al-Andalus: The evidence from the biographical dictionaries», dans Sebastian Günther (éd.), *Ideas, images, and methods of portrayal: Insights into classical Arabic literature and Islam*, Leyde: Brill, 2005, 297-346.

Aḥnāna, Yūsuf, *Taṭawwur al-madhhab al-ashʿarī fī l-gharb al-islāmī*, Rabat: Publications du Ministère Marocain des Affaires Islamiques, 2007.

Allard, Michel, *Le problème des attributs divins dans la doctrine d'al-Ašʿarī et de ses premiers grands disciples*, Beyrouth: Imprimerie Catholique, 1965.

Amharar, Ilyass, *L'ashʿarisme à Fès, des Almohades aux Mérinides (VIIe/XIIIe siècle): La profession de foi d'Abû ʿImrân al-Jawrâʾî*, Paris: Publications du Centre Jacques Berque, 2022.

Amharar, Ilyass, *Langage et théologie chez Ibn al-ʿArabī: Les informations subtiles de la somme de théorie légale (Nukat al-Maḥṣūl fī ʿilm al-uṣūl)*, Piscataway NJ: Gorgias Press, 2023.

Ansari, Hassan et Jan Thiele, «Discussing al-Bāqillānī's theology in the Maghrib: ʿAbd al-Jalīl b. Abī Bakr al-Dībājī al-Rabaʿī's *al-Tasdīd fī sharḥ al-Tamhīd*», dans *al-Qanṭara* 39 (2018), 127-168.

Aʿrāb, Saʿīd, *Maʿa al-Qāḍī Ibn al-ʿArabī*, Tunis: Dār al-Gharb al-Islāmī, 1987.

Arfa Mensia, Mokdad, «Regards d'Ibn Rushd sur al-Juwaynī: questions de méthode», dans *Arabic sciences and philosophy* 22 (2012), 199-216.

Aristotle, *Les politiques*, trad. P. Pellegrin, Paris: Garnier-Flammarion, 1990.

al-Ashʿarī, Abū l-Ḥasan, *Kitāb al-Lumaʿ*, éd. McCarthy, Beyrouth: Imprimerie Catholique, 1953.

al-Ashʿarī, Abū l-Ḥasan, *Maqālāt al-islāmiyyīn*, éd. H. Ritter, Wiesbaden: Franz Steiner Verlag, 1980.

Badawī, ʿAbd al-Raḥmān, *Muʾallafāt al-Ghazālī*, Kuwaït: Wakālat al-Maṭbūʿāt, 1977.

al-Baghdādī, Abū Bakr, *Tārīkh Baghdād*, le Caire: Maṭbaʿat al-Saʿāda, 1931.

al-Bāqillānī, Abū Bakr, *Al-Tamhīd fī al-radd ʿalā l-mulḥida al-muʿaṭṭila wa-l-rāfiḍa wa-l-khawārij wa-l-muʿtazila*, éd. Khuḍayrī & Abū Rayda, le Caire: Dār al-Fikr, 1988.

al-Bayhaqī, Abū Makr *Asmāʾ wa-l-ṣifāt*, éd. M.Z. Kawtharī, le Caire: al-Maktaba al-Azhariyya, 2019.

Benhima, Yassir, «Le mālikisme irakien et le Maghreb: Réseaux, circulations et transferts entre Orient et Occident musulmans», dans M. Meouak et C. de la Puente (éds.), *Connected stories: Contacts, traditions and transmissions in premodern Mediterranean Islam*, Berlin: De Gruyter, 2022, 111-130.

Brown, Jonathan, *The canonization of al-Bukhārī and Muslim: The formation and function of the Sunnī Ḥadīth canon*, Leyde: Brill, 2007.

Bulliet, Richard, *The patricians of Nishapur: A study in medieval Islamic social history*, Cambridge MA: Harvard University Press, 1972.

Cabanelas Rodriguez, Darío, «Notas para la historia de Algazel en España», dans *al-Andalus* 17 (1952), 223-232.

Cano Ávila, P., A. García Sanjuán, et A. Tawfiq, «Ibn al-ʿArabī», dans *Biblioteca de al-Andalus*, Almería: Fundación Ibn Ṭufayl de Estudios Árabes, 2009, ii, 129-158.

Chaumont, Eric, «Encore au sujet de l'ashʿarisme d'Abû Isḫâq al-Shîrâzî», dans *Studia Islamica* 74 (1991), 167-177.

al-Ḍabbī, *Bughyat al-multamis fī tārīkh rijāl ahl al-Andalus*, le Caire: Dār al-Kitāb al-ʿArabī, 1968.

al-Dhahabī, Muḥammad, *Siyar aʿlām al-nubalāʾ*, Beyrouth: Muʾassasat al-Risāla, 2001.

al-Dībājī, ʿAbd al-Jalīl, *Rasāʾil wa-masāʾil*, éd. Tawrātī et ʿAmmūr, Tanger: Dār al-Ḥadīth al-Kattāniyya, 2017.

al-Dībājī, ʿAbd al-Jalīl, *al-Tasdīd fī sharḥ al-Tamhīd*, éd. Nhīrī, Amman: Dār al-Fatḥ, 2021.

Eggen, Nora, «A book burner or not? History and myth: Revisiting al-Qāḍī ʿIyāḍ and the controversies over al-Ghazālī in the Islamic West», dans *Journal of Arabic and Islamic studies* 18 (2018), 87-109.

EI2 = P. Bearman, Th. Bianquis, C.E. Bosworth, E. van Donzel et W.P. Heinrichs (éds.), *The Encyclopaedia of Islam, second edition*, Leyde: Brill, 1954-2004.

EI3 = Kate Fleet, Gudrun Krämer, Denis Matringe, John Nawas et Devin J. Stewart (éds.), *The Encyclopaedia of Islam, Three*, Leyde: Brill, 2007-.

Farrūkh, ʿUmar, *Tārīkh al-adab al-ʿarabī*, Beyrouth: Dār al-ʿIlm li l-malāyīn, 1981.

Fierro, M., «The polemic about the *karāmāt al-awliyāʾ* and the development of Ṣūfism in al-Andalus (fourth/tenth-fifth/eleventh centuries)», dans *Bulletin of the School of Oriental and African Studies* 55 (1992), 236-249.

Fierro, M., «al-Bājī, Abū l-Walīd», dans Delgado et al(éds.), *Enciclopedia de al-Andalus: Diccionario de autores y obras andalusíes*, Grenade: Fundacion El Legado Andalusi, 2002, 118-123.

Fórneas Besteiro, J.M., «*Al-Tamhīd* de al-Bāqillānī y su transmisión en al-Andalus», dans *Miscelánea de estudios árabes y hebraicos* 26-28 (1977-1979), 433-440.

Fórneas Besteiro, J.M., «De la transmisión de algunas obras de tendencia ašʿarī en al-Andalus», dans *Awrāq* 1 (1978), 4-11.

Frank, Richard, «Al-Ustādh Abū Isḥāḳ: An *ʿaḳīda* together with selected fragments», dans *Mélanges de l'Institut dominicain d'études orientales* 19 (1989), 129-202.

Gacek, Adam, *Arabic manuscripts: A vademecum for readers*, Boston: Brill, 2009.

García-Sanjuán, A., «Replication and fragmentation: The Taifa kingdoms», dans Fierro (éd.), *The Routledge handbook of Muslim Iberia*, New York: Routledge, 2020.

Garden, Kenneth, «Al-Māzarī al-Dhakī: al-Ghazālī's Maghribi adversary in Nishapur», dans *Journal of Islamic studies* 21 (2010), 89-107.

Garden, Kenneth, *The first Islamic reviver: Abū Ḥāmid al-Ghazālī and his Revival of the religious sciences*, Oxford: Oxford University Press, 2014.

Garden, Kenneth, "The riḥla and self-reinvention of Abū Bakr Ibn al-ʿArabī", in *Journal of the American Oriental Society* 135 (2015), 1-17.

Gardet, L., et M.M. Anawati, *Introduction à la théologie musulmane*, Paris: Vrin, 1948.

al-Ghazālī, Abū Ḥāmid, *al-Mankhūl min taʿlīqat al-uṣūl*, éd. Hītū, Damas: Dār al-Fikr, 1970.

al-Ghazālī, Abū Ḥāmid, *al-Mustaṣfā min ʿilm al-uṣūl*, éd. ʿAbd al-Shakūr, le Caire: Dār al-Baṣāʾir, 2007.

al-Ghazālī, Abū Ḥāmid, *Ajwibat al-Ghazālī ʿan asʾilat Ibn al-ʿArabī*, éd. ʿAbdū, Beyrouth: Dār al-Kutub al-ʿIlmiyya, 2011.

Gimaret, Daniel, «Un document majeur pour l'histoire du kalām: le *Muǧarrad maqālāt al-Ašʿarī* d'Ibn Fūrak», dans *Arabica* 32 (1985), 185-218.

Gimaret, Daniel, *La doctrine d'al-Ashʿarī*, Paris, Les éditions du Cerf, 1990.

Gimaret, Daniel, «Un extrait de la *Hidāya* d'Abū Bakr al-Bāqillānī: le *Kitāb at-Tawallud*, réfutation de la thèse muʿtazilite de la génération des actes», *Bulletin d'études orientales* 58 (2009), 259-313.

Gimaret, Daniel et Guy Monnot, *Shahrastani: Livre des religions et des sectes, tome 1*, Leuven: Peeters, 2000.

Gómez-Rivas, Camilo, *Law and the Islamization of Morocco under the Almoravids: The fatwās of Ibn Rushd al-Jadd to the Far Maghrib*, Leyde: Brill, 2015.

Griffel, Frank, «Ibn Tūmart's rational proof for God's existence and unity, and His connection to the Niẓāmiyya *madrasa* in Baghdad», dans Patrice Cressier, Maribel Fierro, et Luis Molina (éds.), *Los almohades: Problemas y perspectivas*, Madrid: Consejo Superior de Investigationes Científicas, 2005, ii, 753-813.

Griffel, Frank, *Al-Ghazālī's philosophical theology*, 2009, New York, Oxford University Press.

Griffel, Frank, «Al-Ghazālī at his most rationalist: *The universal rule for allegorically interpreting revelation* (*al-Qānūn al-kullī fī t-taʾwīl*)», dans Georges Tamer (éd.), *Islam and rationality: The impact of al-Ghazālī, Papers Collected on His 900th Anniversary*, Boston: Brill, 2015, i, 89-120.

al-Juwaynī, Abū l-Maʿālī, *al-ʿAqīda al-Niẓāmiyya*, éd. M.Z. al-Kawtharī, le Caire: Maṭbaʿat al-Anwār, 1948.

al-Juwaynī, Abū l-Maʿālī, *al-Irshād ilā qawāṭiʿ al-adilla min ṣaḥīḥ al-iʿtiqād*, éd. Khalāyla et Idrīs, Amman: Dār al-Nūr al-mubīn, 2016.

Ibn al-Abbār, *Takmilat al-Ṣila*, éd. Al-Harrās, Beyrouth: Dār al-Fikr, 1995.

Ibn Abī Zayd, *al-Risāla*, éd. Tawrātī, Tanger: Dār al-Ḥadīth al-Kattāniyya, 2015.

Ibn al-ʿArabī, Abū Bakr, *Qānūn al-taʾwīl*, éd. M. Slīmānī, Tunis: Dār al-Gharb al-Islāmī, 1990.

Ibn al-ʿArabī, Abū Bakr, *ʿĀriḍat al-aḥwadhī bi-sharḥ Ṣaḥīḥ al-Tirmidhī*, Beyrouth: Dār al-Kutub al-ʿIlmiyya, 1997.

Ibn al-ʿArabī, Abū Bakr, *al-ʿAwāṣim min al-qawāṣim*, éd. Ṭālbī, Alger: ENAG Editions, 2013.

Ibn al-ʿArabī, *al-Amad al-aqṣā fī sharḥ asmāʾ Allāh al-ḥusnā*, éd. Tawrātī, Tanger: Dār al-Ḥadīth al-Kattāniyya, 2015.

Ibn ʿAsākir, *Tabyīn kadhib al-Muftarī fīmā nasaba ilā l-imām al-Ashʿarī*, éd. Sharqāwī, Damas: Dār Taqwā, 2018.

Ibn ʿAsākir, *Tārīkh madīnat Dimashq*, éd. ʿAmrāwī, Beyrouth: Dār al-Fikr, 1995-2000.

Ibn ʿAṭiyya, *Fahras Ibn ʿAṭiyya*, Tunis: Dār al-Gharb al-islamī, 1983.

Ibn Bashkuwāl, *al-Ṣila fī tārīkh aʾimmat al-Andalus*, éd. al-ʿAṭṭār al-Ḥusaynī, le Caire: Maktabat al-Khanjī, 1955.

Ibn Faraḍī, *Tārīkh ʿulamāʾ al-Andalus*, le Caire: al-Dār al-Miṣriyya li-l-Taʾlīf wa-l-Tarjama, 1966.

Ibn Fūrak, *Kitāb Mushkil al-ḥadīth*, éd. Gimaret, Damas: Publications de l'Ifpo, 2003.

Ibn Fūrak, *Mujarrad maqālāt al-shaykh Abī l-Ḥasan*, éd. Gimaret, Beyrouth: Dar El-Machrek, 1987.

Ibn al-Jawzī, *al-Muntaẓam fī l-tārīkh*, Beyrouth: Dār Ṣādir, 1934.

Ibn Khaldūn, *al-Muqaddima*, Beyrouth: Dār al-Kutub al-ʿIlmiyya, 1967.

Ibn Khayr, *al-Fahrasa*, éd. Fuʾād Manṣūr, Tunis: Dār al-Gharb al-Islāmī, 2009.

Ibn Ṭāhir, ʿAbd Allāh «Abū Dharr al-Harawī, al-ḥāfiẓ al-muḥaddith al-mālikī, awwal man adkhala al-ʿaqīda al-ashʿariyya li-l-ḥaram al-Makkī wa-awwal man baththahā fī l-Maghāriba", dans *Majallat al-fiqh al-mālikī* 11 (2011).

Idris, H.R., «Essai sur la diffusion de l'ašʿarisme en Ifrîqiya», dans *Les cahiers de Tunisie* 2 (1953), 126-40.

Idris, H.R., «A propos d'un extrait du 'Kitâb al-Mihàd' d'al-Mâzarî al-Iskandarànî», dans *Les cahiers de Tunisie* 2 (1953), 155-159.

Idris, H.R., «Deux juristes kairouanais de l'époque zīrīde: Ibn Abī Zaid et al-Qābisī (Xᵉ-XIᵉ siècle)», dans *Annales de l'Institut d'études orientales* 12 (1954), 122-198.

Idris, H.R., «L'école malikite de Mahdia: l'Imam al-Māzarī (m. 536/1141)», dans *Les études d'orientalisme à la mémoire de Lévi-Provençal*, Paris: Maisonneuve, 1962, 153-163.

Idris, H.R., «L'invasion hilālienne et ses conséquences», dans *Cahiers de civilisation médiévale* 11 (1968), 353-169.

Kaddouri, Samir, «Refutations of Ibn Ḥazm by Mālikī authors from al-Andalus and North-Africa», dans C. Adang, M. Fierro et S. Schmidtke (éds.), *Ibn Ḥazm of Cordoba: The life and works of a controversial thinker*, Leyde: Brill, 2013, 539-599.

Kaddouri, Samir, «Dissimulation des opinions politiques sous contrôle: Le cas d'Ibn Ḥazm à Séville», dans *al-Qanṭara* 35 (2014), 135-150.

Lagardère, Vincent, «À propos d'un chapitre du Nafḫ wal-taswiya attribué à Ġāzalī», dans *Studia Islamica* 60 (1984), 119-136.

Lagardère, Vincent, "Abû Bakr b. al ʿArabi, grand cadi de Seville", dans *Revue des mondes musulmans et de la Méditerranée* 40 (1985), 91-102.

Lagardère, Vincent, «Une théologie dogmatique de la frontière en al-Andalus aux XIe et XIIe siècles: L'ašʿarisme», dans *Anaquel de estudios árabes* 5 (1994), 71-98.

Maḥfūẓ, M., «al-Azdī am al-Adharī?», dans *Majallat al-Fikr* 17 (1971), 56-60.

al-Maqqarī, Aḥmad, *Analectes sur l'histoire et la littérature des Arabes d'Espagne*, trad. Dozy, Amsterdam: Oreintal Press, 1967.

al-Māzarī al-Dhakī, *Īḍāḥ al-maḥṣūl min Burhān al-uṣūl*, éd. Ṭālbī, Tunis: Dār al-Gharb al-Islāmī, 2001.

Naṣīr, Muḥammad, «Dawr al-Bāqillānī fī talqīn al-asānīd al-ashʿariyya bi-bilād al-maghrib», dans Jamāl ʿAllāl al-Bakhtī (éd.), *Al-fikr al-ashʿarī bi-l-maghrib: Khilāl marḥalatay al-taʾsīs wa-l-tarsīm*, Rabat: Publications de la Rābīṭa Moḥammedienne des Savants, 2016, 65-94.

Nguyen, Martin, *Sufi master and Qurʾan scholar: Abū 'l-Qāsim al-Qushayrī and the Laṭāʾif al-Ishārāt*, Oxford: Oxford University Press, 2012.

Ould Cheikh, Abdel Wedoud et Bernard Saison, «Vie(s) et mort(s) de al-Imām al-Ḥaḍramī», dans *Arabica* 34, (1987), 48-79.

al-Qāḍī ʿIyāḍ, *Tartīb al-madārik wa-taqrīb al-masālik li-maʿrifat aʿlām madhhab Mālik*, Rabat: Publications du Ministère des Habous et des Affaires Islamiques, 1965-1983.

al-Qāḍī ʿIyāḍ, *al-Ghunya*, le Caire: al-Maktaba al-Thaqāfiyya, 2002.

Sanseverino, R.V., *Fès et sainteté, de la fondation à l'avènement du proctectorat (808-1912)*, Rabat: Publications du Centre Jacques Berque, 2014.

Schmidtke, S., «Ibn Ḥazm's sources on Ashʿarism and Muʿtazilism», dans C. Adang, M. Fierro et S. Schmidtke (éds.), *Ibn Ḥazm of Cordoba: The life and works of a controversial thinker*, Leyde: Brill, 2013, 375-401.

Serrano Ruano, Delfina, «Los almorávides y la teología ašʿarí: ¿Contestación o legitimación de una disciplina marginal?", dans de la Puente (éd.), *Identitades marginales*, Madrid: CSIC, 2003, pp. 461-516.

Serrano Ruano, Delfina, «ʿIyāḍ b. Musà», dans J. Lirola Delgado (éd.), *Biblioteca de al-Andalus*, Almería: Fundación Ibn Tufayl de Estudios Árabes, 2009, 404-434, no. 1479.

Serrano Ruano, Delfina, «La diffusion de l'ash'arisme et la réforme du crédo malikite à l'époque almoravide: Ibn Rushd al-Jadd, Abū Bakr Ibn al-'Arabī et le Qāḍī 'Iyāḍ», dans C. Aillet et B. Tuil Leonetti (éds.), *Dynamiques religieuses et territoires du sacré au Maghreb médiéval*, Madrid: CSIC, 2015, 79-102.

Serrano Ruano, Delfina, «Later Ash'arism in the Islamic West», dans S. Schmidtke (éd.), *The Oxford handbook of Islamic theology*, Oxford: Oxford University Press, 2016, 514-539.

Serrano Ruano, Delfina, «Entre almorávides y almohades: El concepto de taklīf (responsabilidad individual) y la introducción del imperativo de razonar la fe en el aš'arismo Andalusí», dans D. Villalba Sola (éd.), *Al-Muwahhidūn: El despertar del califato almohade*, Grenade: Patronato Alhambra y Generalife, 2019, 42-59.

Siddiqui, Sohaira, «*Jadal* and *qiyās* in the fifth/eleventh century: Two debates between al-Juwaynī and al-Shīrāzī », dans *Journal of the American Oriental Society* 39 (2019), 923-944.

al-Subkī, Tāj al-dīn, *Ṭabaqāt al-Shāfi'īyya al-kubrā*, éd. Ṭanāḥī et Ḥalū, le Caire: Maktabat al-Bābī al-Ḥalabī, 1963-1976.

Thiele, Jan, «Between Cordoba and Nīsābūr: The emergence and consolidation of Ash'arism (fourth-fifth/tenth-eleventh century)», dans S. Schmidtke (éd.), *Oxford handbook of Islamic theology*, Oxford: Oxford University Press, 2016, 225-241.

Thiele, Jan, «al-Juwaynī», dans H. Lagerlund (éd.), *Encyclopedia of medieval philosophy*, Dordrecht: Springer, 2018, 1028-1032.

Turkī, 'Abd al-Majīd, *Polémiques entre Ibn Ḥazm et Bājī sur les principes de la loi musulmane: Essai sur le littéralisme ẓāhirite et la finalité malikite*, Alger, Etudes et documents, s.d.

al-Ṭurṭūshī, Abū Bakr, *Kitāb al-Ḥawādith wa-l-bida'*, éd. & trad. Maribel Fierro, Madrid : Publications du CSIC, 1993.

8

جهود الإمام أبي بكر بن العربي في النحو واللغة

قراءةٌ فيما كتب عنه

فريد بن عبد العزيز الزامل السُّلَيْم

[Ibn al-ʿArabī's Work on Grammar and Lexicography]

[*A Recent Bibliography*]

[*Fareed Abdulaziz Alzamil Alsulaim*]

إن من نافلة القول الإشارة إلى ارتباط العلوم الإسلامية بعضها ببعض، وتكاملها وتداخلها، ومسائل العلوم يمد بعضها بعضًا، ويُبرهن في بعضها على بعض، فمن جهل فنًّا نقص عليه مادة فن آخر، كما يقول الإمام الطوفي[1]، ويقول الإمام ابن حزم: "فالعلوم كلها متعلق بعضها ببعض، محتاج بعضها إلى بعض، ولا غرض لها إلا معرفة ما أدى إلى الفوز في الآخرة فقط، وهو علم الشريعة"[2].

والإمام أبو بكر بن العربي (543هـ\1148م) رحمه الله تعالى عالم موسوعيٌّ، فألف في التفسير والفقه، والعقائد والأحكام، وما أجمل ما عبر به محمد السليماني في مقدمة تحقيقه لقانون التأويل إذ قال: "أبو بكر بن العربي من أعلام العلماء الذين تعددت مواهبهم... فهو الفقيه البصير الذي جانب التقليد.. وهو المحدث المستنير يعمل عقله وفكره فيما يقرأ أو يسمع، وهو المفسر المقتدر الذي أعد العدة لعمله في التفسير، من تضلع من لغة العرب وأشعارها وروائع نثرها.. وهو الأديب الذي يغوص على المعنى ويفتن في التعبير عنه ... وهو المؤرخ الذي يقارن بين الروايات ويميز حقها من باطلها... وهو المثقف الواسع الثقافة... وهو المتكلم الذي درس عيون كتب الكلام..."[3].

ومن يتعرض لهذه العلوم فلا شك أنه سيتعرض إلى اللغة، استنباطًا واستمدادًا، واستدلالاً واحتجاجًا، وقد ظهر فيما ترك من إرث عظيم علوُّ مرتبته في العربية، وسعة اطلاعه على دقائقها، ومشاركته لعلمائها في الاختيار والترجيح والاستدلال.

1 انظر: شرح مختصر الروضة: 3/586.

2 مراتب العلوم: 4/90.

3 قانون التأويل: 17–18.

 | DOI:10.1163/9789004706859_010

وفي العصر الحديث يمم الباحثون كنوز التراث الإسلامي، فقرؤوها، وحققوها، وأقاموا عليها الدراسات، ونشروا نصوصها، ودرسوا مسائلها، وكان لتراث الإمام ابن العربي نصيبٌ وافر من دراساتهم، ما بين مجيد محسن، ومن هو دون ذلك، وكان النحو واللغة مما تعرضوا له بالدراسة والإشارة، على اختلاف في مستويات ذلك عمقًا وقيمة، وسأعرض في هذه الصفحات المختصرة إلى أبرز معالم ما وقفت عليه من تلك الدراسات، على سبيل الإشارة والإيجاز، دون تفصيل واستقصاء.

ولعل أوَّل ما يلفت نظر الباحث في هذا الباب أن أكثر الدراسات التي تناولت النحو واللغة عند الإمام أبي بكر كانت حول كتابه (أحكام القرآن)، وقد أحصيت في ذلك أكثر من عشر دراسات، ما بين رسالة علمية، أو بحث منشور في دورية محكمة، في حين صدف الباحثون عن كثير من كتبه الأخرى، وخاصة (عارضة الأحوذي)، فلم أقف على دراسة واحدة مختصة بالنحو واللغة في هذا الكتاب، غير رسالة علمية سجلت في جامعة القصيم قبل عدة سنوات، وتعثر صاحبها فلم تنجز.

ومن أوائل الدراسات التي تناولت اللغة والنحو عند الإمام ابن العربي في أحكام القرآن: ابن العربي وجهوده النحوية واللغوية في كتابه أحكام القرآن رسالة ماجستير في جامعة بغداد للباحث ضامن بن محمد الكبيسي عام 1989م.

كما كانت دراسة د. مصطفى إبراهيم المشني: ابن العربي المالكي الإشبيلي وتفسيره أحكام القرآن دراسة وتحليل من الدراسات المتقدمة، فقد كانت الطبعة الأولى عام 1411هـ\1991م، وقد تعرض في الجانب النحوي إلى ما يلي:

1. مصادر ابن العربي من اللغة والنحو، وقد أثبت ما صرح به ككتاب سيبويه، وإصلاح المنطق، ومعاني القرآن للفراء، أو ما نص فيه على عالم معين، كالخليل والفراء والزجاج والمبرد وثعلب، وابن قتيبة وابن الأنباري وغيرهم....[4]
2. بيان منهجه، فبين عنايته بالنحو والتصريف، فكان يذكر أوجه الإعراب، وخاصة إذا ترتب عليه حكم شرعي، وربما نقل عن النحويين إجمالاً، كقوله: قال المعربون، أو النحويون، أو اختلف الناس في إعراب قوله تعالى... وكذا يعزو إلى النحويين عامة، أو البصريين، أو الكوفيين دون تخصيص أو تعيين.[5]

كما كان من منهجه فيما أثبته المنشنى الرد على النحويين، وبيان ما وهموا به.[6]

4 قانون التأويل: 59.

5 قانون التأويل: 153.

6 قانون التأويل: 153.

واستخلص من خلال تتبعه للأمثلة (أن ابن العربي كان يستعرض الأقوال والآراء النحوية في المفردات القرآنية ثم يحكم فيها العقل، ويناقشها مناقشة علمية دقيقة، في ضوء الأدلة اللغوية التي وردت عنده، فيرجح ما وقف الدليل بجانبه، ويرد ما جانبه الصواب، ويضعفه لأنه لم ينهض دليل يدل عليه ويعززه من اللسان العربي أو مما جرى على ألسنة أرباب النحو واهل الصناعة)).[7]

كما فصل في تناول ابن العربي لبعض الموضوعات، كالظروف، والضمائر، وأدوات الاستفهام والشرط، وحروف الجر والعطف.

وكان مما استخلصه أيضًا اعتماده واعتداده بسيبويه، واستشهاده به على أنه الحجة في النحو، وأن قوله هو الدليل الذي يكتفى به ويوقف عنده[8].

وأفرد للصرف مبحثًا ذكر فيه عدة أمثلة.

وتوالت الدراسات اللغوية بعده حول أحكام القرآن، ومن ذلك:

1. منهج ابن العربي في عرض القضايا اللغوية والبلاغية من خلال كتابه أحكام القرآن. د. مصطفى شطة، ود. ناصر عبد العزيز. منشور في مجلة الإمام عبد القادر للعلوم الإسلامية. قسنطينة الجزائر مج 35 ع 1 2021م. ويلحظ أن الباحثين لم يرجعا قط إلى الدراسات السابقة مع كثرتها، وتقدم نشرها.
2. الدلالة اللغوية عند ابن العربي في كتابه أحكام القرآن. أ. بالطير تاج. جامعة معسكر الجزائر. مجلة قراءات للبحوث والدراسات الأدبية والنقدية واللغوية. ع4 جوان 2014م. حتى الأمثلة التي درست من قبل تتكرر.
3. البحث الدلالي في أحكام القرآن لأبي بكر بن العربي الأندلسي. أطروحة دكتوراه في الدلالة وتحليل الخطاب، للباحث حاج إسباغو، جامعة الجيلالي ليابس. الجمهورية الجزائرية. سنة 2016–2017م.
4. أهمية اللغة في الترجيح الفقهي: أبو بكر بن العربي أنموذجا، رسالة ماجستير. للباحث: زكرياء التهامي. وكان البحث من خلال كتابيه أحكام القرآن والقبس. ط: 1435هـ\2013م.

كما تعرض عدد من الدراسات عن الإمام أبي بكر التي لم تخصص للجانب اللغوي لشيء من عنايته في النحو واللغة، وذلك نتيجة متسقة مع الواقع الذي أشرنا إليه من قبل، وهو اعتماد العلوم الإسلامية بعضها على بعض، فكان من أهم نتائج دراسة بعنوان: قواعد الترجيح في التفسير عند القاضي أبي بكر

---

7 قانون التأويل: 155.

8 قانون التأويل: 164.

بن العربي من خلال تفسير الربع الأول من كتابه أحكام القرآن[9] قوله: قواعد الترجيح المتعلقة بالسياق وظاهر القرآن والمتعلقة باللغة ولسان العرب هي أكثر القواعد استعمالاً عند ابن العربي في تفسيره.

أما الدراسات التي أقيمت حول تراث ابن العربي مقرونة بتحقيق نص من نصوصه، فقد تعرض المحققون إلى شيء من عنايته باللغة، إلا أنني لم أجد مَن فصَّل في ذلك تفصيلاً يشفي الغلة، ويشبع النهم، ومن أوفى الدراسات التي اقترنت بتحقيق نص من نصوصه، دراسة الدكتور محمد السليماني لقانون التأويل، إلا أنه لم يتعرض فيها لشيء من جهود الإمام في اللغة والنحو، إلا ما ذكره من مؤلفاته، فقد أثبت له مؤلفين في اللغة: وهما: حواشي على شرح ابن السِّيد لديوان أبي العلاء (سقط الزند)[10]، وملجئة المتفقهين على معرفة غوامض النحويين واللغويين.

كما تعرض الدكتور عبد الله التوراتي في دراسته للأمد الأقصى لمنهج الإمام أبي بكر في تناوله لأسماء الله تعالى، فالفصل الأول في مورد الاسم شريعةً، والثاني في شرحه لغةً، فقال: "ويعول فيه على الاشتقاق مع ضم النظير على النظير، والتمييز لضعيفه من شهيره، مع الاقتدار على الجمع بين المعاني والأقوال، والدلالة على القدر المتفق عليه من المعنى المختلف فيه، ناهجًا نهج ابن فارس في مقاييسه..."[11].

وكان بعض الدراسات مختصرًا جدًّا، لم يتعرض فيه المحققون إلى دراسة النص، ومن ذلك إخراج حسين علي اليدري لكتاب المحصول في أصول الفقه[12]، وقد تعرض فيه ابن العربي لمباحث نحوية ولغوية كثيرة، وهذا هو المنتظر في كتاب في هذا العلم، لكثرة المسائل المشتركة بينهما، غير أن المعتني لم يتناول شيئًا من ذلك.

أما مؤلفات الإمام أبي بكر بن العربي فقد تعرض الدارسون إليها في ترجماتهم، ومنها ما يخص النحو واللغة، فقد أثبتوا:

1. ملجئة المتفقهين إلى معرفة غوامض النحويين واللغويين.
2. إلجاء الفقهاء إلى معرفة غوامض الأدباء.

وهذان كتاب واحد.

3. حواشي على شرح ابن السِّيد لديوان أبي العلاء (سقط الزند).

---

9 أعدها: مسعود عاد. جامعة حما لأخضر – الوايد، 2018م.

10 لابن السيد البطليوسي رد على ما اعترضه فيه ابن العربي، وهو منشور في مجلة الذخائر العددان 16/15 صيف-خريف 1424هـ\2003م. بتحقيق: د. وليد السراقبي.

11 الأمد الأقصى: 41/1.

12 دار البيارق - عمان ط:1 1420هـ\1999م.

4. اختصار إصلاح المنطق.
5. رسالة في شرح قوله ε (لا تُصَرُّوا الإبل).

إضافة إلى شعراء الأندلس، ورسائل في الأدب والشعر.

فأما ملجئة المتفقهين فهو كتاب مفقود، وقد أجمع الدارسون والمترجمون على ذكره، لكثرة ما أحال إليه الإمام ابن العربي في كتبه، ولعل أول من ذكره أبو محمد الأَشِيري (561هـ\1166م)[13] وهو أحد تلاميذه، كما ذكره الذهبي[14] والمَقَّري[15].

وقد جمع كاتب هذه السطور مسائلها ونشرت في بحثين محكمين، ثم جمعت ونشرت في كتاب[16]. وقد فات الجامع بعض المسائل، واستدركها عليه بعض الباحثين، وسيلحقها إن شاء الله في طبعة ثانية.

ولم يزد في خاتمة دراسته عما سبقه، من ذكر لعناية ابن العربي في النحو واللغة، وتضلعه فيها، واعتماده على هذا التضلع في ترجيحه واستنباطه. إلا في مسألة عرض لها ولم يشبعها درسًا، وهي تفرده بآراء نحوية لم يسبق إليها، وهذا ما سأشير إليه بعد قليل إن شاء الله.

وأما حواشيه على ابن شرح ابن السيد ومختصر إصلاح المنطق فمن كتبه المفقودة حسب ما اطلعت عليه، غير أن رد ابن السيد على حواشيه موجود، ونشر مرتين أولاهما عام 1996م.

وأما رسالة شرح (لا تُصَرُّوا الإبل) فقد نشرتها د. حياة قارة في مجلة الدراسات اللغوية الصادرة عن مركز الملك فيصل للأبحاث الإسلامية عام 1421هـ\2000م، كما نشرت في مجلة التاريخ العربي الصادرة في الرباط في العام نفسه، ومع ذلك لم أر من أشار إليها ممن كتب في جهود ابن العربي النحوية غير د. سعيد العمري في رسالة الماجستير: المسائل النحوية في أحكام القرآن لابن العربي، وقد أنجزها عام 1423هـ\2002م.

وفي ختام هذه الورقة أحب أن أشير إلى مسألتين:

الأولى: أن جهود الإمام أبي بكر بن العربي في النحو واللغة لم تجد بعد العناية التي تستحقها، نعم، لقد عُني الباحثون بأحكام القرآن، وتناولوا ما فيه تناولاً كافيًا، إلا أن مؤلفاته الأخرى لا تقل عن الأحكام أهمية في بحث اللغة والنحو، وأذيل هذه المسألة بمسألة منهجية مهمة جدًّا، وهي غياب بناء اللاحق على السابق في كثير من الأبحاث، مع ارتفاع العذر في هذا الزمان الذي تيسرت فيه المعلومة، وقرب

---

13 انظر: المتوسط في الاعتقاد (مقدمة التحقيق): 20.

14 انظر: سير أعلام النبلاء: 199/20.

15 انظر: نفح الطيب: 36/2.

16 دار الحديث الكتانية: طنجة – بيروت: المملكة المغربية، لبنان. ط: 1 1439هـ2018-م.

مأخذها، أوليس من العجيب أن يُنجِز باحثانِ بحثًا بعنوان: منهج ابن العربي في عرض القضايا اللغوية من خلال كتابه أحكام القرآن، عام 2021م لم يعودا فيه إلى أي من الدراسات السابقة، التي نشرت قبل بحثهما بأكثر من عشرين سنة، ثم لا يأتيان بجديد على ما قدم قبلهما؟ فأين تحكيم المحكمين، وأين هيئات التحرير وإدارة تلك المجلات!

الثانية: كان ابن العربي رحمه الله لا يعد نفسه نحويًّا، ويجعل النحويين في كفة وأصحابه في الكفة الأخرى، وعلى إجلاله للنحويين، وخاصة سيبويه فقد كان يقف مواقف المخالفة والنقد، غير محتف بهالات المكانة وشعارات التقديس، يقول معلقًا على قول لسيبويه: "وهذا لا يشبه منصبه، ولا يليق بإدراكه"[17]، ويقول في معرض رده على قول له: وأما سيبويه فلا لعًا لعثرته![18]

وله آراء نحوية لم أقف على من سبقه إليها، ومن ذلك تقديره مبتدأً محذوفًا، في قوله ٤: هل من طعام، يقول: يريد هل مأكول من طعام، وهذا مما وهم فيه رؤساء الصناعة، فجعلوا الجار والمجرور مرفوعًا، فقلبوا القوس ركوة....[19].

ولهذا نظائر من تفرداته، وجدير بها أن تدرس باستصحاب معالم فكره، ومنهجه، فربما كان نواة لنظرٍ جديد، أو نظرية جاء بها من بعده فنسبت إليهم، وما أكثر هذا في تراثنا الفكري، واللغوي على وجه الخصوص.

وصلى الله على نبينا محمد وعلى آله وصحبه أجمعين.

قائمة المصادر والمراجع

ابن العربي المالكي الإشبيلي وتفسيره أحكام القرآن دراسة وتحليل. د. مصطفى إبراهيم المشني. ط:1. بيروت- عمان: دار الجيل- دار عمار، 1411هـ.

ابن العربي وجهوده النحوية واللغوية في كتابه أحكام القرآن. رسالة ماجستير في جامعة بغداد للباحث ضامن بن محمد الكبيسي عام 1989م.

أحكام القرآن. لأبي بكر بن العربي. تحقيق: محمد علي البجاوي. ط:1. بيروت: دار إحياء التراث العربي، 1421هـ.

17 أحكام القرآن: 110/3.

18 انظر: سراج المريدين: 231/1.

19 انظر: عارضة الأحوذي: 37/8.

الأمد الأقصى في شرح أسماء الله الحسنى وصفاته العلى. لابن العربي. ضبط نصه د. عبد الله التوراتي، وخرج أحاديثه أحمد عروبي. طنجة: دار الحديث الكتانية، 1436هـ.

أهمية اللغة في الترجيح الفقهي: أبو بكر بن العربي أنموذجا. رسالة ماجستير. للباحث: زكرياء التهامي.

البحث الدلالي في أحكام القرآن لأبي بكر بن العربي الأندلسي. للباحث حاج إسباغو، أطروحة دكتوراه في الدلالة وتحليل الخطاب، جامعة الجيلالي ليابس. الجمهورية الجزائرية. 2016–2017م.

الدلالة اللغوية عند ابن العربي في كتابه أحكام القرآن. أ. بالطير تاج. جامعة معسكر الجزائر. مجلة قراءات للبحوث والدراسات الأدبية والنقدية واللغوية. ع4 جوان 2014م.

سير أعلام النبلاء. للذهبي. أشرف على التحقيق: شعيب الأرنؤوط. ط:9. بيروت: مؤسسة الرسالة، 1413هـ.

شرح مختصر الروضة. للطوفي. تحقيق: د. عبد الله بن عبد المحسن التركي. ط:1. بيروت: مؤسسة الرسالة، 1407هـ.

عارضة الأحوذي بشرح صحيح الترمذي. لابن العربي. بيروت: دار الكتب العلمية،

قانون التأويل. لابن العربي. تحقيق: محمد السليماني. ط:1. جدة – دمشق: دار القبلة- مؤسسة علوم القرآن، 1406هـ.

القبس في شرح موطأ مالك بن أنس. لابن العربي. تحقيق: د. محمد عبد الله ولد كريم. ط:1. بيروت: دار الغرب الإسلامي، 1992هـ.

الكتاب المتوسط في الاعتقاد. لابن العربي. ضبط نصه: د. عبد الله التوراتي. ط:1. طنجة: المملكة المغربية. دار الحديث الكتانية، 1436هـ.

المحصول في أصول الفقه. لأبي بكر بن العربي. تحقيق: حسين علي اليدري وسعيد فودة. ط:1. عمان: دار البيارق، 1420هـ.

مسائل كتاب ملجئة المتفقهين إلى معرفة غوامض النحويين واللغويين لأبي بكر بن العربي جمعًا ودراسة. د. فرد بن عبد العزيز الزامل السليم. ط: 1. طنجة – بيروت: المملكة المغربية، لبنان: دار الحديث الكتانية، 1439هـ-2018م.

مع لقاضي أبي بكر بن العربي. لسعيد أعراب. ط:1. بيروت: دار الغرب الإسلامي، 1407هـ.

منهج ابن العربي في عرض القضايا اللغوية والبلاغية من خلال كتابه أحكام القرآن. د. مصطفى شطة، ود. ناصر عبد العزيز. مجلة الإمام عبد القادر للعلوم الإسلامية. قسنطينة الجزائر مج 35 ع 1 2021م.

9

Ibn al-ʿArabī, *adīb* et transmetteur de l'*adab* et de la poésie en al-Andalus, d'après le témoignage de quelques contemporains

Bruna Soravia

Parmi tous les aspects de la personnalité intellectuelle d'Abū Bakr Ibn al-ʿArabī[1], celui de son activité de transmetteur de poésie et d'*adab* a été sans doute le moins étudié, tout en ayant été bien présent à l'esprit de ses contemporains, qui connurent et apprécièrent aussi le *qāḍī* en tant qu'*adīb* accompli et connaisseur de la littérature, surtout de celle orientale.

Nous allons présenter, dans ces notes, ce que quatre auteurs de la période almoravide ont témoigné sur ce versant de son œuvre. Ils sont Abū l-Ḥasan Ibn Bassām, Abū Naṣr al-Fatḥ dit Ibn Khāqān, Abū l-Qāsim Ibn ʿAbd al-Ghafūr al-Kalāʿī et Abū Bakr Ibn Khayr : tous ont personnellement connu Ibn al-ʿArabī, chacun d'eux à un moment différent de la biographie de celui-ci. L'ensemble de ces attestations aide à éclairer, nous semble-t-il, aussi bien des aspects moins connus de la personnalité intellectuelle d'Ibn al-ʿArabī que, plus en général, des tendances propres du milieu savant almoravide.

1 Ibn al-ʿArabī dans la *Dhakhīra* d'Ibn Bassām

Né à Santarem, dans le territoire de la *taifa* aftaside de Badajoz, autour de la moitié du v/xi siècle[2], Abū l-Ḥasan ʿAlī Ibn Bassām avait pu entendre parler du père d'Ibn al-ʿArabī, Abū Muḥammad ʿAbd Allāh, bien avant de se rendre à Séville, où Abū Muḥammad avait été *wazīr* de la cour abbadide et aussi *qāḍī* et membre de la *shurā*. Il est, par contre, probable qu'il n'ait pas eu connaissance

1 N'étant pas spécialiste d'Ibn al-ʿArabī, nous nous sommes aidés, pour les repères biographiques, avec l'article sur Ibn al-ʿArabī dans la *Biblioteca de al-Andalus* (dorénavant *BA*), ii, 129-158, no. 298 (P. Cano Avila, A. Garcia Sanjuan, A. Tawfiq).

2 Pour Ibn Bassām, v. désormais l'article dans *BA* ii, 573-592, no. 395 (J. Lirola Delgado). Alors que la datation courante de sa mort est fixée à 542/1147 d'après le seul témoignage tardif d'al-Maqqarī, nous avons proposé de l'anticiper d'une décennie environ, au vu de sa mention dans l'*Iḥkām ṣanʿat al-kalām* d'Ibn ʿAbd al-Ghafūr al-Kalāʿī, dont nous avons fixé la rédaction entre 532 et 536 (1137-1141), v. *infra*.

 | DOI:10.1163/9789004706859_011

de son fils, qui n'était qu'un adolescent lorsque les Almoravides entrèrent dans la ville, en 484. La marque du déplacement forcé qui suivit la conquête du *Gharb* par les armées almoravide et chrétienne, dont pâtit surtout le personnel politique des *taifas*, est bien visible dans les notations autobiographiques disséminées dans la *Dhakhīra* d'Ibn Bassām et dans le *Qānūn* d'Ibn al-ʿArabī. Le premier dit, dans la *captatio benevolentiae* qui préface son anthologie :

> Dieu seul sait que ce livre provient d'un cœur blessé, d'une pensée dont la vivacité s'est éteinte, dans l'étreinte d'un destin changeant comme le caméléon. A l'origine de cela, mon exil forcé de Santarem [en 486, après la prise de la ville par Alfonso VI de Castille], dans l'Occident extrême, ma lame émoussée, mon âme effrayée, après que mes biens se soient évanouis, les anciens comme les plus récents, et avec eux tout le reste, à cause des bandes des *Rūm* qui se sont acharnées contre nous, les unes après les autres, jusque dans le cœur de notre région. Et cela bien que nous nous serions contentés de la noblesse de notre lignage, plutôt que de rechercher un profit ignoble, bien que nous aurions été satisfaits de garder le nécessaire, au lieu de tourner sans but d'un pays à l'autre[3].

Et Ibn al-ʿArabī, avec un bel exemple de sa maitrise du *tağnīs*, après avoir évoqué son enfance gâtée de savant précoce :

> Puis cette condition privilégiée eut fin et devint le sort de tout le monde (*thumma ḥālat hadhihi l-ḥāla al-khāṣṣa bi l-istiḥāla al-ʿāmma*), lorsque les Almoravides envahirent notre pays, en 484 [...]. Obligés de partir, nous sortîmes [d'al-Andalus], tandis que les ennemis se réjouissaient de notre affliction[4].

Au retour d'Ibn al-ʿArabī à Séville, en 495/1102, Ibn Bassām était sans doute encore errant dans le Gharb al-Andalus. Il déclare avoir été à Badajoz en 486/1093 et à Cordoue vers 493/1100[5] ; peu après, il se rend à Séville, qu'il appelle Ḥimṣ, suivant la convention poétique courante à son époque, dans la description très négative qui fait suite au texte cité plus haut :

---

3 Ibn Bassām, *al-Dhakhīra fī maḥāsin ahl al-Jazīra, éd. I. ʿAbbās*, 4 vols en 8 tomes, Beyrouth, 1398/1978, I/i, 19.

4 Ibn al-ʿArabī, Abū Bakr, *Qānūn al-taʾwīl*, éd. M. Sulaymānī, Beyrouth, 1406/1986, 419-420.

5 Ibn Bassām, *al-Dhakhīra*, II/ii, 536; III/ii, 654.

> Je me rendis à Ḥimṣ, mon esprit éteint, presque anéanti par l'affliction, et « plût à Dieu que j'eusse pu vivre avec ce qui m'en restait ! »[6]. Je vécus là-bas en étranger pendant des années, ayant fixé ma demeure à l'ombre du nuage tel une colombe craintive, redoutant de changer de position, et seul, sans le confort d'un ami, obligé de me contenter de ce qui restait de mes provisions de voyage. L'*adab* est en effet ici [à Séville] plus rare que la fidélité au serment, celui qui en possède est plus négligeable que la lune d'hiver. La mesure de la valeur individuelle y est la richesse, les ignorants sont partout le modèle. Il suffit aux hommes que leurs avoirs soient préservés, même aux dépens de leur dignité ; leur seul intérêt est de multiplier l'or et l'argent, même si religion et honneur s'en trouvent diminués[7].

Invective celle-ci quand même curieuse, si l'on pense que le gouverneur almoravide de la ville, entre 484 et 507, était Sīr (ou Sayr) b. Abī Bakr, cousin de Yūsuf b. Tāshfīn et mari de la nièce de celui-ci, Ḥawwā'[8]. C'est aussi à Séville qu'Ibn Bassām acheva probablement la rédaction de son anthologie, en 503/1109-10 : il était alors un homme adulte, sans doute bien avant dans sa quarantaine, alors qu'Ibn al-ʿArabī était encore un jeune homme, si bien qu'Ibn Bassām en parle dans la plupart des cas comme du « fils du *faqīh* Abū Muḥammad ».

Dans la *Dhakhīra*, Ibn al-ʿArabī est d'abord évoqué en tant que transmetteur d'Abū ʿAbd Allāh al-Ḥumaydī (m. 488/1095), qui avait pris une part active à la vie intellectuelle et littéraire en al-Andalus au cours de la première moitié du V/XI siècle, dans l'entourage d'Ibn Ḥazm, avant de s'installer à Bagdad jusqu'à la fin de sa vie[9]. Puisqu' Ibn al-ʿArabī n'atteignit la capitale abbaside que vers la fin de sa *riḥla*, en 490/1096, on peut penser qu'il avait rencontré al-Ḥumaydī

6 Citation d'al-Mutanabbī, *Dīwān al-Mutanabbī bi-sharḥ al-ʿUkbarī*, éds. M. al-Saqqā', I. al-Ibyārī, ʿA. H. Shalabī, Beyrouth, 1978, iii, 171, v. 25.

7 Ibn Bassām, *al-Dhakhīra*, 1/i, pp. 19-20. Traduction espagnole par E. García Gomez, *Un éclipse de la poésia en Sevilla: La época almorávide*, Madrid: Real Academia Española, 1945, 304; trad. italienne par Bruna Soravia, « L'introduzione d'Ibn Bassām al *Kitāb al-Dhakhīra fī maḥāsin ahl al-Ǧazīra*: Presentazione e traduzione », dans F. Díaz Esteban (éd.), *Bataliús II: Nuevos estudios sobre el reino Taifa de Badajoz*, Madrid: Letrúmero, 1999, 253-271, à pp. 66-67.

8 Sīr b. Abī Bakr apparaît dans un caméo assez curieux dans le *Sirāj al-muridīn*, masquant sans doute une dynamique de patronage. Tout en soulignant le manque d'éducation formelle (*ʿujma*, ignorance de l'arabe) de Sīr, Ibn al-ʿArabī y fait l'éloge de son *fiqh* et de sa *maʿrifa* innés, après que celui-ci avait apparemment accepté de satisfaire sa demande d'aide (Ibn al-ʿArabī, Abū Bakr, *Sirāj al-muridīn fī sabīl al-dīn*, éd. A.A. al-Tawrātī, 4 vols., Tétouan, 1436-38/2014-16, iv, 262-263). Sur Ḥawwā', v. Manuela Marin, « The Princess and the Palace. On Ḥawwa' bin Tashufin and Other Women from the Almoravid Royal Family », dans M. Hamilton, N. Silleras-Fernandes (éds), *In and Of the Mediterranean: Medieval and Early Modern Iberian Studies*, Nashville (TN): Vanderbilt U.P., 2015, 29-47.

9 V. l'article dans *BA* i, 460-464, no. 162 (G. Rossello Bordoy; J. Hareska).

lors d'une différente étape de son itinéraire, car Ibn Bassām rapporte que ce dernier l'avait personnellement informé d'un fait dont il sera question plus avant. De toute manière, Ibn al-ʿArabī acquit l'ouvrage que nous connaissons comme *Jadhwat al-muqtabis*, et qui était également connu en Orient comme « Histoire d'al-Andalus » (*Taʾrīkh al-Andalus*)[10], par l'enseignement d'Abū Bakr Muḥammad Ibn Ṭarkhān, l'un des élèves orientaux d'al-Ḥumaydī. En effet, la *Jadhwa* s'avère la source d'une partie des notices qu'Ibn al-ʿArabī avait transmises à Ibn Bassām au sujet d'Ibn Ḥazm en tant que poète et *adīb*, mais aussi sur quelques poètes et *udabā kuttāb* du milieu cordouan de celui-ci[11], à savoir Abū ʿUbāda Ibn Māʾ al-Samāʾ, Abū Marwān Ibn al-Ṭubnī (l'un des trois frères Banū al-Ṭubnī), Abū Ḥafṣ Ibn al-Shahīd et l'historien Ibn al-Faraḍī.

Dans la troisième partie de la *Dhakhīra*, portant sur le Sharq al-Andalus, Ibn al-ʿArabī est cité comme témoin au sujet d'une histoire d'un grand intérêt[12], qu'Ibn Bassām déclare avoir trouvée dans quelques notes de la main d' Ibn Ḥazm, et qui est ensuite corroborée par Ibn al-ʿArabī, d'après al-Ḥumaydī[13]. Ibn Ḥazm aurait entendu ce récit de son maître, le cordouan Ibn al-Kattānī, soit, Abū ʿAbd Allāh Muḥammad b. al-Ḥasan al-Madhḥijī (m. 420/1029-30)[14], médecin célèbre, ainsi qu'*adīb* et marchand d'esclaves. Ibn al-Kattānī avait pris part à une réunion de la cour de la fille de « Sanche roi des Basques » (*Shanj malik al-Bashkuns*), à savoir Urraca Gomez[15], épouse de « Sanche fils de García fils de Ferdinand », c'est à dire, Sancho García (r. 995-1017/385-407), pour des

10 Al-Ḥumaydī, *Jadhwat al-muqtabis*, éd. I. al-Abyārī, 2 vols., le Caire, 1403/1982; al-Dhahabī, *Siyar aʿlām al-nubalāʾ*, plusieurs éds., 23 vols. + 2 vols. d'indexes, Beyrouth 1406/1986, xix, 123. On le trouve en effet sous le titre *Jadhwat al-muqtabis fī Taʾrīkh al-Andalus*, dans la liste des livres qu'Ibn al-ʿArabī avait appris au cours de sa *riḥla*, dans le *Sirāj al-muridīn* (iv, 399-411). ʿAmmar Ṭālibī, qui cite cette liste dans l'appendice de son édition des *ʿAwāṣim*, semble indiquer qu'Ibn ʿArabī avait matériellement ramenés (*atā bi-hā*) ces ouvrages (Ibn al-ʿArabī, Abū Bakr, *al-ʿAwāṣim min al-qawāṣim*, éd. ʿAmmar Ṭālibī, le Caire, 1394/1974, 377-379: p. 377). En réalité, il s'agit de *riwāyāt*, comme ʿAbd Allah al-Tawrātī, l'éditeur du *Sirāj*, l'indique, d'après Ibn Khayr dans la plupart des cas.

11 Sur le cercle des *ikhwān* d'Ibn Ḥazm, avant sa transformation définitive en *ʿālim*, v. Soravia, Bruna, « A portrait of the *ʿālim* as a young man: The formative years of Ibn Ḥazm, 404/1013-420/1029 », dans C. Adang, M. Fierro et S. Schmidtke (éds.), *Ibn Ḥazm of Cordoba: The life and works of a controversial thinker*, Leyde: Brill, 2013, 25-50, à pp. 27-31.

12 Cette anecdote est connue grâce à la traduction de Henri Pérès, *La poésie andalouse en arabe classique au XIe siècle : ses aspects généraux, sa valeur documentaire*, Paris : Maisonneuve, 1937, 386-387.

13 Ibn Bassām, *al-Dhakhīra*, III/i, 318-319.

14 Sur lui, *BA* iii, 735-738, no. 726 (S. Sadiq ; J. Lirola Delgado).

15 La présentation de cette princesse chrétienne semble renforcer l'hypothèse qu'elle soit la fille de Sancho Garcés de Pamplona et non du comte Gómez Díaz (v. de Salazar y Acha, Jaime, « Urraca: Un nombre egregio en la onomástica altomedieval », dans *En la España medieval* 1 (2006), 29-47, à pp. 37-38).

affaires entre la *taifa* de Saragosse et les Chrétiens. Un grand nombre d'esclaves musulmanes, offertes par le marwanide Sulaymān b. al-Ḥakam lors de l'un de ses deux courts califats[16], assistaient au *majlis*. À la demande de la princesse chrétienne, une esclave chanteuse avait entonné un poème d'amour, faisant pleurer une des esclaves qui se tenaient debout tout autour de leur maîtresse. À Ibn al-Kattānī qui, ému, s'enquiert à son sujet, l'esclave révèle que les vers sont de son père, Sulaymān b. Mihrān al-Saraquṣṭī, et lui avoue sa grande nostalgie pour sa famille.

Ibn Bassām questionne la véracité du récit contenu dans l'écrit autographe d'Ibn Ḥazm, car il s'étonne qu'Ibn al-Kattānī ait pu laisser la fille avec sa maîtresse chrétienne, sans chercher à la racheter. Toutefois, il mentionne tout de suite après Ibn al-ʿArabī, lequel, citant al-Ḥumaydī, confirme le récit par la description circonstanciée de l'occasion dans laquelle Ibn Ḥazm aurait entendu le poème :

> Abū Bakr Ibn al-ʿArabī m'a également informé de ces vers, en me disant : « C'est al-Ḥumaydī qui m'en a parlé, d'après le *faqīh* Abū Muḥammad Ibn Ḥazm qui lui avait dit : 'Muhammad b. al-Hasan al-Madhḥijī [Ibn al-Kattānī] les a récités devant moi dans le *majlis* du vizir Abū l-Aṣbagh ʿIsā b. Saʿīd, vizir d'al-Muẓaffar b. al-Manṣūr Ibn Abī ʿĀmir, et ce sont les quatre vers susmentionnés' »[17].

Le fait qu'Ibn al-ʿArabī n'apparaît pas comme informateur au sujet des biographies des *udabaʾ* gharbiens, dans la seconde partie de la *Dhakhīra*, semble indiquer que la référence hazmienne était alors pour lui bien plus significative que les relations qu'il avait pu avoir à Séville, avant et tout après sa *riḥla*. Par contre, on le retrouve dans la quatrième et dernière partie de l'anthologie, celle concernant « un groupe de personnages éminents de ces contrées, parmi ceux qui surgirent à notre époque en Ifriqiya, en Syrie et en Iraq »[18]. Il y est la source de la citation de quelques *udabāʾ shuʿarāʾ* qu'il avait connus

16 Sulaymān b. al-Ḥakam fut calife pour quelques mois en 400/1009-10, et une deuxième fois entre 403/1013 et 407/1016, soutenu par les troupes berbères de Zāwī b. Zīrī et par Sancho García.

17 Ibn Bassām, *al-Dhakhīra*, III/i, 319. La déclaration d'Ibn Ḥazm est reprise, presque dans les mêmes termes, dans la biographie de Sulaymān b. Mihrān al-Saraquṣṭī, le père de l'esclave, dans al-Ḥumaydī, *Jadhwat al-muqtabis*, i, 349, no. 458.

18 *Dhakhīra* I/1, p. 30. Ce sont les personnages recensés dans la seconde partie du quatrième tome de l'édition courante. Bien qu'ils n'aient pas eu de rapport avec al-Andalus, Ibn Bassām en parle « pour faire pièce avec Abū Manṣūr al-Thaʿālibī, qui avait mentionné dans la *Yatīma* un groupe de gens d'al-Andalus » (*ibidem*, IV/1, 8).

ou dont il avait reçu l'œuvre lors de son séjour à Ascalon, en Palestine, où il avait passé six mois en 491/1098, en rencontrant « un mar de literatura que se bebió a borbotones »[19]. Ibn Bassām ne conteste guère le jugement critique d'Ibn al-ʿArabī, car il présente sur sa seule parole ces personnages, au demeurant plutôt obscurs, à savoir, Abū l-Fityān al-ʿAsqalānī, le *qāḍī* Abū Muḥammad b. Niʿma b. Khalīl et al-Mujīd b. Abī al-Shakhbāʾ al-ʿAsqalānī[20]. Le chapitre au sujet de ce dernier *adīb*, qui est aussi est le seul à être recensé par d'autres auteurs, est assez long mais lacunaire dans le début, où Ibn Bassām aurait sans doute identifié son informateur comme étant également Ibn al-ʿArabī, comme il sera montré dans les sections suivantes.

2 Ibn al-ʿArabī dans la *Maṭmaḥ al-anfus*[21] d'al-Fatḥ Ibn Khāqān

Abū Naṣr al-Fatḥ b. Muhammad, dit Ibn Khāqān (m. 529/1134), originaire de la région de Jaén mais sévillan d'adoption[22], est l'auteur, entre autres ouvrages, des *Qalāʾid al-ʿIqyān* et du *Maṭmaḥ al-anfus*, florilèges littéraires écrits quelques années après la *Dhakhīra* d'Ibn Bassām et en concurrence directe avec celle-ci. Ibn al-ʿArabī y est mentionné dans le plus tardif des deux, le *Maṭmaḥ*, écrite depuis 517/1123, bien que la même *tarjama*, à quelques variantes près, ait été incluse aussi dans la tradition textuelle des *Qalāʾid*, écrits entre 503/1110 et 517/1123[23]. Cette mention témoigne ainsi d'une période de la biographie d'Ibn al-ʿArabī postérieure à celle attestée par la *Dhakhīra*, correspondant à sa cinquantaine, ce qui explique pourquoi il y est surtout décrit dans sa qualité de *faqīh ḥāfiẓ* mais non encore comme *qāḍī*, charge qu'il n'obtiendra qu'en 528/1134.

Après une ouverture louant le savoir d'Ibn al-ʿArabī, lequel avait « fait oublier la perspicacité d'Iyās, abandonné le *taqlīd* pour le *qiyās*, fait fructifier le droit appliqué (*farʿ*) à partir de ses principes (*aṣl*) »[24], la *tarjama* continue en faisant

19 *BA* ii, 132 no. 298.

20 Ibn Bassām, *al-Dhakhīra*, IV/ii, 615-625 et 627-661. Ibn al-ʿArabī est probablement aussi la source des quelques extraits attribués à l'*adīb* et *qāḍī* fatimide Jalāl al-Dawla Ibn ʿAmmār, qu'Ibn Bassām transcrit sans commentaire (*ibidem*, 625-627).

21 Abū Naṣr al-Fatḥ, dit Ibn Khāqān, *Maṭmaḥ al-anfus*, éd. M. ʿA. Shawābkah, Beyrouth, 1403/1983.

22 V. l'article dans *BA* iii, 622-626, no. 692 (W. Saleh al-Khalifa).

23 Ibn Khāqān, *Maṭmaḥ al-anfus*, 297-300; Id., *Qalāʾid al-ʿiqyān*, éd. H.Y. Kharyūsh, 2 vols, Irbid : Maktabat al-Manār, 1989, 692-694. Pour la datation, v. l'introduction de l'éditeur de la *Maṭmaḥ*, pp. 114-115.

24 Ibn Khāqān, *Maṭmaḥ al-anfus*, 297. Iyās est Iyās b. Muʿāwiya, que ʿUmar b. ʿAbd al-ʿAzīz avait nommé *qāḍī* de Basra, et dont l'esprit vif devint proverbial.

l'éloge de son père Abū Muḥammad, « pleine lune parmi les étoiles, cœur du *majlis* de son souverain. Al-Muʿtamid b. ʿAbbād le choisit comme al-Maʾmūn avait choisi Ibn Abī Duʾād (Abū ʿAbd Allāh Aḥmad al-Iyādī) [pour sa *shurā*][25], lui conférant les plus hautes fonctions et le distinguant par les charges les plus élevées ».

Le texte décrit ensuite l'exile forcée de père et fils et leur *riḥla* orientale, au cours de laquelle un Abū Muḥammad craintif avait renoncé à réclamer ce dont il avait été dépossédé, car « il n'avait pas trouvé un [autre] al-Muʿtamid prêt à combler ses besoins et à lui accorder sa faveur » et s'était replié sur « la transmission et l'écoute [du savoir] », suivi en cela par son fils, qui ne tarda pas à briller de sa propre lumière. A la mort de son père, Abū Bakr Ibn al-ʿArabī avait fait retour en al-Andalus, en y trouvant un accueil chaleureux et des auditeurs avides de connaître ce qu'il avait rapporté. La *tarjama* s'achève sur un poème d'Ibn al-ʿArabī au sujet de son amour nostalgique pour Bagdad, qu'il adresse à ses amis chéris (*ahl al-widād*), dont voici notre tentative de traduction :

> Est-ce de toi qu'il provient, tandis que la nuit est trompée par l'aurore,
> Le fantôme d'un aimé (*ṭayf ḥabīb*), portant avec soi le bâton de la gloire,
> Dont la lumière resplendissante dévoile l'obscurité ténébreuse ?
> Il n'a pas frappé les ténèbres avec les astres luisantes (soleil et lune),
> Il n'a pas été content de laisser sa trace sur la terre plate,
> Mais il est monté jusqu'aux Jumeaux, vers une étoile qui avance dans la nuit,
> Poussant ses montures, qu'il avait déjà incité avec sa puissance,
> Les forçant à marcher sur le sommet de l'Aigle,
> Alors qu'elles sont alourdies par la housse qui les couvre.
> Elles ont marché rapidement, par crainte d'entendre le cri perçant [du conducteur],
> Trainant leur queue sur la queue de la Voie Lactée,
> Raison pour laquelle les étoiles qui s'y trouvent sont visibles au voyageur de nuit,
> Puis sont allés au-delà des Jumeaux, qui se trouvent au-dessus d'elle,
> Et les traces qu'elles ont lassées sont les taches de la pleine lune,
> Qui répandent le parfum éternel du Jardin céleste.
> Oublie donc les sables qui s'étendent à l'ombre d'al-Unayʿim[26]

25 Sur Aḥmad b. Abī Duʾād (ca. 160-240/776-854), savant mutazilite et inspirateur de la *miḥna*, v. *EI2*, « Aḥmad b. Abī Duʾād » (K.V. Zettersteen ; Ch. Pellat).

26 Yāqūt mentionne cette localité, apparemment désertique, sans la localiser (*Muʿjam al-buldān*, éd. Beyrouth : Dār Ṣādir 1397/1977, i, p. 273.

Car elles [tes montures] n'ont pas cherché à éviter Qays ni le cheval de ʿĀmir,
Ni se sont-elles cachées par crainte de rencontrer les Banū Ḍamr.
Puisse Dieu abreuver l'Égypte, l'Iraq et ses gens,
Bagdad et les deux Syries, avec une pluie abondante ![27]

Il est sans doute risqué de vouloir proposer une interprétation ponctuelle de ces vers hermétiques et fascinants, déployant une maîtrise parfaite du discours par énigmes, dans le sillon d'al-Maʿarrī. Il est tout de même possible de lire, entre les lignes du *fakhr,* une allégorie de la *riḥla* de leur auteur, décrite comme un voyage à travers les constellations du ciel nocturne, avant d'en désigner les destinations dans l'invocation finale.

3 Ibn al-ʿArabī dans l'*Iḥkām ṣanʿat al-kalām* d'Ibn ʿAbd al-Ghafūr al-Kalāʿī[28]

Le troisième témoin que nous allons considérer est Abū l-Qāsim Muḥammad Ibn ʿAbd al-Ghafūr al-Kalāʿī (dorénavant Ibn ʿAbd al-Ghafūr), né sans doute à Séville vers la fin du v/xi siècle et mort après 537/1143, auteur de l'*Iḥkām ṣanʿat al-kalām*, la *Maîtrise de l'art de la prose*. De souche yéménite, comme les Maʿafirites Banū l-ʿArabī, aussi bien le grand-père et homonyme d'Ibn ʿAbd al-Ghafūr, Abū l-Qāsim Muḥammad, que son père Abū Muḥammad ʿAbd al-Ghafūr, avaient vécu à la cour abbadide, où ils avaient sans doute côtoyé le père d'Ibn al-ʿArabī. Le grand-père avait connu une mort prématurée dans la première partie de l'émirat d'al-Muʿtamid b. ʿAbbād (r. 461-484/1069-1091), autour de la naissance d'Abū Bakr Ibn al-ʿArabī. Le père, ʿAbd al-Ghafūr, né vers la moitié du v/xi siècle, n'avait pas encore atteint l'âge pour entrer à son tour au service d'al-Muʿtamid, lorsque la *taifa* abbadide tomba aux mains des Almoravides.

Abū Muḥammad ʿAbd al-Ghafūr dut certainement connaître les mêmes péripéties que les Banū al-ʿArabī et Ibn Bassām, après la conquête de Séville et du Gharb al-Andalus, mais il ne quitta apparemment pas la ville. Dans le

27 Cité aussi par al-Maqqarī, *Nafḥ al-ṭīb min ghuṣn al-Andalus al-raṭīb*, éd. I. ʿAbbās, 7 vols. + indexes, Beyrouth, 1388/1968, ii, 34-35.

28 Édité par R. al-Dāya, Beyrouth 1966 (les numéros de page entre parenthèses se réfèrent à cette édition); trad. française par B. Soravia et M. Zekraoui, *La maîtrise de l'art de la prose*, Cordoue, University of Cordoba Press, 2023. Sur Ibn ʿAbd al-Ghafūr, v. *BA* i, 587-591, no. 184 (B. Soravia) ; *EI3*, « al-Kalāʿī, Ibn ʿAbd al-Ghafūr » (B. Soravia); HATA : https://www.eea.csic.es/red/hata/autor.php?idg=4315 (consulté le 6/12/2022). Ni sa date de naissance ni celle de sa mort ne sont connues.

chapitre de la *Dhakhīra* qui lui est consacré, Ibn Bassām, qui fut son ami, déplore que les vicissitudes du destin l'aient empêché de suivre les traces de son père, avant de présenter sa production, principalement en prose, de manière ample et élogieuse. Cette introduction très favorable eut probablement l'effet de signaler ʿAbd al-Ghafūr à l'attention de ʿAlī b. Yūsuf b. Tāshfīn (r. 500-537/1106-1143), qui venait de succéder à son père comme *amīr al-muslimīn*, dans les mêmes années où le projet de la *Dhakhīra* allait s'achever. Aussi, c'est sans aucun doute à cette même époque que pourrait dater la relation entre Abū Bakr Ibn al-ʿArabī et Abū Muḥammad ʿAbd al-Ghafūr (vraisemblablement plus âgé que l'autre d'une quinzaine d'années), avant que leurs carrières respectives ne s'écartassent, celle d'Ibn al-ʿArabī dans la direction du *fiqh* et du *ʿilm al-ḥadīth*, celle d'Ibn ʿAbd al-Ghafūr vers la *kitāba*. En effet, avec l'avènement de ʿAlī b. Yūsuf, les *kuttāb udabāʾ* qui avaient fait partie du *dīwān* abbadide et de celui aftaside de Badajoz furent recrutés dans le *dīwān* almoravide à Marrakech, la nouvelle ville capitale de la dynastie. Ce fut notamment le cas d'Ibn ʿAbdūn de Evora, Ibn al-Qaṣīra (m. 508/1114), Abū Bakr Ibn al-Qabṭurnuh (m. ap. 520/1126) et Abū l-Qāsim Ibn al-Jadd (m. 515/1121), auxquels s'ajouta, dans la mouvance du nouvel émir, ʿAbd al-Ghafūr, portant sans doute avec lui son enfant, Abū l-Qāsim Muḥammad, l'auteur de l'*Iḥkām*.

L'*Iḥkām* montre l'empreinte des événements liés à la première période du règne de ʿAlī, et celle des personnages qui, comme Ibn al-Qabṭurnuh et Ibn ʿAbdūn (ce dernier appelé dans le traité *kabīrī*, mon ainé), avaient contribué à la formation d'Ibn ʿAbd al-Ghafūr. Ibn Bassām y est souvent mentionné avec sa *Dhakhīra*, comme source des longues citations des *udabāʾ* andalous, mais aussi comme informateur direct de l'auteur, au sujet de quelques récits de grand intérêt. Bien sûr, on y trouve aussi le père de l'auteur, ʿAbd al-Ghafūr, décrit dans sa relation amicale avec Ibn ʿAbdūn, ou lors d'une attaque vicieuse que lui porta un contemporain, peut-être al-Fatḥ Ibn Khāqān, ou vis-à-vis de son fils, qui lui témoigne dans ces cas une affection respectueuse et nostalgique.

Nous avons montré ailleurs que l'*Iḥkām* a été vraisemblablement rédigé à Séville dans les dernières années de la domination almoravide, entre 532/1137 et 536/1141, lorsque son auteur avait la quarantaine et tous les personnages susmentionnés étaient décédés. Seule exception, celle d'Abū Bakr Ibn al-ʿArabī, qui était encore vivant lors de la fin probable de sa rédaction, bien qu'il semble décrit de mémoire, sur la base d'interactions qui durent avoir lieu avant son départ de Séville, en 529/1134-5. La première de ces mentions décrit la réception d'une *risāla* dans laquelle Ibn ʿAbd al-Ghafūr avait placé de grands espoirs, la *Risālat al-Sājiʿa wa-l-ghirbīb*, imitation assez médiocre de la *Risālat al-Ṣāhil wa-l-shāḥij* d'al-Maʿarrī, auteur pour qui Ibn ʿAbd al-Ghafūr avoue une admiration sans bornes. Le jeune auteur avait envoyé le texte de l'épitre, peut-être de Marrakech, à un ami de famille, l'*adīb* sévillan Abū Ayyūb Ibn Abī Umayya (m. 522/1138). C'est al-Fatḥ Ibn Khāqān, qui, en mentionnant cette même histoire, nous apprend qu'Ibn Abī Umayya avait tardé à accomplir son devoir d'amitié et qu'Ibn ʿAbd al-Ghafūr s'en était impatienté et lui avait réclamé son ouvrage. Et l'autre de se tirer d'embarras en lui envoyant un billet amical, reproduit dans l'*Iḥkām*, où l'épitre qu'il retourne, sans l'avoir peut-être lue, est personnifiée en une jeune vierge que l'on rend à son légitime amant.

Si l'on comprend pourquoi Ibn ʿAbd al-Ghafūr avait pu omettre les circonstances peu flatteuses de cet échange, l'on se demandera par contre pourquoi il a voulu rappeler, dans une autre partie du traité, la réaction d'Ibn al-ʿArabī à ce même texte, qu'il avait écouté, peut-être lors d'une lecture publique dans un *majlis* à Séville, sans doute vers 520/1136, lorsque son auteur était dans sa vingtaine. Il s'agit, en effet, d'une réponse dans la forme d'une *risāla* « au titre insultant de *Lamḥat al-bāriq fī taqrīẓ lawāhiẓ al-sābiq* (L'éclat de l'éclair : éloge des yeux du vainqueur) », titre qui reprend celui d'un *radd* qu'Ibn Abī l-Khiṣāl avait opposé à la célèbre épitre *shuʿūbīya* d'Ibn Gharṣiya[29]. Ibn ʿAbd al-Ghafūr cite un extrait de son ouverture, dans le cadre du chapitre sur le double sens (*al-muwarrā*) (pp. 190-1) :

> Il y a eu récemment en Orient un homme qui fit de ce style (*fann*, i.e., le double sens) son point d'honneur. Alors, bravo pour mon petit frère, qui jouta si bien de patience avec lui qu'il l'enterra ! Tandis que celui-là se borna à donner une question suivie d'une réponse, l'autre se plut à imaginer toute sorte d'arguments, dont le grand nombre compense ce qui manque en justesse et exactitude.

La raison pour laquelle ce texte dénigrant a été quand même cité, est probablement qu'il montre Ibn al-ʿArabī, appelé ici et ailleurs « vizir *faqīh* », s'intéresser à cette *risāla*, au point de l'écouter et de la critiquer, si vrai qu'Ibn ʿAbd al-Ghafūr continue en faisant l'éloge de son détracteur, comme si rien n'était. L'on rappellera ici qu'Ibn al-ʿArabī avait transmis en al-Andalus la plupart des ouvrages d'al-Maʿarrī, comme il sera dit mieux dans la section suivante, et qu'il avait été, avec son rival Ibn al-Sīd al-Baṭalyawsī, le catalyseur de la passion dévorante que les Andalous de son temps nourrissaient pour le poète aveugle[30]. Ibn ʿAbd

29 Cf. Ibn Khayr, *Fahrasa*, éd. F. Codera et J. Ribera, Bagdad, 1382/1963, 419 ; éds. B. ʿAwād Maʿrūf, M. Bashshār ʿAwād, Tunis, 2009, 514, no. 1224. Ce *radd*, qui n'a pas survécu, n'est pas mentionné dans l'étude de G. Larsson, *Ibn García's shuʿūbiyya letter: Ethnic and theological tensions in medieval al-Andalus*, Leyde: Brill, 2003.

30 V. *Taʿrīf al-qudamāʾ bi-Abī l-ʿAlāʾ*, éd. M. al-Saqqāʾ, I. al-Abyārī, H. ʿAbd al-Maǧīd, supervision de Ṭaha Husayn, Caire 1944/1363 (réimp. 1385/1965), 439 ss. Aussi, Blankinship, Kevin,

al-Ghafūr n'étant pas une exception, non seulement il égrène ses sélections des ouvrages d'al-Maʿarrī, suivies par ses propres imitations maladroites, mais il arrive à présenter le poète syrien comme le modèle inégalé d'auteur original, après avoir recensé une vingtaine de ses ouvrages qu'il avait connus, sans doute à travers Ibn al-ʿArabī[31].

Ibn al-ʿArabī est également évoqué au sujet de l'autre auteur dont l'*Iḥkām* s'est idéalement inspiré, c'est à dire, Abū Manṣūr al-Thaʿālibī. Ibn Bassām avait raconté à Ibn ʿAbd al-Ghafūr qu'Ibn al-ʿArabī lui avait dit « avoir eu connaissance (*saqaṭa ilay-hi*) » de vingt et un ouvrages d'al-Thaʿālibī, sans lui dire lesquelles. Notre auteur avait ensuite trouvé un autographe d'Ibn Bassām, après la mort de celui-ci, contenant cette liste[32], qui représente une ultérieure et significative confirmation de la popularité de cet auteur oriental à l'époque almoravide, bien qu'il ne soit pas précisé si ces ouvrages avaient été physiquement introduits en al-Andalus, tous ou en partie, ni, en l'absence d'autres témoins, si Ibn al-ʿArabī en avait transmis quelques-uns[33]. Cette histoire est importante aussi car elle remet en question la date courante de la mort d'Ibn Bassām, que le seul al-Maqqarī fixe en 542/1147[34], alors qu'elle devrait être avancée

« Al-Maʿarrī's esteem in the Islamic West: A preliminary overview », dans *Al-Masāq* 31 (2019), 253-271, à pp. 1-19.

31 Ibn ʿAbd al-Ghafūr al-Kalāʿī, *Iḥkām ṣanʿat al-kalām*, éd. R. al-Dāya, Beyrouth, 1966, 231-232. Ce sont, parmi les ouvrages en prose : le *Kitāb al-Qāʾif*, le *Kitāb al-Ṣāhil wa-l-shāḥij* et son commentaire linguistique, *Lisān al-Ṣāhil*; le *Kitāb al-Fuṣūl wa-l-ghāyāt fī tamjīd Allāh wa-l-ʿiẓāt*; le *Kitāb al-Sajʿ al-sulṭānī*; la *Khutba al-faṣīḥ li-Thaʿlab* et son commentaire. Parmi les épîtres (*rasāʾil*): la *Risālat al-falāḥa*; la *Risālat al-ghufrān*; la *Risālat al-jinn*; la *Risālat al-nikāḥ*; la *Risālat al-ighrīḍ* et la *Risālat al-manīḥ*. Parmi ses ouvrages poétiques : le *Kitāb Saqṭ al-Zand* et son commentaire linguistique, le *Ḍawʾ al-Saqṭ*; le *Kitāb Luzūm mā lā yalzam*, le *Kitāb al-Istighfār* et le *Kitāb Jāmiʿ al-Awzān*. Les ouvrages sur d'autres auteurs : le *Kitāb Dhikrā Habīb*, puis un livre sur la poésie d'Abū l-Ṭayyib al-Mutanabbī « que je n'ai pas appris ni vu » (sans doute le *Muʿjiz Aḥmad*) « et d'autres encore qui ne nous sont pas parvenus et dont nous n'avons même pas la mention ».

32 Ibn ʿAbd al-Ghafūr al-Kalāʿī, *Iḥkām*, 232. Ce sont : *Fiqh al-lugha*; *Yatīmat al-dahr*; *Thimār al-qulūb fī l-muḍāf wa-l-manṣūb*; *Bard al-akbād fī l-aʿdād*; *Ḥall al-ʿaqd*; *Mirʾāt al-murūʾa*; *Aḥsan mā samiʿtu*; *Aḥāsin al-maḥāsin*; *Ghurar al-maḍāḥik*; *al-Farāʾiḍ wa-l-qalāʾid*; *al-Tamthīl wa-l-muḥāḍara*; *Ajnās al-tajnīs*; *al-Mubhij*; *al-Ṭarāʾif wa-l-laṭāʾif*; *al-Kifāya wa-l-nihāya*; *al-Thalj wa-l-maṭar*; *al-Siḥr wa-l-balāgha* (sic); *Sajʿ al-manthūr*; *al-Lumaʿ al-ghaḍḍa*; le *Kitāb Alf ghulām*; la *Tatimmat al-yatīma*.

33 En effet, il ne figure pas dans l'*isnād* des ouvrages d'al-Thaʿālibī mentionnés dans la *Fahrasa* d'Ibn Khayr.

34 Al-Maqqarī dut avoir lui-aussi des doutes, car il dit que le décès d'Ibn Bassām « tarda » jusqu'à cette date (*Nafḥ al-ṭīb*, iii, 458). Il est vraisemblable qu'Ibn Bassām, déjà actif à la fin de l'époque des *taifas*, ait été du même âge que le père de notre auteur, alors qu'une interprétation superficielle de la notice biobibliographique d'Ibn Abbār relative à Ibn

d'une décennie au moins, étant donné que tous les indices chronologiques conduisent à penser que la rédaction de l'*Iḥkām* s'acheva avant cette date.

Dans la troisième et dernière mention, Ibn al-ʿArabī est indiqué comme source d'une *khuṭba* d'al-Mujīd al-ʿAsqalānī, qu'Ibn ʿAbd al-Ghafūr présente comme modèle du « style équivoque (*al-mushkil*) » du *sajʿ* (pp. 247-8). Ailleurs, (p. 62) après avoir cité le début d'une épître d'al-Mujīd, mentionnée aussi par Ibn Bassām dans le *faṣl* de la *Dhakhīra*, notre auteur saisit cette occasion pour surenchérir sur sa propre admiration pour al-Mujīd, en présentant un dispositif de style que celui-ci avait conçu comme l'exemple de la « maîtrise de l'art de la prose » (*iḥkām ṣanʿat al-kalām*), qui est le sujet propre de son traité. Dans ce cas aussi, il est vraisemblable que le parrainage influent d'Ibn al-ʿArabī ait pu justifier, chez Ibn ʿAbd al-Ghafūr comme chez Ibn Bassām, la surestimation d'un *adīb* dont la réputation ne repose que sur la trace que ses écrits ont laissé en al-Andalus.

Finalement, c'est Ibn al-Abbār qui, dans la notice biographique sur Ibn ʿAbd al-Ghafūr dans la *Takmila*, confirme, s'il en était besoin, le lien avec Ibn al-ʿArabī, en disant :

> Dans certains de ses ouvrages, il [Ibn ʿAbd al-Ghafūr] a rapporté le meilleur des traditions d'Abū Bakr Ibn al-ʿArabī, et il y eut entre eux une correspondance écrite.

Puisque l'*Iḥkām* est le seul ouvrage survécu d'Ibn ʿAbd al-Ghafūr, nous ne savons pas à quelles traditions Ibn al-Abbār est en train de se référer. Néanmoins, il est donné de penser il s'agit d'une transmission de poésie et d'*adab*, compte tenu du fait qu'Ibn ʿAbd al-Ghafūr, qui arborait aussi le titre de *faqīh*, était surtout connu comme *adīb* et que, pour citer Ibn al-Abbār, « les matières de l'*adab* étaient sa marchandise à lui ».

4 Ibn al-ʿArabī dans la *Fahrasa* d'Ibn Khayr

Qu'Ibn al-ʿArabī ait été aussi la source de quelques autres ouvrages de l'*adab* mentionnés dans l'*Iḥkām* est, à son tour, confirmé par le quatrième auteur que nous allons considérer, Abū Bakr Muḥammad Ibn Khayr (m. 575/1179). Né à Fès mais naturalisé andalou, Ibn Khayr fut le seul parmi eux à survivre à l'écroulement du régime almoravide en al-Andalus et à poursuivre son activité

ʿAbd al-Ghafūr (citée tout après), qui le décrit comme ami d'Ibn Bassām et des *udabāʾ* « de sa *ṭabaqa* », en a fait le contemporain de ceux-ci.

savante sous le nouveau gouvernement almohade. Son ouvrage le plus connu, la *Fahrasa*, répertorie les environ 1400 ouvrages qu'il avait appris, de différentes manières, de ses 103 maitres, andalous et orientaux[35]. Elle représente, à bien des égards, le témoin principal et le plus complet de la vie intellectuelle en al-Andalus sous la domination almoravide, dont elle décrit les chaînes de transmissions, les trajectoires et les lieux de l'enseignement, les réseaux des maitres et, ainsi faisant, les courants et les tendances dominants. Abū Bakr Ibn al-ʿArabī, déjà décédé au moment de la rédaction de l'ouvrage[36], y apparaît souvent au nombre des maîtres révérés d'Ibn Khayr, auquel il transmit un grand nombre d'ouvrages de sujets divers.

Parmi ceux-ci, on ne considérera ici que ceux afférant aux champs de l'*adab*, de la poésie et de la lexicographie, cette dernière se superposant, à bien des égards, à la pratique des deux autres, en commençant par l'auteur oriental dont on a suivi la trace de sa réception en al-Andalus, c'est à dire al-Mujīd al-ʿAsqalānī. Ibn Khayr atteste la transmission de son œuvre complète par Ibn al-ʿArabī – poésie, sermons et épîtres -, en ajoutant que ce dernier l'avait reçue de ʿAlī b. Ḥarīz al-Shahīd, au sujet duquel nous n'avons pas trouvé de renseignements. Ce que ce témoignage confirme, une fois pour toutes, c'est qu'Ibn al-ʿArabī est la source de l'admiration, peut-être excessive, que les Andalous de son époque avaient nourri pour cet *adīb* et *kātib* fatimide.

Parmi les grands auteurs dont la connaissance est faite remonter à Ibn al-ʿArabī, c'est al-Maʿarrī qui prime avec son œuvre plus ou moins complète, qu'Ibn al-ʿArabī avait reçue de l'un des disciples directs du poète syrien, Abū Zakarīyāʾ al-Tibrīzī. Ibn Khayr enregistre aussi séparément le *Saqṭ al-zand*, qu'il avait également reçu non seulement à travers Ibn al-ʿArabī mais aussi par deux *riwāyāt* concurrentes, remontant à Ibn al-Sīd al-Baṭalyawsī ; et la *Khuṭbat al-Faṣīḥ*, qu'il avait aussi reçue par une *riwāya* andalouse remontant à un autre disciple d'al-Maʿarrī, Abū ʿAmr al-Safāqusī. À remarquer qu'aussi bien le *Saqṭ* que la *Khuṭbat al-faṣīḥ*, à savoir, le *Prologue* au *Livre de l'Éloquent* du grammairien de l'école de Kufa Abū l-ʿAbbās Aḥmad b. Yaḥyā, dit Thaʿlab, (200-291/815-904), avaient fait l'objet d'une tentative d'émulation de la part de l'auteur de l'*Iḥkām*, résultant dans la conservation de deux extraits de la *Khuṭba*, laquelle aurait autrement disparu sans laisser de trace[37]. Ibn Khayr attribue aussi à Ibn al-ʿA-

35 Sur Ibn Khayr et la *Fahrasa*, v. BA iii, 719-723, no. 712 (J. Vizcaino Plaza).

36 La mention de son nom est souvent accompagnée par la formule *raḥimahu Llāh*, p.ex. dans l'*isnād* de la *Jadhwat al-muqtabis* d'al-Ḥumaydī, p. 281.

37 Ibn ʿAbd al-Ghafūr al-Kalāʿī, *Iḥkām*, 38 et 180-181, où il explique que « c'est un noble prologue, renfermant beaucoup de science et d'*adab*, qui intègre les vocables recensés par Thaʿlab dans son *Faṣīḥ* », ce qui est confirmé par Ibn al-ʿAdīm, pour qui cette *khuṭba* aurait intégré chaque terme du *Faṣīḥ* « dans un morceau plus étendu de prose littéraire » : *Inṣāf*,

rabī, par deux fois, la transmission du *Faraj baʿd al-shidda* d'Abū ʿAlī al-Tanūkhī, auteur que les Andalous avaient connu grâce à la *Yatīma* d'al-Thaʿālibī[38].

Ibn al-ʿArabī avait transmis à Ibn Khayr deux autres ouvrages mentionnés dans l'*Iḥkām*, le *Mujmal al-lugha* et les *Futyā faqīh al-ʿarab*, du grammairien et lexicographe du IV/X siècle Abū l-Ḥusayn Ibn Fāris. C'est notamment d'Ibn Fāris, en tant qu'auteur du *Mujmal*, qu'Ibn ʿAbd al-Ghafūr dérive, dans le chapitre le plus important de son traité, l'argument en défense des Modernes contre la supériorité des Anciens[39] et l'autorisation à se poser lui-même en auteur à part entière.

Puisque la grande poésie orientale était déjà arrivée en al-Andalus entre la fin de l'époque émirale et les califats d'al-Nāṣir et d'al-Mustanṣir, on ne trouvera pas de *dīwān*-s classiques parmi les *riwāyāt* attribuées à Ibn al-ʿArabī, sinon une sélection (*mukhtār*) d'Abū l-ʿAtāhiya, auteur apprécié par al-Maʿarrī, faite par un Abū l-Ḥasan ʿAlī b. Ṭalḥa, autrement inconnu. D'al-Tibrīzī, mentionné plus haut comme lien avec al-Maʿarrī, Ibn al-ʿArabī avait aussi transmis le commentaire au *dīwān* d'al-Mutanabbī, qui avait été sans doute connu et utilisé par Ibn ʿAbd al-Ghafūr, lequel se vantait à son tour d'être spécialiste d'al-Mutanabbī.

En revanche, Ibn Khayr attribue à Ibn al-ʿArabī la transmission du *dīwān* d'Ibn Darrāj al-Qashṭallī, poète andalou célèbre de l'époque amiride, dans deux *riwāyāt* distinctes, remontant l'une à Ibn Darrāj lui-même, à travers le père d'Ibn al-ʿArabī, l'autre à Ibn Ḥazm, à travers al-Ḥumaydī et son disciple oriental Ibn Ṭarkhān[40]. Abū Bakr Ibn al-ʿArabī est également mentionné, s'agissant de la célèbre *qaṣīda rāʾiya* d'Abū Marwān Ibn al-Jazīrī, poète et vizir amiride lui aussi, mort dans la prison d'al-Manṣūr b. Abī ʿĀmir[41]. Ibn Khayr déclare avoir reçu ce poème (dont il cite, cas rare, le vers initial) d'Abū l-Ḥasan ʿAbbād b. Sirḥān, qui déclare à son tour l'avoir reçu par deux filières distinctes, incluant Abū Bakr Ibn al-ʿArabī, le père de celui-ci et, peut-être, le grand-père, jusque

---

apud *Taʿrīf al-qudamāʾ*, 530, ajoutant que cet ouvrage aurait consisté de quinze fascicules ; aussi p. 201, p. 449 ; Smoor, P., « Enigmatic allusion and double meaning in Maʿarrī's newly-discovered *Letter of a horse and a mule* », dans *Journal of Arabic Literature* 12 (1981), 49-73, à p. 61.

38 Ibn Bassām, *al-Dhakhīra*, IV/ii, 576-577. Le sujet de la délivrance après l'épreuve était bien présent à l'esprit du *qāḍī*, qui lui consacre une section du Ibn al-ʿArabī, *Sirāj al-muridīn* iv, 311-313.

39 Cette tirade provient de la *risāla* d'Ibn Fāris en défense de Abū l-Ḥasan Muḥammad b. ʿAlī al-ʿIjlī, auteur d'un ouvrage au sujet de la *ḥamāsa*, contre son critique Abū ʿAmr Muḥammad b. Saʿīd.

40 Ibn Khayr, *Fahrasa*, éds. B. ʿAwād Maʿrūf, M. Bashshār ʿAwād , 508, no. 1205. Le *dīwān* d'Ibn Darrāj et la *qaṣīda* faisaient sans doute partie du *Kitāb Shuʿarāʾ al-Andalus* qui est attribué à Ibn al-ʿArabī (v. BA ii, 156, no. 91).

41 Ibn Khayr, *Fahrasa*, éds. B. ʿAwād Maʿrūf, M. Bashshār ʿAwād, 503, no. 1189.

des membres de la *khidma* califale cordouane du calibre d'Ibn al-Ḥayy et Ibn al-Qūṭiyya, l'historien califale auteur de l'*Iftitāḥ al-Andalus*. Notons qu'Ibn Khayr s'avoue surpris par cet *isnād*, car Ibn al-ʿArabī ne lui avait pas parlé d'Ibn al-Qūṭiyya, mais seulement d'Ibn al-Ḥayy[42].

Pour conclure, on pourra ajouter à ces témoignages d'auteur ce qu'Abū l-Ḥasan Ibn Saʿīd al-Maghribī (m. 685/1286) a inclus dans ses deux anthologies, *al-Mughrib fī ḥulā al-Maghrib* et *Rāyāt al-mubarrizīn*[43], citant d'autres auteurs contemporains d'Ibn al-ʿArabī, dont l'œuvre n'a pas survécu. Le premier est Abū Muḥammad ʿAbd Allāh al-Ḥijāri, vécu entre la fin du V et la première moitié du VI siècle[44], auteur du *Mushib fī faḍā'il al-Maghrib*. Dit al-Ḥijārī, avec une hyperbole alambiquée, que « si on ne pouvait attribuer à Séville personne d'autre que cet *imām* illustre, cela serait tout de même pour elle une source de fierté telle que même les yeux fatigués ne s'en détourneraient guère ». Ensuite, Abū ʿAmr ʿUthmān Ibn al-Imām (m. ap. 550/1155-6), de Silves, auteur du *Simṭ al-jumān*, lequel, après avoir fait l'éloge de l'immense savoir d'Ibn al-ʿArabī, fait allusion au poème de nostalgie pour Bagdad traduit plus haut. Ibn Saʿīd transcrit aussi dans le *Mughrib* trois courts poèmes d'occasion de sujet érotique, deux desquels sont également compris dans les *Rāyāt*. Al-Maqqarī (m. 1041/1632), qui reprend l'ensemble de ces poèmes dans la longue section sur Ibn al-ʿArabī dans le *Nafḥ*, en y ajoutant d'autres, observe prudemment que « certains pensent que ces versicules ne sont pas de lui, mais qu'on les a façonnés de manière à le faire croire ». En tout état de cause, cela signalerait un aspect de la réputation du *qāḍī* auprès de ses contemporains[45], suffisant à justifier leur mention, du moins dans le cas des trois cités par Ibn Saʿīd.

Le premier de ces petits poèmes, qu'on ne trouve que dans le *Mughrib*, est adressé au « fils d'un émir almoravide (*ibn amīr min umarā' al-mulaththimīn*) », qui, l'ayant défié à la course à cheval, avait agité sa lance sur lui pour plaisanter :

42 Ibn al-ʿArabī cite six vers de la *qaṣīda* dans le *Sirāj*, sous le titre de *Qaṣīdat al-ādāb wa-l-sunna*, d'après un *isnād* comprenant, en plus de son père, le seul Ibn al-Ḥayy (*Sirāj al-muridīn* iii, 200-201).

43 Ibn Saʿīd al-Maghribī, *al-Mughrib fī ḥulā al-Maghrib*, éd. Sh. Ḍayf, 2 vols., le Caire, 1953-1955, i, 254-255, no. 177, (l'ensemble de cette notice est repris par al-Maqqarī, *Nafḥ al-ṭīb*, ii, 26-27) ; *Rāyāt al-mubarrizīn wa ghāyāt al-mumayyazīn*, éd. M.R. al-Dāya, Damas, 1987, 60-61; trad. esp. par E. García Gomez, *El libro de las banderas de los campeones*, Barcelone, 1978, 139.

44 Sur lui, l'article biographique dans BA i, 454-456, no. 157 (J. Lirola Delgado).

45 Ibn al-ʿArabī n'avait pourtant pas hésité à définir la poésie érotique (*al-ashʿār al-ghazaliyya*) un "instrument du diable" (*min silāḥ al-shayṭān*) (dans *Sirāj al-muridīn* ii, 223).

Une svelte gazelle fait vibrer sa lance sur moi,
Joueuse, se moquant du cœur des simples mortels.
Si ce n'était qu'une lance, alors tu ne devrais pas t'en garder,
Mais c'est une lance, deux, trois même !

L'occasion de l'autre distique, provenant d'Ibn al-Imām, est la rencontre avec un jeune et bel esclave aux vêtements rêches :

Il s'est habillé de laine rugueuse (*al-ṣūfa*)[46] pour me déplaire,
Et est venu chez nous, tout maigre, le visage sévère.
J'ai dit : « C'est bien lui, on t'a reconnu, voyons !
Une housse disgracieuse n'est pas un défaut du cheval.
Tu es beau dans n'importe quelle robe,
Comment la beauté s'habille, cela nous est égal !

Le troisième morceau, qui, d'après les *Rāyāt*, avait été rapporté par al-Shaqundī (m. 629/1231) dans son *Ẓarf* aujourd'hui perdu[47], s'adresse à quelqu'un qui avait suggéré à Ibn al-ʿArabī de répandre du sable sur la feuille d'une lettre qu'il était en train d'écrire[48]:

Pas la peine de le défigurer en lui jetant du sable,
Car pour cela il lui suffit le souffle ardent[49] de cette passion.
Le sable qu'on disperserait sur lui
Ne serait qu'un furoncle sur sa belle joue arrondie.

5 Quelques considérations finales

On distinguera trois filières principales dans l'ensemble des témoignages apportés au sujet de l'activité d'Ibn al-ʿArabī comme *adīb* et transmetteur de l'*adab* et de la poésie. La première, liée à son histoire familiale et à la conservation de l'héritage du siècle des *taifas*, recoupe le sillon hazmien, à travers

46 Ibn al-ʿArabī avait consacré un chapitre du *Sirāj* à la question de savoir si le *ṣūf*, soit le vêtement de laine grossière, était en relation avec les mystiques soufis, et avait porté quelques citations poétiques à l'appui (Ibn al-ʿArabī, *Sirāj al-muridīn* i, 110-113).

47 Sans doute, le *Kitāb ṭuraf al-ẓurafā*ʾ, pour lequel v. HATA : https://www.eea.csic.es/red/hata/obra.php?ido=21000 (consulté en ligne le 9 décembre 2022).

48 C'est à dire, pour faire sécher l'encre ; la réponse est un double sens érotique (lettre = jeune homme), permis par le genre masculin du mot *kitāb*.

49 Le mot utilisé, *habūb*, indique le vent qui soulève la poussière du désert.

al-Ḥumaydī principalement, et cela bien qu'Ibn al-ʿArabī eusse notoirement émis un jugement très négatif sur Ibn Ḥazm, dont son père avait été l'un des derniers élèves[50].

La deuxième porte sur la connaissance et diffusion de l'œuvre d'al-Maʿarrī en al-Andalus, initiative à laquelle Ibn al-ʿArabī consacre un effort considérable, sans reculer devant le combat critique, comme celui qu'il engage avec Ibn al-Sīd al-Baṭalyawsī. L'appréciation pour al-Maʿarrī, frôlant le fanatisme chez un auteur mineur comme Ibn ʿAbd al-Ghafūr, semble quand même poser un problème. Il est évident que notre lecture d'al-Maʿarrī, soulignant la profondeur de sa vision morale, ne coïncide pas avec celle qu'en ont donné les lettrés andalous du VI/XII siècle, pour qui al-Maʿarrī était surtout le maître inégalé du haut style et des acrobaties linguistiques les plus audacieuses. Si, pour connaître vraiment les enjeux de l'opposition entre Ibn al-ʿArabī et Ibn al-Sīd au sujet d'al-Maʿarrī, il faudrait lire entre les lignes du *Radd*, on peut tout de même se référer à l'influence que le *qāḍī* eut sur Ibn ʿAbd al-Ghafūr pour saisir certains aspects de son jugement critique, car, si l'on peut dire que l'*Iḥkām* a été conçu sous le signe d'al-Maʿarrī, il s'agit sans doute, pour paraphraser le titre de l'ouvrage pionnier de Salvador Peña[51], d'al-Maʿarrī selon Ibn al-ʿArabī.

Finalement, l'importance primordiale des faits de style chez Ibn al-ʿArabī est confirmée par sa transmission et diffusion de l'œuvre du *kātib adīb* fatimide al-Mujīd al-ʿAsqalānī, dans lequel il dut voir le porte-parole du style recherché dans la prose rimée, le *sajʿ*, qu'il souhaitait proposer comme modèle aux *kuttāb* andalous de son époque.

Nul doute qu'Ibn al-ʿArabī aurait pu être compté parmi les grands *udabāʾ kuttāb* de son temps s'il avait choisi cette carrière, au demeurant périlleuse comme l'expérience de son père lui avait montré, au lieu de celle dans le champ juridico-religieux, plus valorisante et prestigieuse. Ses œuvres savantes contiennent plusieurs morceaux poétiques, qu'il serait intéressant de recenser, et elles trahissent parfois une sensibilité inattendue, notamment dans les nombreuses digressions autobiographiques. Les quatre auteurs qu'on vient de considérer, attestent d'ailleurs des talents littéraires du *qāḍī* et de sa réputation de connaisseur et de juge de la littérature, mais aussi de sa personnalité charismatique et impérieuse, capable de lancer des auteurs dans les milieux lettrés andalous, ou alors, de tâcher des réputations. Pour mieux comprendre

---

50 V. pour tous, Ibn al-ʿArabī, *al-ʿAwāṣim*, 249: « à mon retour [en al-Andalus] je pus constater que l'Occident avait été saturé par le verbe zahirite, par œuvre d'un être insignifiant basé dans le pays de Séville, connu comme Ibn Ḥazm » (cité aussi par al-Dhahabī, *Siyar*, xviii, 188).

51 Peña, Salvador, *Maʿarrī según Baṭalyawsī: Crítica y poética en al-Andalus, siglo XI*, Grenade: Universidad de Granada, 1990.

ce personnage plus grand que nature, mais aussi, plus en général, la mentalité des élites savantes andalouses à l'époque almoravide, il serait alors intéressant d'intégrer dans sa biographie intellectuelle ces aspects apparemment problématiques, mais qui ne présentaient aucune contradiction pour ses contemporains, vis-à-vis de son identité principale de *ʿālim* et de *faqīh*.

Bibliographie

al-Dhahabī, *Siyar aʿlām al-nubalāʾ*, plusieurs éds., 23 vols. + 2 vols. d'indexes, Beyrouth : Mu'assasat al-Risāla, 1406/1986.

BA = *Biblioteca de al-Andalus*, dirigée et éditée par J. Lirola Delgado et J.M. Puerta Vilchez, Almeria : Fundación Ibn Tufayl de Estudios Arabes, 2004-2012, 7 vols. + 3 appendices.

Blankinship, Kevin, «Al-Maʿarrī's esteem in the Islamic West: A preliminary overview», *Al-Masāq* 31 (2019), 253-271.

EI2 = *The Encyclopaedia of Islam, second edition*, P. Bearman, Th. Bianquis, C.E. Bosworth, E. van Donzel et W.P. Heinrichs (éds.), Leyde: Brill, 1954-2004.

EI3 = *The Encyclopaedia of Islam, Three*, Kate Fleet, Gudrun Krämer, Denis Matringe, John Nawas et Devin J. Stewart (éds.), Leyde: Brill, 2007-.

García Gomez, Emilio, *Un éclipse de la poésia en Sevilla: La época almorávide*, Madrid: Real Academia Española, 1945.

HATA = *Historia de los autores y transmisores andalusíes*, éd. par M. Fierro, accessible en ligne : https://www.eea.csic.es/red/hata/index.html

al-Ḥumaydī, *Jadhwat al-muqtabis* , éd. I. al-Abyārī, 2 vols., Le Caire : Dār al-kitāb al-miṣrī, 1403/1982.

Ḥusayn, Ṭaha (dir.), *Taʿrīf al-qudamāʾ bi-Abī l-ʿAlāʾ*, éd. M. al-Saqqāʾ, I. al-Abyārī, H. ʿAbd al-Majīd, Le Caire : Dār al-qawmīya li-l-ṭibāʿa wa-l-našr, 1385/1965 (réimp. de l'éd. 1944/1363).

Ibn ʿAbd al-Ghafūr al-Kalāʿī, *Iḥkām ṣanʿat al-kalām*, éd. M.R. al-Dāya, Beyrouth : Dār al-thaqāfa, 1966.

Ibn al-ʿArabī, Abū Bakr, *al-ʿAwāṣim min al-qawāṣim*, éd. ʿAmmar Ṭālibī, Le Caire : Maktabat al-sana, 1394/1974.

Ibn al-ʿArabī, Abū Bakr, *Qānūn al-taʾwīl*, éd. M. Sulaymānī, Beyrouth : Mu'assasat ʿulūm al-Qur'an, 1406/1986.

Ibn al-ʿArabī, Abū Bakr, *Sirāj al-muridīn fī sabīl al-dīn*, éd. A.A. al-Tawrātī, 4 vols., Tétouan : Dār al-ḥadīth al-kattāniyya, 1436-38/2014-16.

Ibn Bassām, *al-Dhakhīra fī maḥāsin ahl al-Jazīra*, éd. I. ʿAbbās, 4 vols en 8 tomes, Beyrouth: Dar al-thaqāfa, 1398/1978.

Ibn Khāqān, Abū Naṣr al-Fatḥ (dit), *Maṭmaḥ al-anfus*, éd. M. ʿA. Shawābkah, Beyrouth: Dār ʿammār, 1403/1983.

Ibn Khāqān, Abū Naṣr al-Fatḥ (dit), *Qalāʾid al-ʿiqyān*, éd. H.Y. Kharyūsh, 2 vols, Irbid : Maktabat al-Manār, 1989.

Ibn Khayr, *Fahrasa*, éd. F. Codera et J. Ribera, Bagdad, s.l., 1382/1963, (réimp. de l'éd. de Saragosse : Comas, 1893); éds. B. ʿAwād Maʿrūf, M. Bashshār ʿAwād, Tunis : Dār al-gharb al-islāmī, 2009.

Ibn Saʿīd al-Maghribī, *al-Mughrib fī ḥulā al-Maghrib*, éd. Sh. Ḍayf, 2 vols., Le Caire : Dār al-maʿārif, 1953-1955.

Ibn Saʿīd al-Maghribī, *Rāyāt al-mubarrizīn wa ghāyāt al-mumayyazīn*, éd. M.R. al-Dāya, Damas : Tlasdar, 1987 ; trad. esp. de E. García Gomez, El libro de las banderas de los campeones, Barcelone : Seix Barral, 1978 (facs. de l'édition de 1942).

Larsson, Göran, *Ibn García's shuʿūbiyya letter: Ethnic and theological tensions in medieval al-Andalus*, Leyde: Brill, 2003.

al-Maqqarī, *Nafḥ al-ṭīb min ghuṣn al-Andalus al-raṭīb*, éd. I. ʿAbbās, 7 vols. + indexes, Beyrouth: Dār al-thaqāfa, 1388/1968.

Marín, M., « The Princess and the Palace. On Ḥawwaʾ bin Tashufin and Other Women from the Almoravid Royal Family », dans M. Hamilton, N. Silleras-Fernandes (éds), *In and Of the Mediterranean : Medieval and Early Modern Iberian Studies*, Nashville (TN) : Vanderbilt U.P., 2015, p. 29-47.

al-Mutanabbī, *Dīwān al-Mutanabbī bi-sharḥ al-ʿUkbarī*, éds. M. al-Saqqāʾ, I. al-Abyārī, ʿA. H. Shalabī, Beyrouth: Dār al-maʿrifa, 1978.

Peña, Salvador, *Maʿarrī según Baṭalyawsī: Crítica y poética en al-Andalus, siglo XI*, Grenade: Universidad de Granada, 1990.

Henri Pérès, *La poésie andalouse en arabe classique au XIe siècle : ses aspects généraux, sa valeur documentaire*, Paris: Maisonneuve, 1937.

de Salazar y Acha, Jaime, «Urraca: Un nombre egregio en la onomástica altomedieval», dans *En la España medieval* 1 (2006), 29-47.

Smoor, P., «Enigmatic allusion and double meaning in Maʿarrī's newly-discovered *Letter of a horse and a mule*», dans *Journal of Arabic Literature* 12 (1981), 49-73.

Soravia, Bruna «L'introduzione d'Ibn Bassām al *Kitāb al-Dhakhīra fī maḥāsin ahl al-Ǧazīra*: Presentazione e traduzione», dans F. Díaz Esteban (éd.), *Bataliús II: Nuevos estudios sobre el reino Taifa de Badajoz*, Madrid: Letrúmero, 1999, 253-271.

Soravia, Bruna, «A portrait of the *ʿālim* as a young man: The formative years of Ibn Ḥazm, 404/1013-420/1029», dans C. Adang, M. Fierro et S. Schmidtke (éds.), *Ibn Ḥazm of Cordoba: The life and works of a controversial thinker*, Leyde: Brill, 2013, 25-50.

Yāqūt, *Muʿjam al-Buldān*, s.n., 5 vols, Beyrouth: Dār Ṣādir, 1397/1977.

10

أبو بكر ابن العربي المعافري

رائد التأليف في « فن المسلسلات الحديثية » بالغرب الإسلامي

نورالدين الحميدي

[Ibn al-ʿArabī, the Pioneer of Writings on *musalsalāt ḥadīthiyya* in the Islamic West]

[*Nourredine Elhmidy*]

1 مقدمة

أبدى المغاربةُ فرْطَ عنايةٍ بالحديث الشريفِ وعلومه، وجارَوْا المشارقةَ في تصانيفهم المُخْتَرعةِ في هذا المضمار، وحاولوا رغم النأيِ عن مهبط الوحي ومنبع الرواية، أن يُسْفِروا عن البَذِّ والشفوف فيما جمعوه وتصرَّفوا فيه، فألَّفُوا في جمع متون السنة وانتقائها، إذ هي القصد من رواية الحديث، والثمرةُ المرجوَّةُ من التأصيل لعلومه وقواعده، فبها يقع التدين والتعبد.

ثم أمعنوا في باب التصنيف تفننا، فراعَوْا معانٍ زائدة في حاليْ المروي والراوي، غير الصحة والضعف، مثل: العلو والنزول، والتسلسل وصفا أو وقولا، ورواية الأكابر عن الأصاغر أو التدبيج، وغيرها من الألوان المُسْتَلطفة، والنِّكات المُسْتَمْلَحة في الرواية، مما عُني أرباب النقل والحديث بجمعها وحشدها، وأثني عِنان القولَ أَحَصُره على فن المسلسلات الحديثية، إذ هو مَعْقِدُ هذا البحثِ ومَحَزُّهُ. فالإمام أبو بكر ابنُ العربي المعافري الإشبيلي ؛ يعدُّ رائد المغاربة المؤلفين في هذا الفن، إذ تمحَّضتْ له الأولية في هذه الحلبة رواية وتأليفا ونشرا[1]، ولم أجد فيما عُنيتُ به من التراث الحديثي المغربي - بعُدوتيه الأندلس والمغرب - ما يُسْقط هذه المنقبةَ عن هذا الإمام، ويبقى قرينه الحافظ أبو علي ابنُ سُكَّرَة الصدفي السرقسطي نظيره في بثِّ المسلسلات وإذاعتها بالأندلس والمغرب، ومن تتبع ما رُوِيَ عنه من المسلسلات، يُلْفيها مُشْعِرَةً بأن أبا علي الصدفي جمعها في مجموع واحد، لكن لا نجزم بأن جمعه

1 وقد تهيأ لي - بفضل الله تعالى - إعادة بناء كتابه « المسلسلات »، ضمن المجموع الذي أصدرته بعنوان (الإمام ابن العربي، فقيه المحدثين وشيخ المسندين بالأندلس) عن دار الغرب الإسلامي، سنة: 2020م.

 | DOI:10.1163/9789004706859_012

لها كان على سبيل التأليف المستقل، فأبو علي الصدفي لا نعرف له كبير تواليف سوى ما خرَّج له تلاميذه أو من جاء بعدهم[2].

وأقدم مغربي - أوقفنا عليه البحث - يُروى عنه حديث مسلسل، هو: أبو مروان عبد الملك الطُبُنِّي (457هـ\1064م)، فهو تحمَّل « مسلسلات » أبي العباس ابنِ بُندار الرازي (409هـ\1018م)، وأدخلها إلى الأندلس[3]، ومن هذا الوجه خرَّج ابنُ الطيلسان جملة من المسلسلات في كتابه. وممن أدخل هذه المسلسلات وحدث بها في الأندلس؛ أبو العباس العذُري (478هـ\1085م)[4].

وممن وردت بعضُ الأحاديث المسلسلة من طريقه؛ أبو بكر جُماهر بن عبد الرحمن الحَجْري الطُّلَيْطُلي (466هـ\1073م)، عن شيخه أبي عبد الله القُضاعي (454هـ\1062م)، لكن هي مدرجة ضمن كتابه « مسند الشهاب »، ولم تقع ضمن تواليف مقصود بالتسلسل ومُفْرَدٍ في هذا الفن، وقد أخرج ابنُ العربي بعضَها في « مسلسلاته » بإسناده المشرقي.

وإن تأكدتْ عنايةُ أبي مروان الطبني وأبي العباس العذريّ برواية المسلسلات وإدخالها، فهذا لا يدفع عن المغاربة إبطاءهم في العناية بهذا الفنِّ رواية وتصنيفا، إذ إن أوَّلَ وضعٍ مشرقي مذكور في المسلسلات، يُنمى للإمام أبي حاتم ابن حبان البُسْتي (354هـ\962م)[5]، وإزاء ذلك ظل الغرب الإسلامي خِلوًا من التأليف في « المسلسلات » أزمانا، إلى أن هتَكَ الإمام ابنِ العربي سِتْرَ ورود هذا المشرع البُهْم، فتوارد على نهجه جماعة من كبار حفاظ الأندلس ومحدثي المغرب، وتعاطوا رُقيَّ مرقبه السامق، خاصة من طبقةِ تلاميذِ تلاميذِهِ، فكان مُعَوَّلهم في روايتها، ومُعْتَمَدَهم في سلسلتها: كتابُ ابنِ العربي.

وممن زخرت كتبهم في هذا الفن بمروياته؛ أبو الربيع ابن سالم الكَلاعي البلنسي (634هـ\1237م)، وأبو القاسم ابن الطيلسان القرطبي (643هـ\1246م)، وأبو بكر ابنُ مَسْدي الغرناطي (663هـ\1264م)[6]، وأبو علي ابن أبي الأحوص الغرناطي (679هـ\1280م)[7]، وغيرهم ممن لا أطيل بذكرهم.

2 أستثني من ذلك تواليف لِطاف، لا تتعدى خمسة، إحداها في جمع طرق حديث « ما من عبد مسلم يذنب ذنبا، ثم يتوضأ فيحسن الوضوء »، والآخر ثَبَتٌ بمشيخته المشارقة الذين استجازهم لابن غشليان، ومشيخة ابن الجارود، وغيرها.

3 الفهرسة: 148، لابن خير الإشبيلي.

4 مل العيبة - الجزء الثاني -:186، لابن رشيد السبتي.

5 سير أعلام النبلاء: 452/20.

6 صدر بتحقيقنا ودرسنا كتاب عن هذا الإمام، نشرته دار البشائر الإسلامية ومكتبة نظام يعقوبي الخاصة.

7 صدر بتأليفي وعنايتي كتاب عن هذا الإمام والتعريف به وبتراثه، ذكرت في تمهيد تحقيقي لـ« مسلسلات ابن سالم » ستة عشر تأليفا مغربيا في المسلسلات، كلها خرَّجت عن مسلسلات ابن العربي على الأرجح.

2 مصادر ابنِ العربي في « مسلسلاته »

اتَّكأ ابنُ العربي في هذا السبيل على جملة من مشيخته المشرقية، ومنهم من ألف في هذا الضرب مثل: قوَّام السنة أبو القاسم التَّيْمي الأصبهاني (535هـ\1140م)، وتوجد نسختان من « مسلسلاته » [8]، والظاهر بعد تصفحها أنها مختصرة ومنتخبة.

كما صدر عن « مسلسلات » عبد العزيز الكتاني الدمشقي (466هـ\1073م) بعد طول ورود، فبث منها في توليفِهِ مروياتٍ كثيرة، تحصَّلتْ له روايةً عن شيخه أبي القاسم ابن أبي الجن الدمشقي النسيب (508هـ\1114م).

وممن اعتنى بتحمُّلِ « مسلسلاته »؛ ما وقع مسلسلا للحافظ أبي عبد الله القُضاعي المصري (454هـ\1159م) في « مسند الشهاب »، وهو يحملها بشرطها عن شيخه عبد الجليل الساوي (493هـ\1099م)، والظاهر أن ابنَ العربي حرص على سماع مسلسلات الكتاب بشرطها دون غيرها، فأبو بكر ابنُ خير يفيد أن ابنَ العربي لا يحمل الكتاب كاملا بالسماع [9]، وهذا يومئ إلى أن نشأة مشروع كتاب « المسلسلات » عند ابنِ العربي كانت أيام الطلب والتحصيل.[10]

وأكثرَ ابنُ العربي كذلك عن « مسلسلات » أبي محمد الخلال البغدادي (439هـ\1047م)، وطريقه إليها عن شيخه ابنِ الطُّيُوري، وتُعَدّ مصدرا أصيلا في كتابه لوفرة ما أسند من جِهَتِها.

وممن متح من معينه؛ أبو بكر الخطيب البغدادي (463هـ\1071م) في « مسلسلاته »، لكن عن قُلٍّ ونَزْرٍ.

والمستغرب أن هذه الكتب في هذا الفن المستطرف لا توجد ضمن مرويات المغاربة من طريق ابنِ العربي، ولم يعرِّجْ على إحداها ابنُ خير ومن أتى بعده على قلتهم، فأرجح - من غير جزم - أن ابنَ العربي لم يُدْخِلْ هذه الكتب إلى المغرب، بل اكتفى بإدخال ما انتخب منها واستصفى، مما يستحق أن يُنَسَّقَ في مصنف مفرد، والسبب كما نصصت عليه في تقدمة « مسلسلات ابنِ سالم الكَلاَعي » أن هذا الفن

---

8 محفوظة في المكتبة الظاهرية، ضمن مجاميع العمرية رقم: (34) ورقم: (51).

9 الفهرسة:156.

10 وهناك عبارات كثيرة لابن العربي تفيد أن بعض تواليفه كانت مجاميع مسودة محفوظة عنده، انتهض لتنقيحها وتصحيحها لما تصدى للإفادة والتدريس حتى تكون تواليف تُحْمل عنه وتعزى إليه، يُنظر مقدمة « جزء ليس من البر » حيث ذكر أن طرق الحديث كانت مجموعة عنده قِدما، صدر بتحقيقنا ضمن مجموع عن دار الغرب الإسلامي، سنة: 2020م.

من مُلح العلم وليس من متينه، فلا يُشْتَغل به عن أصول العلم ولبابه، ومما يؤيد هذا أن ابنَ العربي أغفل جملة كثيرة من المسلسلات، وإنما خرّج لنفسه ما رآه صحيحا وحقيقا بالرواية.

3 مُضمَّن « المسلسلات » ومنهجُ ابنِ العربي في تأليفها

أُبدي بدءًا التعجب من إغفال القاضي ابنِ العربي لحديث الرحمة المسلسل بالأولية، وأكاد أجزم أنه لم يخرجه في كتابه « المسلسلات »، مع أنه من أصحِّ المسلسلات وأجلِّها وأشهرها، فهذا أبو الربيع ابنُ سالم شديد الحرص على أن يُخَرِّج الحديثَ المسلسل من طريق ابنِ العربي، لأنه يصل إليه من قِبَلِ أجلِّ مشيخته وأعلاهم ؛ أبو القاسم ابنُ حُبيش، فلما ابتدأ تصنيفَه بحديث الأولية، لم يخرجه من طريق ابنِ العربي، وكذا ابنُ الطيلسان ضمَّن كتابَهُ جُلَّ « مسلسلات » ابن العربي لكن لم يسند حديث الأولية عنه، بل هذا تلميذه ابنُ بُشْكُوال لما خرج المسلسل بالأولية في « الفوائد المنتخبة »، لم يسلسلها عن ابنِ العربي مع احتفاله بمروياته ومُسنداته، وهؤلاء كلهم نوَّعوا في أسانيدهم في المسلسل بالأولية وعدَّدوا في طرقه، ورغم ذلك لم يأت لابن العربي في جميعها ذكر ولا إيراد، ولا ندري لم أخلى ابنُ العربي « مسلسلاته » من ذلكم الحديث.

وأفيض القول في مُضمَّنِ « مسلسلات » ابنِ العربي وما اشتملت عليه؛ فأكاد أجزم أن ما وقفت عليه ضمن مسلسلات ابنِ سالم وابنِ الطيلسان وغيرهما، محتوٍ على كلِّ كتاب ابنِ العربي، وهذا يخفِّفُ أسى فقد وضياع هذا السفر الأندلسي المحظيِّ بالأسبقية في هذا الباب، ويسلي الوالهَ المتشوق لمرآه إلى حين إقرار العين بالكشف عنه، وجلوه من خبئه للناظر المرتقِب.

وقد قرن ابنُ العربي روايته للمسلسلات بإيراد ما يشهد لها من الأحاديث مما عَرِيَ من وصف التسلسل، تعضيدا لأصل تلك المسلسلات وتقوية لمعناها. وقد أفادنا هذا صنيعه في المسلسل رقم: (2)[11] حيث أردفه بما يشهد لمعناه، ونصَّ عليه الحافظ أبو عبد الله العبدري المراكشي، حيث قال بعد أن أورد المسلسل بالسؤال عن الاسم والكنية والبلد: وذكر ابنُ العربي بعَقِبِ هذا الحديث حديثين في معناه، أحدهما قوله صلى الله عليه وسلم: « استكثروا من الإخوان، فإن لكل مؤمن شفاعة »، والآخر قوله صلى الله عليه سولم: « إذا آخى الرجل الرجل فليسأله عن اسمه، واسم أبيه، وممن هو فإنه أوصل للمودة »[12].

---

11 الإمام أبو بكر ابن العربي، فقيه المحدثين وشيخ المسندين بالأندلس.

12 الجواهر المكللة: 285.

وهذه الخصلة المفترعة انتهض لمحاكاة حسنها، كل من حذا حذوَ ابنِ العربي في التصنيف في المسلسلات من المغاربة.

ووقفت على نصوص قد يستفاد منها أن ابنَ العربي أورد في كتابه بعض الإنشادات والأشعار[13]، وهذا ما درج عليه جلُّ المصنفين في « المسلسلات »، حيث أوردوا فيها مُنشدات مسلسلة وغير مسلسلة، إذ منهم من راعى تسلسل صيغة الإنشاد، ومنهم من اعتبر وصفا زائدا على مجرد صيغة الإنشاد.

ومما نُسب لـ« مسلسلات » ابن العربي؛ الحديث المسلسل بالسبحة، قال السخاوي: وهكذا رواه القاضي عياض في « مشيخته » عن القاضي أبي بكر ابن العربي، فيما كتبه له سمعت أبا إسحاق الحبَّال به[14].

وقد دخل الوهم على السخاوي (902هـ\1496م) في هذا الكلام من جهتين، الأولى: أن القاضي عياض إنما أسند هذا المسلسل عن شيخِه أبي علي ابنِ الأنماطي المصري كما في « الغنية في مشيخته »[15]، ولا أدري كيف تطرَّق الوهم إلى السخاوي في هذا، ولعله اعتمد على نسخة سقيمة من « الغنية »، ومما يُقْطَعُ به على خطأ السخاويِّ في هذه النسبة، أن الفقيه العلامة أبا الفضل قاسم العُقباني التلمساني (845هـ\1441م)، نقل هذا السندَ عينَهُ من « الغنية » في رسالة له في حكم الذكر جماعة[16]، وأورد الإسناد على الجادة.

الثانية: أن ابنَ العربي ليس له رواية عن أبي إسحاق الحبال المصري (482هـ\1089م) قطعا، فهو لا يروي عنه إلا بواسطة، فهو لم يحمل عنه بأي وجه من وجوه التحمل، ولا أدري كيف غاب هذا عن السخاوي. ويُعْطَفُ على ما سبق أن القاضي عياض انخرم شرط التسلسل بينه وبين ثلاثة رجال ممن فوقه في هذا الإسناد، فهو لم يقع له بشرط التسلسل، إنما دخل في عموم إجازة ابن الأنماطي له من مصر. وتسلسل لابن الطيلسان من جهة القاضي عياض على الوجه المذكور[17].

وأوغل المتأخرون ممن ألف في « المسلسلات » في الوهم، وهم ممن يُعَوِّلُ على السخاوي في هذا الفن، فزعموا أن ابنَ العربي أثبت هذا المسلسل في « مسلسلاته »، قال ابنُ الطيب الشرقاوي

---

13 أفاد ابنُ عبد الملك المراكشي في (الذيل والتكملة،223/5) أن ابنَ حوط الله أبا محمد روى إنشادة في « مسلسلاته » من طريق ابن العربي.

14 الجواهر المكللة: 285.

15 الغنية: 180.

16 هذه الرسالة هي جواب على سؤال، أثبتها بتمامها الونشريسي في المعيار: 58/1.

17 الجواهر المفصلات: 62.

(1170هـ\1756م): مُسَلْسَلُ هَذَا الأثر أَجَلُّ المسلسلات، وأورده القاضي عياض في « مشيخته »، وإمام الحفاظ القاضي أبو بكر ابن العربي في « مسلسلاته »، كالكتاني، والسلفي، وأبي الحسن الأنماطي.[18]

وهو إبعاد في الوهم وإيغال في الخطأ، اتبعه فيه من جاء بعده، باعثه التقليد من غير تحر وتثبت. ومما أكاد الجزم بخروجه عن شرط كتاب ابن العربي في التسلسل، الأثر المسلسل بحذو مثال النعل النبوية الشريفة، وهو مما رواه عنه ابنُ خير الإشبيلي وابنُ بشكوال وابن عبد الملك المراكشي[19]، ولكني أثبته لأنه داخل في هذا الفن، لا أن ابنَ العربي أورده ضمن ما انتخبه من المسلسلات في كتابه.

ومما يتصل بما سلف؛ الحديث المسلسل بآل البيت الشرفاء؛ وهو مما تفرد بسلسلته ابنُ الطيلسان من طريق ابنِ العربي، وقد أشعر ابنُ الطيلسان أنه ليس مما أخرجه ابنُ العربي في « المسلسلات »، حيث اعتذر عن إخراجه بكونه من رواية حَفَدةِ بيتٍ واحد، ولو أخرجه ابنُ العربي لما احْتِيجَ إلى هذا الاعتذار، ومما يدل على ذلك أن ابنَ الطيلسان لم يروه من طريق ابن علوش وابن وهبون الإشبيليين، وهما معتمده في الرواية عن « مسلسلات ابن العربي »، وإنما رواه عن أبي الحسن الشَّقُّوري، وليس له في « مسلسلات ابنِ الطليسان » سوى هذا الحديث الفرد.

وقد أخرج ابنُ العربي أثرا مسلسلا بآل البيت، لكن لم يعتبر في تسلسله رواية آل البيت عن بعضهم البعض، لأن هذا الشرط ينخرم فيما بينه وبين ست طبقات ممن فوقه، إنما اعتبر فيه التسلسل بالسماع، إذ هو مسلسل بالسماع من مُبْتدئه إلى منتهاه، والأثر مما تواطأ على إدراجه في المسلسلات جميعُ من جاء بعد ابنِ العربي اتِّكاءًا عليه.

ومما هو أدخل وأمكن في خُلُوِّ كتاب ابن العربي منه، المسلسلات الصوفية التي رُوِيتْ من طريق ابن العربي، عن طريق تلميذ واحد وهو أبو علي ابن حِرْزِهِم، وهي كالتالي:

المسلسل بالتلقيم، المسلسل بتلقين الذكر، المسلسل بالمصافحة من طريق الصوفية، المسلسل بلُبس الخرقة.

وهذه المسلسلات قَصَدَ بها من أثبتها وأوردها، إثبات تلقِّي القوم لأصول التصوف وآدابه بإسناده عن أحد أربابه بالمغرب الإسلامي، هو الولي الصالح أبو مدين الإشبيلي ثم التلمساني (594هـ\1198م)، كما يفيده صنيع المؤرخ العدل ابنِ قنفذ القسنطيني[20]، وقد يُراد بها - على بُعد - إثبات سندهم من طريق

18 عيون الموارد السلسلة: 101.

19 يُنظر توثيق رواياتهم في موطنها من كتابي (الإمام ابن العربي، فقيه المحدثين وشيخ المسندين بالأندلس)، وكذا غيرها مما أغفلت توثيقه في هذا التصدير.

20 أنس الفقير وعز الحقير: 40 و42، ويُنظر: 43، حيث أومأ إلى ما في السند من وهن من غير تصريح، وأعاده من وجه آخر في ص93.

أبي حامد الغزالي، ولا شك أن ابنَ العربي تأدب بالغزالي وأخذ من علمه وزهده وهديه، ولكن مثل ابنِ العربي في تحفظه وتحرزه لا يحفل بمثل هذه الروايات، ولكنني أوردتها لأن أئمة كبار أثبتوها على رأسهم ؛ الإمام أبو عبد الله المقري (759هـ\1358م)، والإمام أبو عبد الله الساحلي المالَقي (754هـ\1353م)، والمؤرخ ابنُ قنفذ القسنطيني (810هـ\1407م) وغيرهم من أعيان الحفاظ والفقهاء المتأخرين، فعلى إثباتي لها فهي ليست مما أخرجه ابنُ العربي في « مسلسلاته ».

ولا أطيل القول في أسانيد الصوفية، فتُطلب في مظانها على أن متأخري الفقهاء والمحدثين حاولوا تصحيح هذه الأسانيد، لكن ذلك بعيد، ولكني أحصر القول في هذا الإسناد بخصوصه، فهذا السند الذي من طريق ابن العربي، فيه رواية أبي محمد الجويني (438هـ\1046م) عن أبي طالب المكي (386هـ\1482م)، فهو من طبقة شيوخ أبي محمد الجويني، فروايته عنه ممكنة، ولا تُدفع إلا بحجة، لكن رواية أبي طالب المكي عن الجريري فيها بُعد، فالجريري توفي سنة: 311هـ\923م، بل هي مستحيلة، إذ أعلى شيخ عند أبي طالب المكي هو عبد الله بن جعفر ابن فارس توفي سنة: 346هـ\957م، ولا يروي عنه إلا بالإجازة في « أربعينه »، كما أفاده الذهبي[21]، فمن نسب هذا الإسناد إلى ابن العربي تساهل تساهلا بيِّنًا، ويُنظر الرواية رقم: (34) [22].

4 ثَبَتٌ بالأحاديث المسلسلة التي خرَّجها ابنُ العربي في كتابه

استقصيتُ في هذا الموطن ألقاب الأحاديث المسلسلة التي خرَّجها ابن العربي في كتابه، معوِّلا على المصادر التي روتْ من طريقه أو نقلت عنه، وأُخُصُّ بالتنصيص منها مصدرين، هما: « الجواهر المفصلات في الأحاديث المسلسلات » لابن الطيلسان القرطبي و« المسلسلات والإنشادات » لابن سالم البلنسي، فهما اشتملا على جلِّ الكتاب إن لم أقل كلَّه، واقتصرت على الأحاديث المقطوع بكونها داخلة في الكتاب، أما ما أوردته على الاحتمال في كتابي عن الإمام ابنِ العربي، فإني حذفته، ويظهر من خلال هذا الثَّبَتِ المَسُوقِ حفالةُ كتاب ابنِ العربي وثراءه، وهذه هي المرويات المسلسلة الواردة في كتابه:

[المسلسل بالسؤال عن الإخلاص]

[المسلسل بالسؤال الاسم والكنية والنسب والبلد]

21 تاريخ الإسلام: 8/599.

22 الإمام ابن العربي، فقيه المحدثين وشيخ المسندين بالأندلس.

[المسلسل بالمصافحة]

[المسلسل بأخذ فلان بيدي]

[المسلسل بالعد في اليد]

[المسلسل بأخذ باللحية]

[المسلسل بالعض على الأصبع]

[المسلسل برحم الله فلانا لو أدرك أهل زماننا]

[المسلسل بشهدت على فلان]

[المسلسل بقوله: سمعت فلانا يحلف بالله]

[المسلسل بذاك]

[المسلسل بأوصى بها وبإني أحبك]

[المسلسل بقول: أنظر في المصحف]

[المسلسل بصُمَّتْ أُذُنَاي]

[المسلسل برأيتُ بعينيَّ هاتين وإلا فَعُمِيَتَا، وسمعتُ بأُذُنَيَّ هاتين وإلا فصُمَّتا]

[المسلسل بأصمَّ الله هاتين]

[المسلسل بصفة التيمم]

[المسلسل بالعيدين]

[المسلسل بالدعاء في المُلتزَم]

[المسلسل بالحَسَن]

[المسلسل بالنعوت]

[المسلسل بالتحديث والكنى]

[المسلسل بآل البيت عليهم السلام]

[المسلسل بالسماع]

[حديث آخر: مسلسل بالسماع]

[حديث آخر: مسلسل بالسماع]

[حديث آخر: مسلسل بالسماع]

[حديث آخر: مسلسل بالسماع]

[حديث آخر: مسلسل بالسماع]

[حديث آخر: مسلسل بالسماع]

[حديث آخر: مسلسل بالسماع]

[حديث آخر: مسلسل بالسماع]

[حديث آخر: مسلسل بالسماع]

[حديث آخر: مسلسل بالسماع وآل البيت، موقوف على علي - رضي الله عنه-]

[آثار مسلسلة بالسماع]

فهذه أزيد من خمسة وثلاثين رواية، احتوى عليها كتاب ابنِ العربي، وسلف التنبيه على خلوه من المسلسل بالأولية، وهذا الاحتشاد يفيد حرص ابنِ العربي على أصول الأحاديث المسلسلة وعيونها.

5 رواة « المسلسلات » وأثرها في كتب الفن اللاحقة

أما رواة « المسلسلات » عن مؤلفها، والمُخَرِّجُون من طريقها والناقلون عنها؛ فعدد كثير أعرِّج على ما وقفت عليه من ذلك، **فأما الرواة:**

فأُصدِّر بقرينه في الجلالة والأصالة القاضي أبي الفضل عياض السبتي (544هـ\1149م)، فهي مما دخلت في مروياته ومرويات ابنِه أبي عبد الله محمد[23].

وأُردف بأبي بكر ابنِ خير اللمتوني، إذ قال: حدثني بها - رضي الله عنه - قراءة مني عليه في المسجد الجامع بإشبيلية - عمره الله بالإسلام -، بين المغرب والعشاء، والعيديات منها في يوم عيد الأضحى [24].

وممن سمعها عليه سنة: 540هـ\1145م، أبو عبد الله محمد بن الحسن ابن بداوة المرسي الطبيب (605هـ\1208م) [25]، ومن طريقه تحصَّلتْ للحافظ أبي بكر ابنِ مَسْدي الغرناطي، وهذا أعلى إسناد انتهى إلينا بالسماع وشرط التسلسل في هذا الكتاب، ومن جهة ابنِ مسدي وابن العربي الحاج (617هـ\1221م) انتشرت « مسلسلات ابن العربي » بالمشرق، وعلى ابنِ مسدي عوَّل ابنُ حجر في « معجمه » [26].

---

23 التعريف بالقاضي عياض: 25.

24 الفهرسة: 147.

25 مشيخة ابن مسدي: 114، من جمعنا وترتيبنا.

26 المعجم المفهرس: 161.

والحافظ المؤدب أبو محمد عبد الله بن أحمد ابن عَلُّوش الإشبيلي، ثم المراكشي توفي بعد: 599هـ\1203م [27]، وهو ثاني اثنين عوَّل عليهما ابنُ الطيلسان في روايتها وسلسلتها، وكان سماعُ ابنِ الطيلسان عليه حين وفوده على قرطبة.

والثاني؛ أبو الحجاج يوسف بن إبراهيم ابن وهبون الكَلاعي الإشبيلي (609هـ\1212م) [28]، وإنما تحصَّلَ لابن وهبون الإشبيليِّ سماعُ جزءٍ من « مسلسلات ابنِ العربي » لا جميعِها، وهذان العَلمان رَسَنُ ابنِ الطيلسان الوثيقُ في الرواية عن ابنِ العربي.

ومن طريق ابنِ وهبون؛ تحصَّلتْ للحافظ الرُّحَلة الحاج أبي بكر ابن العربي المعافري (617هـ\1220م)، وكان سماعه عليه سنة:611 هـ بإشبيلية [29].

ومن قِبَل ابنِ الطيلسان تهيأ وَصْلُها لابن جابر الواداياشي (749هـ\1348م) في « مسلسلاته » [30].

ومن الجلة الأعيان الذين رووها عنه؛ أبو القاسم ابن حُبيش ألمروي ثم المرسي (581هـ\1185م) [31]، وعنه يروي الحافظ ابنُ سالم البلنسي « مسلسلات ابن العربي »، وكذا الحافظ أبو جعفر ابن عميرة الضبي المرسي (599هـ\1203م) [32] في « مسلسلاته المبوبة »، والحافظ أبو محمد ابن حوط الله البلنسي في « مسلسلاته » [33].

ومن جهة ابنِ سالم البلنسي، يرويها الحافظ أبو علي ابن أبي الأحوص الغرناطي (679هـ\1280م) في « المشرع السلسل في الحديث المسلسل » [34]، ومن هذا السبيل كذلك تأتَّتْ للحافظ ابن جابر الواداياشي (749هـ\1348م) في « مسلسلاته » [35].

ويرويها ابنُ أبي الأحوص بأعلى من هذا من غير شرط التسلسل، بالإجازة عن أبي القاسم ابن عمر التاجر القرطبي ثم المكناسي (616هـ\1219م)، وهو آخر الرواة عن ابنِ العربي بالإجازة، وابن أبي الأحوص آخر الرواة عن أبي القاسم التاجر بالإجازة، إلا أن هذا ينخرم فيه شرط التسلسل كما ترى، وهو ليس من شرطي.

---

27 له ترجمة في التكملة: 221/4.

28 يُنظر مشيخة ابن مسدي: 114، من جمعنا وترتيبنا.

29 جزء فيه خمسة أحاديث من روايته، ل 2/ب.

30 الفهرسة: 531–532 للسراج، والفهرسة:89 لابن غازي.

31 أفردته بدراسة مطولة في العدد 4 من مجلة مرآة التراث الصادرة عن الرابطة المحمدية للعلماء.

32 صاحب « بغية الملتمس ».

33 ملء العيبة: 186/2.

34 أفردت له كتابا عن حياته وتراثه، نشر عن دار المقتبس ومركز الدبوس الأدبي سنة: 2020م.

35 الفهرسة:531–532، للسراج، والفهرسة: 89، لابن غازي.

وتتحصل روايتها للحافظ المؤرخ أبي القاسم الملاَّحي من جهة أبي جعفر ابن عميرة المرسي، وهما تاريخيان.

غريبة: ومما يستغرب منه، أن الحافظ أبا القاسم الملاحي الغرناطي (619هـ\1222م)، لم تتحصل له رواية « مسلسلات » ابن العربي إلا بواسطتين مع تقدم وفاته [36]، فهو لا يرويها إلا عن ابن عميرة الضبي عن ابن حُبيش عن المؤلف، بينما تلميذه ابنُ مسدي الغرناطي (663هـ\1264م) توفي بعده بأزيد من أربعين سنة يرويها بواسطة واحدة، وهذا ابنُ سالم البلنسي (634هـ\1237م) وابنُ الطيلسان القرطبي (642هـ\1244م) توفيا بعده بأمد، ويرويانها بواسطة واحدة كما تقدم، ومن عَلِمَ سَعَةَ مشيخةِ أبي القاسم الملاحي وغزارةَ مروياته وتفنُّنَهُ في هذا الشأن، استغرب من نزوله في كتاب ابن العربي هذا النزول البيِّن.

ومن مشاهير رواتها؛ الحافظ التاريخي أبو القاسم ابن بَشْكُوال القرطبي (578هـ\1182م)، وقد بثَّ كثيرا منها في « مسلسلاته » بلا شك، وأورد بعضها في « الفوائد المنتخبة »، ومن جهته كذلك يرويها الحافظ أبو محمد ابنُ حوط الله (612هـ\1215م) في « مسلسلاته » [37]، ثم الحافظ أبو عبد الله ابن حيان الشاطبي ثم التونسي (718هـ\1318م) [38]، وكذا الحافظ ابنُ رشيد السبتي (721هـ\1321م).

وقد تفنَّن الحافظ أبو عبد الله ابنُ حيان الأوسي بإيراد أسانيده في المسلسلات، حيث اتَّسق له روايتها من طريق أربعة تلاميذ لابن العربي [39]، وهم: ابن حُبيش وابنُ بشكوال وقد تقدما، والثاني: أبو الحسن علي بن محمد بن فَيْد الفارسي القرطبي (567هـ\1172م)، والثالث: أبو محمد ابن الفرس الغرناطي (ت: 567هـ\1171م).

وممن تحملها عن المؤلف من صفوة أصحابه وكُبرائهم؛ أبو محمد ابن عُبيد الله الحجري السبتي (591هـ\1194م)، وانبرى لإسماعها وإقرائها بمعقل درسه من سبتة، وممن قرأها عليه أبو العباس العَزَفي السبتي (633هـ\1236م)، نصَّ عليه في « برنامج مشيخته »، كما أفاده الحافظ أبو الحسن الرُّعيني الإشبيلي [40].

---

36 لمحات الأنوار في فضائل القرآن: 328/1، وأخذ الملاحي على جماعة كاثرة يبلغون العشرين رجلا من تلاميذ ابنِ العربي من الشرقيين والغربيين، وهذا يدل على أنه لم يتهيأ له سماع « المسلسلات » عليهم بشرطها.

37 ملء العيبة: 186/2.

38 ملء العيبة: 178/2.

39 ملء العيبة: 180/2–181.

40 برنامج شيوخه: 44.

وقد حُفِظت إحدى المسلسلات من روايته عند ابنِ غريون البجائي (731 هـ\1331م) في « جزء من حديثه »[41]، وقد رواها عنه الحافظ المفيد أبو الحسن الشارِّي (649هـ\1251م)، وهو آخر من حدث عنه بالسماع.

وخالف ابنُ سالم البلنسي عادته، فسلسل حديثا مسسلسلا واحدا من جهة ابنِ عبيد الله الحجري سمعه عليه بسبتة[42].

ومن العِلية المسندين المتأدِّبين؛ أبو إسحاق ابنُ قُرقُول الحمزي ألمروي (569هـ\1174م) صاحب « مطالع الأنوار »، فهو ممن سمعها على المؤلف، ومن جهته يرويها ابنُ عبد الملك المراكشي[43]، وكان سماع ابنِ عبد الملك لها - أو لبعضها - بثغر آسفي من غرب العدوة على الحافظ الزاهد أبي علي ابن حسون الماجري الآسفي.

وأختم بتلميذه الأخص وراويته الألزم؛ الحافظ أبو عبد الله ابن الفخار المالقي ثم المراكشي (590هـ\1194م)، فقد روى عنه ابنُ سالم البلنسي حديثا مفردا في المسلسلات[44]، وقد أفادنا الحافظ أبو الحسن المَطْماطي أن ابنَ الفخار روى حكاية مسلسلة بالسماع[45]، لكن رواها بواسطة ابنِ بشكوال عن ابنِ العربي، هذا مع أن ابنَ الفخار كان شديد اللزوم لابنِ العربي مكثرا عنه، وكل من أسند تلك الحكاية رواها من طريق ابنِ بشكوال.

وإن كان ما سبق مُغْنيا في بيان رواة « المسلسلات » والمخرِّجين من طريقها، فإني أسوق مسردا بمن ذُكر عنه أَخْذُهُ لـ« المسلسلات » عن المؤلف، وبعضهم لم يسمع عليه سواها:

أبو القاسم محمد ابن غلبون التُّجيبي اللورقي[46]، رحل إلى قرطبة فسمع من ابن العربي « المسلسلات » وغيرها.

أبو عبد الله محمد بن مالك بن محمد بن مالك الغافقي المرسي توفي في حدود:590هـ\1194م، ليس له رواية إلا عن ابن العربي، روى عنه « مسلسلاته »[47].

---

41 (ص:216) صدر بتحقيقنا ضمن مجموع عن دار البشائر الإسلامية، سنة: 2018م.

42 المسلسلات والإنشادات: الرواية:11.

43 الذيل والتكملة: 38/4.

44 المسلسلات والإنشادات، الرواية: 11.

45 يُنظر توثيق هذه الرواية ومن رواها في الرواية رقم: (53) من كتابنا (أبو بكر ابن العربي، فقيه المحدثين وشيخ المسندين بالأندلس).

46 التكملة: 128/2، لم تُذكر سنة وفاته.

47 التكملة: 66/2.

أبو الحجاج يوسف ابن غصن التجيبي الإشبيلي (597هـ\1201م) [48]، سمع « المسلسلات » على مؤلفها.

أبو محمد عبد الله بن يحيى الأنصاري اللُّرْيِّي البلنسي (550هـ\1155م)، سمع على ابن العربي « مسلسلاته »، وعنه يرويها المؤرخ الحافظ أبو عمر ابن عياد (575هـ\1179م) [49].

وممن انتصب لإسماع هذه « المسلسلات » وسلسلتها للقاصدين والواردين؛ أبو عبد الله ابن حيان الشاطبي (718هـ\1124م) بتونس، وممن تحملها عنه بشرطها، الحافظ ابنُ رشيد السبتي، فأخبر عن ذلك بقوله: وقرأت عليه من مسلسلات القاضي أبي بكر ابن العربي رحمه الله، وسمعت منه ما افتقر إلى السماع إلى قوله: بأي كف صافحت كف رسول الله صلى الله عليه وسلم. وعاق السفر عن إتمام ذلك...، ولم أسلسل مع صاحبنا أبي عبد الله الحديث الذي فيه « ذاك الأبلح ذاك الأصلع ». وإن كنت قرأت المتن مع سنده من ابن العربي، ولم أترك أبا عبد الله أن يقول لي: صفات من بينه وبين ابن العربي، وسلسلت معه الحديث الذي قبله على شرطه [50].

ومن الذين تصدوا لبثِّ هذه « المسلسلات » ونشرها من مهاجِرة الأندلس؛ أبو عبد الله محمد بن صالح الكناني الشاطبي (699هـ\1300م) ببجاية، فهذا أبو القاسم التجيبي السبتي يتحملها عنه حين وفوده على بجاية، وذكر ذلك قائلا: كتاب في مسلسلات القاضي أبي بكر محمد بن عبد الله بن العربي المعافري، رحمه الله تعالى، قرأت بعضه ببجاية على الخطيب أبي عبد الله بن صالح، وتناولت سائره من يده بسنده فيه. [51]

وتقدمهما في هذا المضمار جلالة وزمنا؛ الإمام الحافظ المعمَّر أبو محمد ابن هارون الطائي القرطبي، نزيل تونس (702هـ\1303م)، وممن احتفى بأخذها عنه، الحافظ الأديب الرُّحلة أبو عبد الله محمد ابن سُعُود الحيحي المراكشي توفي بعد سنة720 هـ\1320م [52].

وقد تبَدَّى مما سقناه وتتبعناه جملة من المُخَرِّجين من طريق المسلسلات والمفيدين منها رواية وتخريجا، فيغني ما سلف من ذكرهم عن تكريرهم، ويتخلص من هذا؛ سؤالٌ مفاده، هل التواصي بهذه « المسلسلات » عناية ورواية وانتساخا امتد إلى العصور المتأخرة؟

48 التكملة: 217/4.

49 التكملة: 261/2.

50 ملء العيبة: 179/2–180.

51 برنامجه: 173.

52 برنامجه: 553–554.

والذي تحصَّل لدينا من متأخري المسندين المغاربة الذين رووا من طريق هاته المسلسلات؛ الحافظ المسند أبو عبد الله ابنُ خطيب ألمرية ألمريي توفي بعد سنة:777هـ[53]، والحافظ أبو زكريا السراج الفاسي (803هـ\1400م)، والحافظ أبو عبد الله المِنْتَوري (834هـ\1430م)، ويتحقق عندي رواية الحافظ أبي عبد الله ابن مرزوق الخطيب (781هـ\1379م) لها.

وهؤلاء كلهم إنما يصلون إلى « مسلسلات ابنِ العربي » من طريق أربعة غالبا، وهم: أبو الربيع ابن سالم البلنسي وابن الطيلسان القرطبي وابن أبي الأحوص الغرناطي وابن جابر الوادياشي، وعليه فالجزم بأن سندهم مستوعب لجميع كتاب ابن العربي في المسلسلات متعذر، والصحيح أنه إسناد لهم في بعض كتاب « المسلسلات » لابن العربي التي خرّجها أولئك المؤلفون من طريقه. صحيحٌ أنها محتويةٌ لأغلب الكتاب، لكن الغرض تحقيق بقاء سند الكتاب كاملا بالشرط المعتبر في الرواية المسلسلة، لا بعض الكتاب أو ما وقع من مروياته ضمن مؤلفات أخرى.

وبهذا تعلم أن الحافظ ابنَ حجر العسقلاني (852هـ\1448م) [54] توسَّع عندما جعل إسناده في كتاب « المسلسلات » من طريق ابنِ مَسدي، إذ هذا ليس سوى إسناده فيما خرجه ابنُ مسدي من كتاب « المسلسلات » لابن العربي ضمن ديوانه في « المسلسلات »، نعم هو داخل في عموم الرواية، إلا أن المسلسلات ميزتها أن لها شروطا في التحمل والرواية، إذا لم تُراعَ سقط معها التسلسل وخرجت عن حده، ولم يُقْطَعْ لصاحبها بروايتها إلا على وجه يُذْهِبُ فائدتها، وروايتها بشرطها هو ما تحرَّينا تحقيق القول فيه، فحاصل ما سبق أن رواية كتاب « المسلسلات » لابن العربي بشروطه انقطع في حوالي منتصف القرن التاسع الهجري.

ونقف على القُرَأة الضابط ابنِ غازي المكناسي (919هـ\1513م) [55] يروي من طريقها من غير التزام شرط التسلسل بينه وبين رجلين فوقه، ولم يستقم له التسلسل إلا في الرجل الثالث، وهو: الحافظ أبو زكريا يحيى السراج الفاسي (803هـ\1400م)، على أنه يروي هذا الحديث مسلسلا من طريق ابن الطيلسان. فهذا الحافظ المقرئ ابنُ غازي على تقدم زمنه بالنسبة لمن يأتي بعده، لم يتصل له الحديث مسلسلا إلى ابن العربي، فإن سُدَّتْ سُبُلُ ذلك دونه رحمه الله تعالى، فتحصيلها عن غيره أبعد وأعسر.

إلا أن امتداد تأثير « المسلسلات المعافرية » على الدرس الحديثي بالمغرب لم يغب، وإن انقطع إسناد الكتاب بشرطه، فما احتوته الكتب اللاحقة والناهلة من نبعه، نبهت على جليل قدره ومدَّتْ

53 فرائد المرويات: 127.

54 المعجم المفهرس: 161.

55 فهرسته: 90.

الخالفين بغزيرِ فوائده ودِرَرِ عوائده، فمسلسلات ابن الطيلسان وابن أبي الأحوص وابن جابر الوادياشي وهي مما اعتمد عليه متأخرو المغاربة في هذا الشأن، مشحونة بالرواية عن ابن العربي، وأستشهد بالعلامة أبي عبد الله الفاسي الصغير (1134هـ\1721) فهو في « معجم مشيخته ومروياته »[56]، أفرد بابا بوصل المسلسلات، ساق فيه أزيد من عشرة مسلسلات من طريق ابن العربي، إلا أن شرط التسلسل ينخرم في روايتها، وهي لا تخلو من انقطاع بين ابن غازي وابن جابر الوادياشي.

وأُقَفِّي بالعلامة ابنِ الطيب الفاسي (1170هـ\1756)، إلا أن اعتماده في الرواية من طريق « مسلسلات ابن العربي » على شيخه أبي عبد الله الفاسي الصغير المتقدم[57]، وما ذكرته عنه يشمله أيضا.

ومن متأخري المشارقة الذين خرجوا من طريق مسلسلات ابن العربي؛ الحافظ المسند المرتضى الزبيدي (1205هـ\1791)، الذي روى بعضها في « أماليه »[58] عن شيخه ابن الطيب الشرقي الفاسي المذكور.

وأما في المشرق ؛ فالذي شهَّر كتاب ابنِ العربي هو الحافظ ابنُ مَسدي الغرناطي على الأرجح، فمن قِبَلِه يرويه ابنُ حجر العسقلاني، وعن « مسلسلاته » تحصل لتلاميذ ابن مسدي وغيرهم من أعلام المشرق وحفاظه روايةُ بعض « مسلسلات ابن العربي »، نعم سبقه إلى التحديث ببعضها شيخه الحاج أبو بكر ابن العربي (617هـ\1220م) في جزء انتصب لإسماعه بمصر سنة:614هـ[59]، إلا أننا لا نتيقن تحديثه بالكتاب كاملا، لأن شيخه في الكتاب ابنَ وهبون الإشبيلي إنما سمع بعضا منه على مؤلفه لا جميعه كما سبق ذكره، واشتمل ذاك الجزء على حديثين مسلسلين من طريق ابن العربي، وهما: مسلسل الأخذ باليد، ومسلسل المصافحة، وممن روى هذا الجزء عن تلاميذ الحاج ابن العربي، الحافظ النَّسَّابة شرف الدين الدمياطي (703هـ\1303م)، وممن حرص على قراءة الجزء الحافظ السخاوي (909هـ\1503م)، وقد أثبت ذلك على ظهر الجزء بخطه، ونص قيد القراءة: "الحمد لله،

قرأته على المسندة الأصيلة سارة بنت عمر بن عبد العزيز ابن جماعة الحموية الأصل، بإجازتها من الحلاوي محمد بن يوسف عن الدمياطي بسنده.

---

56 (المنح البادية،2/26-27-28-29-30-31-34-48-49)، وإسناده هو: أخبرنا به أبو المكارم، عن أبي المعارف، عن المنجور، عن سُقِّينْ، عن ابن غازي، عن ابن جابر، عن ابن الغماز، عن ابن سالم، عن ابن حبيش، عن ابن العربي به.

قلت: بين ابن غازي وابن جابر ثلاث وسائط.

57 عيون الموارد المسلسلة، ل 35/ب.

58 الأمالي، ق29/أ.

59 جزء فيه خمسة أحاديث من روايته، ل 2/ب.

فسمعه محمد بن محمد بن محمد السنباطي، وصح وثبت في يوم الأحد 13 ربيع الأول سنة 851هـ\1447م بمنزلها، وأجازت جميع مروياتها.

قاله: أبو الخير محمد بن عبد الرحمن السخاوي، وكتب حامدا مصليا مسلما محسبلا محوقلا". [60]

وهذا سند ينخرم فيه شرط التسلسل بين السخاوي والدمياطي كما هو ظاهر.

وممن نبه على « مسلسلاته » الحافظ الذهبي [61]، ونقل منها ابنُ حجر مما يفيد وقوفه على نسخة منها [62]، وذكر السخاويُّ ابنَ العربي ضمن المؤلفين في هذا الفن في سرده لأسمائهم [63].

وهذا الاحتفال بـ« المسلسلات المعافرية » والاحتفاء بها، قاطعٌ بكثرةِ نُسَخِها الخطية ووفرتها، فهي سُمعت وانْتُسِخَتْ في حواضر الأندلس؛ مثل: قرطبة وإشبيلية وبلنسية ومرسية وغرناطة ومالقة وغيرها، وتم التحديث بها وإملاؤها اعتمادا على نسخ متقنة وأصول معتمدة من غير شك بسبتة وفاس ومراكش وبجاية وتونس وغيرها، لكن خلو الخزائن والمكتبات من نسخة واحدة وحيدة منها مستدع للعجب وقاض بالنُّكرة، على أن أمل الكشف عن إحداها لا زال متقد الجذوة، وبقاء بعضها محفوظا منزويا في إحدى الأركان المحجوبة والزوايا المستورة، تمنٍّ ينعقد عليه قلب المتشوق المتشوف.

خاتمة

تغيّتْ هذه المقالة الكشفَ عن إمامةِ ابنِ العربي في فنٍّ لطيف لدى المحدثين، وأبدى سبقَه في هذا المضمارِ بالغرب الإسلامي، حيث غدا أسوةً لمن أتى بعده، وتقصَّى أثرَ هذه الإمامةِ في المسار التاريخيِّ لهذا العلمِ بحواضر الأندلس والمغرب، وعرَّف بمحتوى الكتاب ومسلك مؤلفه فيه، وجلَبَ فوائدَ حافَّةً بالموضوع تُوسع فناءه وتُغني مادته، وهذا التقري المبذول في هذا الموضوع، دليل على أن مكامن النبوغ لدى الإمام ابن العربي متعددة الأنحاء، مترامية الأرجاء، ولا شك أن جلو تراث ابن العربي المحكوم بالفقد وإبانته، سيسعف بمظاهر مستجدة لشفوف هذا الإمام وريادته في شتى أنواع المعرفة وضروب العلم، وإظهار مدى تأثيره على المسار الفكري والثقافي بالغرب الإسلامي.

60 جزء فيه خمسة أحاديث من روايته، ل 2/ب.

61 سير الأعلام:199/20.

62 فتح الباري:109/4.

63 الجواهر المكللة:46.

المصادر والمراجع

المخطوطة

أمالي في الحديث والأدب. لمرتضى الزبيدي، محفوظة بمكتبة الأزهر، رقم: (1243 – خاص).

جزء فيه خمسة أحاديث مسلسلة وعوالي. للحافظ أبي بكر ابن العربي الحاج المتأخر، محفوظة بالخزانة المحمودية بالمدينة المنورة.

الجواهر المفصلات في الأحاديث المسلسلات. لأبي القاسم ابن الطيلسان، نسخة المكتبة الوطنية بالرباط.

الرحلة المغربية. لأبي البركات ابن الحاج النُّميري، محفوظة بدير الاسكوريال تحت رقم: (1734).

عيون الموارد السلسلة في الأحاديث المسلسلة. لابن الطيب الشرقي الفاسي، نسخة خزانة الحرم المكي، رقم: (4339).

المسلسلات. لأبي القاسم التيمي الأصبهاني، ضمن مجموع (34) من مجاميع العمرية.

المطبوعة

الإمام أبو بكر ابن العربي، فقيه المحدثين وشيخ المسندين بالأندلس. جمع وتحقيق: نورالدين الحميدي الإدريسي، ط: دار الغرب الإسلامي، سنة: 2020م.

الإمام أبو بكر ابن مسدي الغرناطي حياته وآثار. من جمعي وتحقيقي، ط: دار البشائر الإسلامية، سنة: 2018م.

أنس الفقير وعز الحقير. لأبي العباس ابن قفنذ القسنطيني، تـ: محمد الفاسي ومن معه، ط: جامعة محمد الخامس الرباط، سنة: 1965م.

البرنامج. لأبي عبد الله ابن جابر الوادياشي، تـ: محمد محفوظ، ط: دار الغرب الإسلامي، سن: 1400هـ.

البرنامج. للقاسم التجيبي السبتي، تـ: عبد الحفيظ المنصور، ط: الدار العربية للكتاب، سنة: 1981م.

برنامج الشيوخ. لأبي الحسن الرعيني، تـ: إبراهيم شبوح، ط: المطبعة الهاشمية بدمشق، سنة: 1381هـ.

التعريف بالقاضي عياض. للقاضي محمد بن القاضي عياض السبتي، تـ: محمد بنشريفة، ط: وزارة الأوقاف.

جزء فيه أحاديث مسلسلة وحكايات. لأبي عبد الله ابن غريون البجائي، بتحقيقي، ضمن مجموع، عن دار البشائر الإسلامية، سنة: 2018م.

الجواهر المكللة في الأخبار المسلسلة. لشمس الدين السخاوي، تـ: محمد بن إبراهيم الحسين، ط: دار الحديث الكتانية، سنة: 2013م

الذيل والتكلمة لكتابي الموصول والصلة. لابن عبد الملك المراكشي، تـ: محمد بنشريفة وإحسان عباس وبشاو عواد، ط: دار الغرب الإسلامي، سنة: 2012م.

الرحلة. لأبي عبد الله العبدري المراكشي، تـ: علي إبراهيم كردي، ط: دار سعد الدين، سنة: 2005م.

عيون الموارد السلسلة في الأحاديث المسلسلة - جزء منه. لابن الطيب الشرقي الفاسي، تـ: أيوب أصيل، بحث لنيل شهادة الماستر، بكلية الآداب والعلوم الإنسانية بجامعة المولى سليمان – مدينة بني ملال المغرب، سنة: 2018م.

الغنية أو الفهرسة. للقاضي عياض السبتي، تـ: ماهر زهير جرار، ط: دار الغرب الإسلامي، 1402هـ.

فتح الباري بشرح صحيح البخاري. لأبي الفضل ابن حجر العسقلاني، ط: دار المعرفة، سنة: 1379هـ.

فهرس المشيخة والمرويات. لأبي عبد الله ابن غازي المكناسي، تـ: محمد الزاهي، ط: دار بوسلامة بتونس، سنة: 1984م.

الفهرسة. لأبي بكر ابن خير اللمتوني الإشبيلي، تـ: محمد فؤاد منصور، تـ: دار الكتب العلمية، سنة: 1998م.

الفهرسة. لأبي زكريا السراج، تـ: نعيمة بنيس، ط: دار الحديث الكتانية، سنة: 2013م.

فهرس الفهارس والأثبات ومعجم المعاجم والمشيخات والمسلسلات. لعبد الحي الكتاني، تـ: إحسان عباس، ط: دار الغرب الإسلامي، سنة: 1982هـ.

لمحات الأنوار في فضائل القرآن. لأبي القاسم الملاحي، تـ: رفعت فوزي عبد المطلب، ط: دار البشائر الإسلامية، سنة: 1997م.

مذكرات رحلة ابن الحاج النميري سنة: 745هـ. تحقيق ودراسة المستشرق ألفريد لويس دي بريمار، وهي أطروحة دكتوراه، بجامعة ليون الفرنسية، سنة: 1978م.

المسلسلات والإنشادات. لأبي الربيع ابن سالم البلنسي، بدرسي وتحقيقي، قيد الطبع.

المعجم المفهرس. لأبي الفضل ابن حجر العسقلاني، تـ: محمد شكور المياديني، ط: مؤسسة الرسالة، سنة: 1998م.

معجم أصحاب القاضي أبي علي الصدفي. لأبي عبد الله ابن الأبار، ط: دار الثقافة الدينية، سنة: 2000م.

المعيار المعرب والجامع المغرب. لأبي العباس الونشريسي، تـ: محمد حجي ومن معه، ط: وزارة الأوقاف المغربية، سنة: 1401هـ.

ملء العيبة بما جمع بطول الغيبة في الوجهة الوجيهة إلى الحرمين مكة وطيبة - الجزء الثاني. لأبي عبد الله ابن رشيد السبتي، تـ: محمد الحبيب ابن الخوجة، ط: الدار التونسية للنشر، سنة:1402هـ.

المنح البادية في الأسانيد العالية. لأبي عبد الله الفاسي الصغير، تـ: محمد الصقلي، ط: وزارة الأوقاف المغربية، سنة: 2005م.

المناهل السلسلة في الأحاديث المسلسلة. لمحمد عبد الباقي الأيوبي، ط: دار الكتب العلمية، سنة: 1403هـ.

11

مخطوطات القاضي أبي بكر ابن العربي المعافري بخزائن سوس

محمد علوان

[Manuscripts of Ibn al-ʿArabī's Texts Preserved in the Private Libraries of Souss (Morocco)]

[*Mohammed Aalouane*]

لا تقل الخزائن السوسية بين مثيلاتها من الخزائن المغربية منزلةً ومكانةً في الحفاظ على نفائس الأعلاق المخطوطة؛ التي تتحقق فيها معايير النّفاسة المعتبرة عند أهل التخصص، بل بعضها متميز ونادر؛ مما يدل على غنى خزائنها.

هذا؛ ونتوخى من هذا المقالة لفت انتباه الباحثين وأنظار المختصين في التراث الإسلامي الأندلسي المخطوط بما تزخر به الخزائن السوسية من نفائس ونوادر المخطوطات الأندلسية، وما تحتضنه من أعلاق قلّ نظيرها في المكتبات المشهورة؛ خاصة ما للإمام القاضي ابن العربي المعافري -رحمه الله- موضوع كلمتي.

وسأعرض مقالتي في محورين:

المحور الأول: نوادر أندلسية في الخزائن السوسية:

المحور الثاني: مخطوطات ابن العربي المعافري في الخزائن السوسية.

1 نوادر أندلسية في الخزائن السوسية

آثرت قبل البدء بعرض ما تختزنه الخزائن السوسية من تراث الإمام ابن العربي المعافري الأندلسي-رحمه الله-؛ أن أذكر بعض نوادر الأعلاق الأندلسية التي وقفنا عليها بالصُّقع السوسي، مما لم نكشفه قبل؛ معضداً ذلك بصورة لكل مخطوط تجلي المقصود، وتكشف عن الوجه الحقيقي للمخطوطات بالمنطقة، والمكانة التي تحتلها الخزائن السوسية في الحفاظ على التراث المغربي الأندلسي المخطوط؛ ونشرع الآن بذكر نماذج من هذه النفائس النادرة؛ فمنها:

 | DOI:10.1163/9789004706859_013

- نسخة أندلسية من كتاب في « المناجاة »، من النسخ الفريدة UNICUM، مكتوبة على ورق شاطبي، خطها أندلسي جميل، مبتورة الأول والأخير، عدد لوحاتها (141)، مسطرتها(15) سطراً، عليها تعقيبة، والنسخة مضبوطة بالشكل، وعليها رموز المقابلة والتصحيح، ميِّزت فيها الفصول بلون أسود كبير، ومحتوى الكتاب حول مناجاة الأنبياء والتابعين والعباد والزهاد..، والقرائن من المخطوط تدل على أنه أندلسي من القرن السابع من الفترة الموحدية؛ لأنه يذكر كاتب الخلافة العلية أبا الحسن الرعيني(666هـ\1268)، والأمير أبا إبراهيم إسحاق بن يوسف بن عبد المؤمن الموحدي(610هـ\1213م) بما يدل على معاصرته له أثناء كتابته للمخطوط، وكل هذه القرائن تدل أنه كتاب: « المناجاة » للإمام أبي محمد حسن بن علي بن عبد الملك الكتامي المعروف بابن القطان(ت: في العقد السابع من القرن السابع)[1]، وقد ذكره ابن القطان في كتابه « السِّلك المُثَنَّى النِّظام »(ص:408)، وأبو العباس ابن عذاري في كتاب « البيان المغرب »(584/3)، وهو من كُتُبِه التي ألّفها للخليفة المرتضَى الموحدي.

وهذا الكتاب قال عنه الرهوني في نظم الدرر (649-650/2):

وكصنَاعَةِ المُنَاجَاةِ التي　　لم تُدرَ فيما قَبْلُ فاسْتَقَلَّتِ
وَبرزَت في عِدَّةِ أَسفَاراً　　تُسْفِرُ عن عَلائِهِ إِسْفَارَا

- رسالة في شرح الأبيات الكندية؛ للخطيب الأديب فَضْل بن محمد بن عليّ بن إبراهيمَ بن فَضِيلَة المَعافِرِيّ، من أهل أوْرِيُولِيٍّ؛ سكَنَ غَرْناطةَ، أبو الحَسَن، ويعرف بابن فَضِيلة(696هـ\1296م)[2]، هي ضمن مجموع، يحتوي على رسائل للفقيه أبي محمد عبد العزيز بن محمد البوفرحي(899هـ\1493م)[3]، في مجلد متوسط الحجم، عدد لوحاتها 28، ومسطرتها 21، خطها مغربي مقروء، جاء في بدايته بعد البسملة والتصلية:

« قال الشيخ الأجل الفقيه الصوفي السّني المتفنن العارف المدرك الخطيب الفاضل أبو الحسن؛ فضل بن محمد بن علي بن إبراهيم بن فضيلة المعافري-رحمه الله-:

سألتَ – أخذ الله بأيدينا في مواطن الحيرة؛ أيها الأخ الكريم؛ وسلك بنا في منهج المشكلات صراطه المستقيم- عن بيان معانٍ تضمّنتها قطعة أنشدتنيها، وذكرت أنّه أشكل عليك شيءٌ من معانيها،

1 ينظر ترجمته في مفاخر البربر:171، والمنتوري في فهرسته :355.

2 تنظر ترجمه في صلة الصلة لابن الزبير: 4/الترجمة: 371.

3 تنظر ترجمته في: جذوة الاقتباس: 452/2، وسلوة الأنفاس: 127/3.

[illegible]
وان سأله لقضيه وان استغفره غفر له وما
ترددت عن شيء انا فاعله ترددي عن موت
المومن يكره الموت وانا اكره مساءته ولابد له منه
وروي عنه صلى الله عليه وسلم انه قال
ان الله يحب كل قلب حزين
وقال صلى الله عليه وسلم عن الله تعالى
وجبت محبتي للمتحابين في [illegible]
وقال ابو بكر الصديق رضي الله
تعالى عنه من ذاق خالص محبة الله عز وجل منعه
ذلك من طلب الدنيا واوحشه من جميع البشر
وقال الحسن البصري رحمه الله تعالى من
عرف الله تعالى احبه ومن عرف الدنيا زهد
فيها والمومن لا يلهو حتى يغفل واذا تفكر حزن
[illegible]

صورة 11.1 لوحة من كتاب مناجاة الأبرار لابن القطان، خزانة خاصة بأكادير

وأنّه وقع بينك وبين إخوانك خلاف فيها، في لفظة: « حُلتَ »، وأنهم قالوا فيها خلافَ ما قلت، فزعموا أنها بالخاء المعجمة أصوب، وزعمتَ أنها بالحاء المهملة أشبه بالمعنى وأقرب، واستشهد كل لمقالته، واستدل على ما ذهب إليه من حُولَته أو إخالته، ولم يرجع أحد إلى ما ظهر لأخيه، ورأى أن الأولى ما ظهر له فيه، وبقيت المسألة في غطاء من الإشكال، مفتقرة إلى أوضح مما ظهر من الاستدلال، و

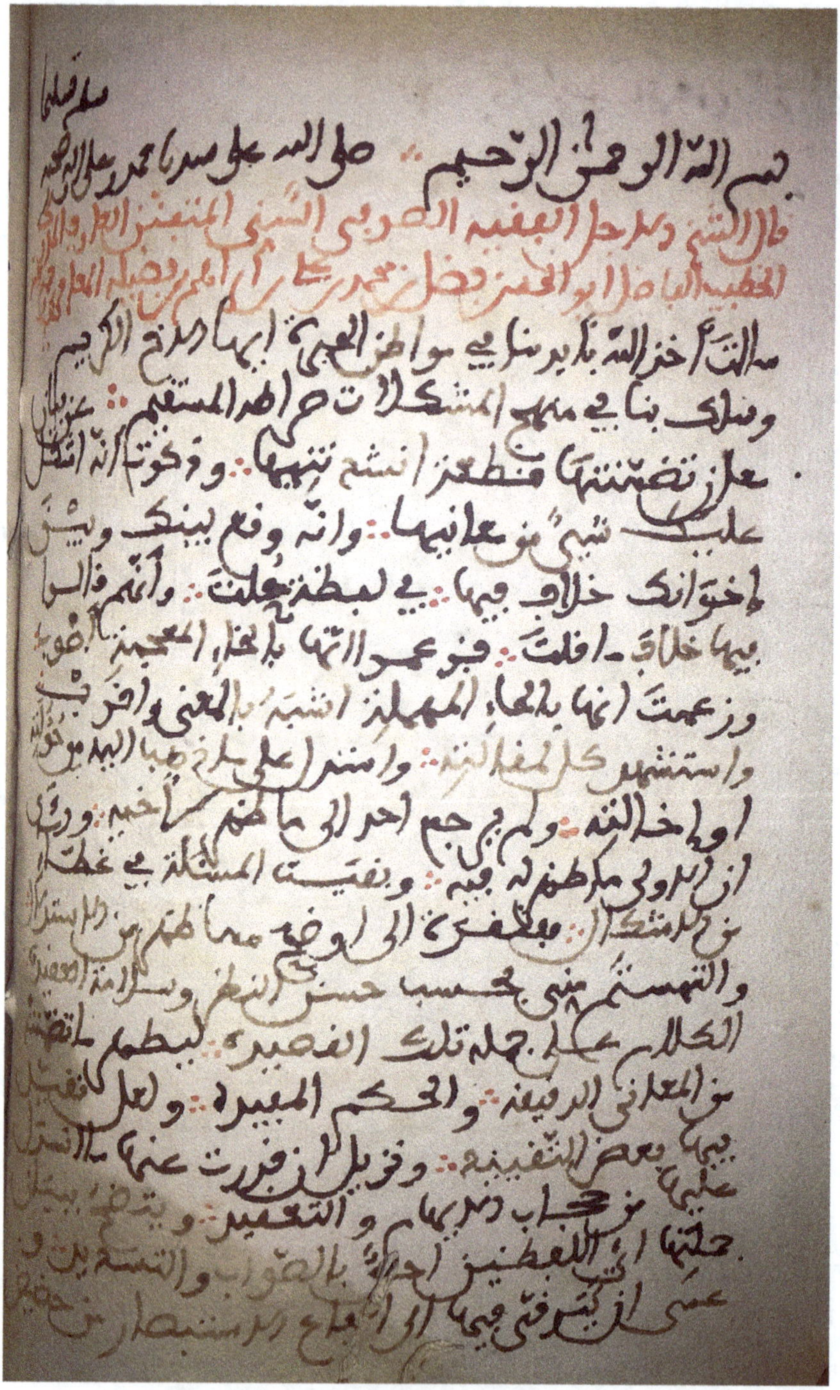

صورة 11.2 لوحة من أول كتاب شرح أبيات الكندي لابن فضالة الغرناطي، خزانة خاصة بضواحي أكادير

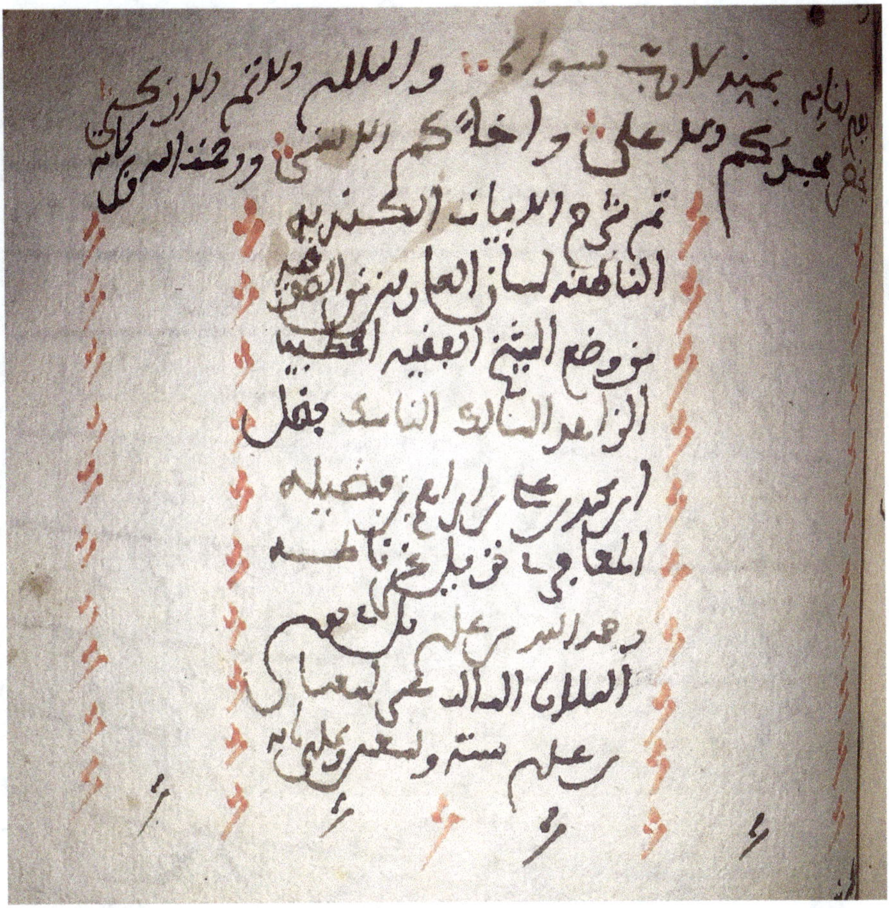

صورة 11.3 لوحة من آخر كتاب شرح أبيات الكندي لابن فضالة الغرناطي، خزانة خاصة بضواحي أكادير

التمستم مني بحسب حسن النظر وسلامة العقيدة الكلام على جملة تلك القصيدة؛ ليظهر ما تضمنته من المعاني الدقيقة، والحكم المفيدة، لعل نقيّد فيها بعض التّقييد، ونزيل إن قدرت عنها ما انسدل عليها من حجاب الإيهام والتعقيد، ويتضح ببيان جملتها؛ أي اللفظين أحرى بالصّواب والتسديد، وعسى أن يترقى فيها إلى ارتفاع الاستبصار من حضيض التقليد ...».

وجاء في قيد نسخها: « تم شرح الأبيات الكندية الناطقة لسان العارفين من الصوفية، من وضع الشيخ الفقيه الخطيب الزاهد السالك الناسك؛ فضل بن محمد بن علي بن إبراهيم بن فضيلة المعافري؛ نزيل غرناطة –رحمه الله- من عام بل في يوم الثلاثاء الثالث شهر شعبان من عام ستة وتسعين وثمانمائة ».

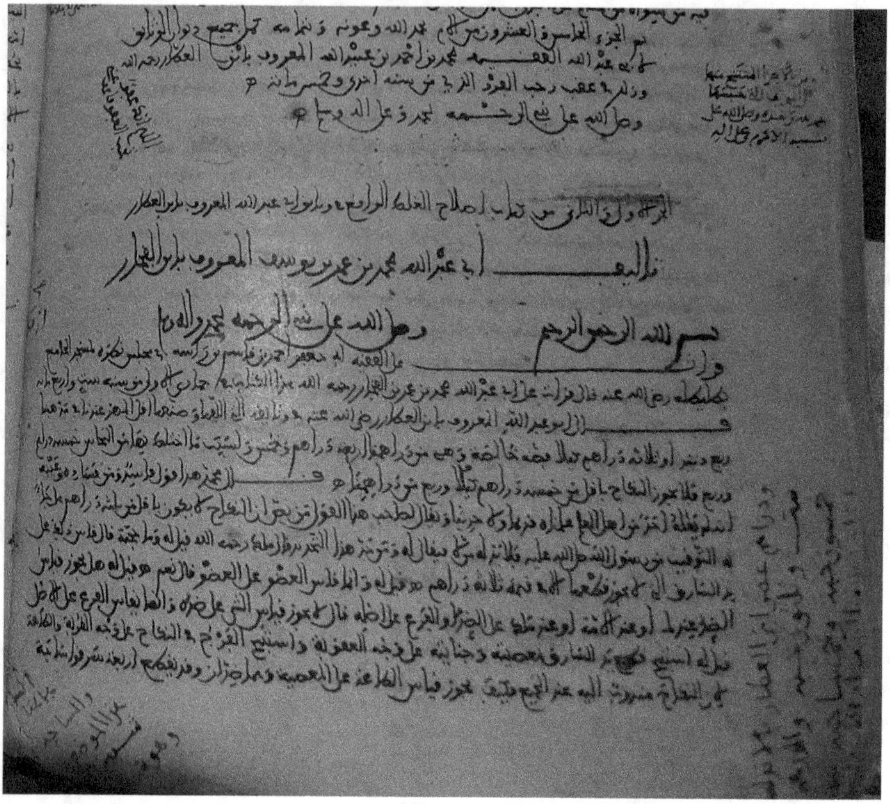

صورة 11.4 لوحة من أول كتاب إصلاح الغلط لابن العطار، خزانة تدسي

– إصلاح الغلط الواقع في وثائق ابن العطار(399هـ\1008م)؛ لأبي عبد الله محمد بن عمر بن يوسف القرطبي؛ المعروف بابن الفخار(419هـ\1028م)[4]، نسخة أندلسية عتيقة، بخط أندلسي دقيق، مكتوبة على ورق شاطبي ثخين، عدد لوحاتها 10، مسطرتها 33، خالية من التعقيبة، وهو الثاني ضمن مجموع، بعد كتاب "ديوان الوثائق" لابن العطار(399هـ\1008م)؛ الذي جاء في قيد نسخه: « تم الجزء الخامس والعشرون من الأم بحمد الله وعونه وبتمامه كمل جميع ديوان الوثائق لأبي عبد الله الفقيه؛ محمد بن أحمد بن عبيد الله، المعروف بابن العطار –رحمه الله-، وذلك في عقب رجب الفرد الذي من سنة إحدى وخمسمائة ».

وجاء في آخر الجزء الثاني من إصلاح الغلط:

4 ينظر ترجمته في ترتيب المدارك: 289/7–286، والصلة: 138/2–138.

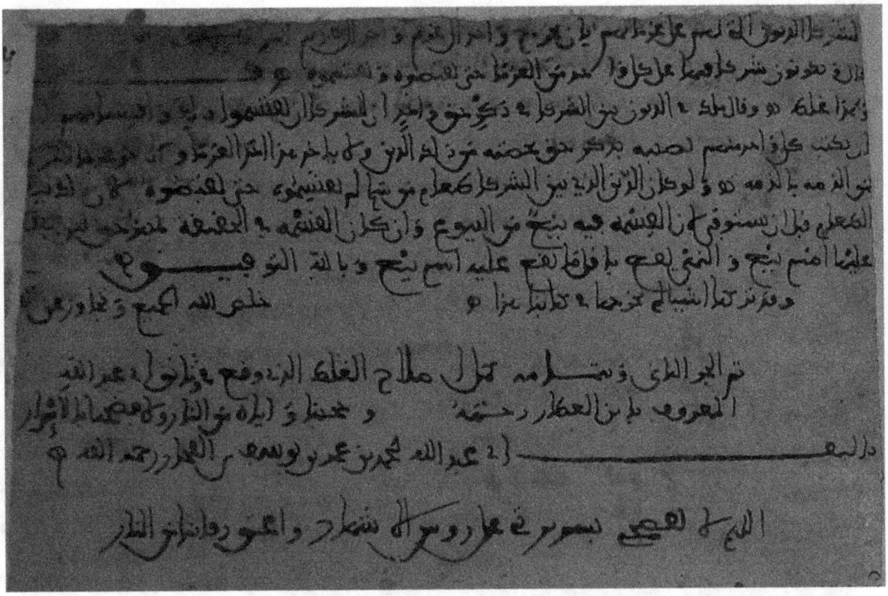

صورة 11.5 لوحة من آخر الجزء الثاني من كتاب إصلاح الغلط لابن العطار، خزانة تدسي

«تم الجزء الثاني وبتمامه كمل إصلاح الغلط الذي وقع في وثائق أبي عبد الله المعروف بابن العطار –رحمه الله- وحجبنا وإياه من النار ولا فضحنا بالأسرار، تأليف أبي عبد الله محمد بن عمر بن يوسف بن الفخار- رحمه الله- ».

- شرح الجمل؛ لأبي عبد الله المرسي؛ محمد بن أحمد بن إسماعيل الأنصاري[5]، عدد لوحاته 100، ومسطرتها 26، خطها أندلسي، مكتوبة على ورق شاطبي ثخين، خالية من التعقيبة، عارية من التسفير.

جاء في بدايته بعد البسملة والتصلية:

«يقول محمد بن أحمد بن إسماعيل الأنصاري المرسي –وفقه الله وسدده وأعانه وأرشده-:

الحمد لله الذي رفع حجاب ظلمة الجهل عن قلوب العارفين، ونصب على وحدانيته دلائل [...]عبداً [...]، وتقدس عن مضارعة صفات المخلوقين وسمات المحدثين، وعلم ما كان وما يكون وما هو كائن من جميع العلمين، أحمده على ما علم من البيان [...] ما فهم من العلوم وأبان، وأشهد أن لا إله إلا الله وحده لا شريك له شهادة [...] اللسان ونفاق الجنان، وأشهد أن محمدا عبده ورسوله النبي العربي

5 لم أقف على ترجمة له.

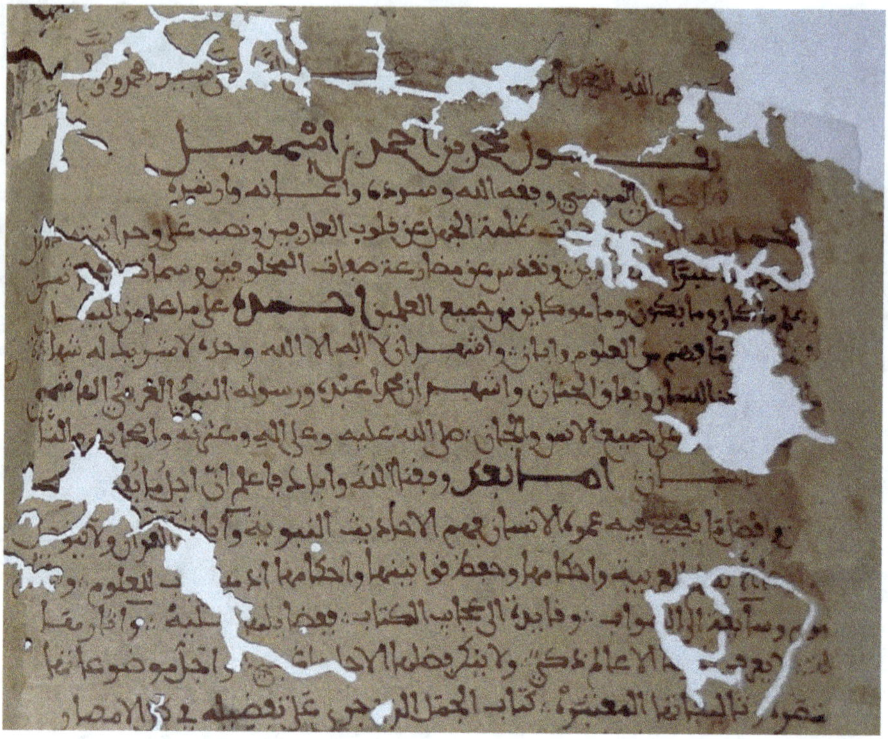

صورة 11.6 لوحة من أول كتاب شرح الجمل لأبي عبد الله المرسي، خزانة خاصة بسوس

الهاشمي [...] على جميع الأنس والجان، صلى الله عليه وعلى آله وعشرته وأصحابه والتا[بعين لهم] بإحسان ...».

وجاء في قيد نسخها:

« فتم بحمد الله له ما طلب وحصل له بعون الله ما رغب نفعنا وإياه، ومن قرأه أو كتبه أو رآه، وجعل ثوابه دخراً ليوم نلقاه بمنه وكرمه [...]، والحمد لله رب العالمين والصلاة على سيدنا ومولانا محمد خاتم النبيين الطيبين الطاهرين وصحبه والتابعين لهم بإحسان إلى يوم الدين، وسلم تسليماً، [تم الـ[ـفرغ من تأليفه يوم الجمعة غرة رجب الفرد من عام عشرين وسبعمائة، بمدينة تدنيست عمرها الله وحفظها، آمين.....، وكان الفراغ من كتبه بعيد الظهر من يوم الخميس ...عام اثنان وعشرون وسبعمائة، والحمد لله رب العالمين ».

ومدينة تدنيست المذكورة في قيد النسخ من مدن حاحا بضواحي مدينة الصويرة، ذكرها الوزاني في وصف إفريقيا (1/100–98)؛ فقال في وصفها: « تَدْنِسْتْ: مدينة من حَاحَا، تدنست مدينة قديمة

صورة 11.7 لوحة من آخر كتاب شرح الجمل لأبي عبد الله المرسي، خزانة خاصة بسوس

بناها الأفارقة في سهل على جانب كبير من الرونق والجمال، يحيط بها من كل جانب سور من الآجر والطين، وتوجد في داخله الدور والدكاكين. يوجد فيها خمسمائة كانون فأكثر، ويسيل على مدى السور جدول ماء ينبع بالقرب من ذلك المكان. وفي هذه المدينة بعض دكاكين التجار الذين يبيعون القماش المستعمل في تلك الناحية، ونسيج الكتان المستورد من البرتغال. وليس هناك من الصناع غير الاسكافيين والحدادين والخياطين وبعض الصاغة اليهود. ولا يوجد في أي مكان من هذه المدينة فندق

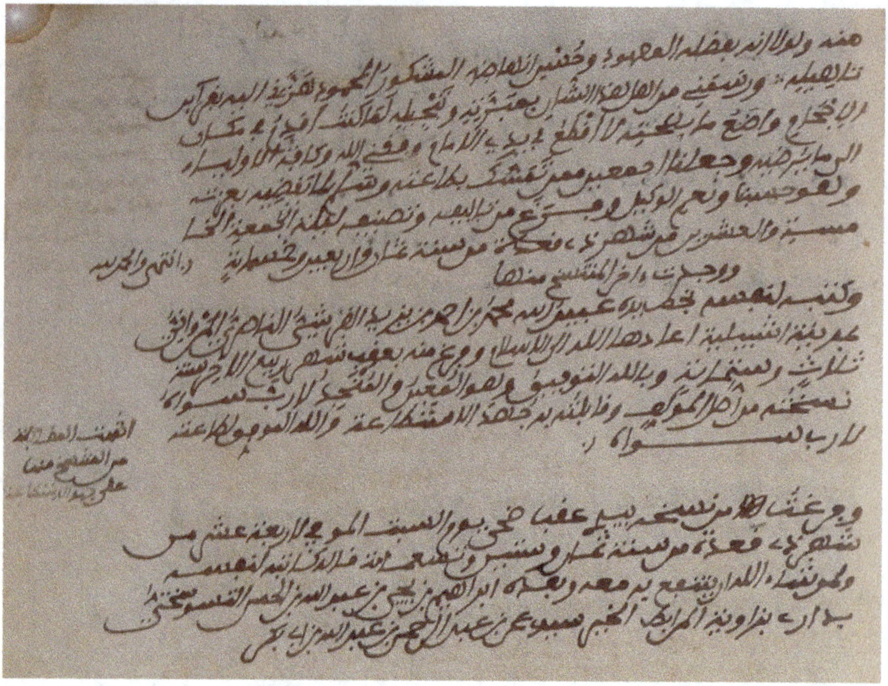

صورة 11.8 لوحة من كتاب غريب الموطأ لأبي علي الأشيري، خزانة خاصة بسوس

ولا حمام ولا دكان حلاق...وفي وسط المدينة جامع عظيم مبني بالحجر والجير في غاية الحسن، وهو عتيق شيد في الوقت الذي كانت البلاد خاضعة لسلطة ملوك مراكش....».

– غريب الموطأ؛ تصنيف أبي علي حسن بن عبد الله بن حسن الكاتب؛ المعروف بالأَشِيرِي من أهل تلمسان (بعد سنة 569ه ـ\1171م)[6]، وهي في مجلد ضمن مجموع عدد لوحاته 180 لوحة، فيه: كتاب تفسير القرآن لعبد الرحمن الثعالبي، وغريب الحديث له أيضاً، وبينهما غريب موطأ المهدي، عدد لوحاتها 40 لوحة، كتبت بخط مغربي سوسي، والمجموع غير مسفر، محفوظ بخزانة خاصة، وجاء في قيد الختام:

« وفرغ من تأليفه وتصنيفه ليلة الجمعة الخامسة والعشرين من شهر ذي القعدة من سنة ثمان وأربعين وخمسمائة. وانتهى والحمد لله.

ووجدت آخر المنتسخ منها:

6 ينظر ترجمته في: التكملة لكتاب الصلة: 412/1.

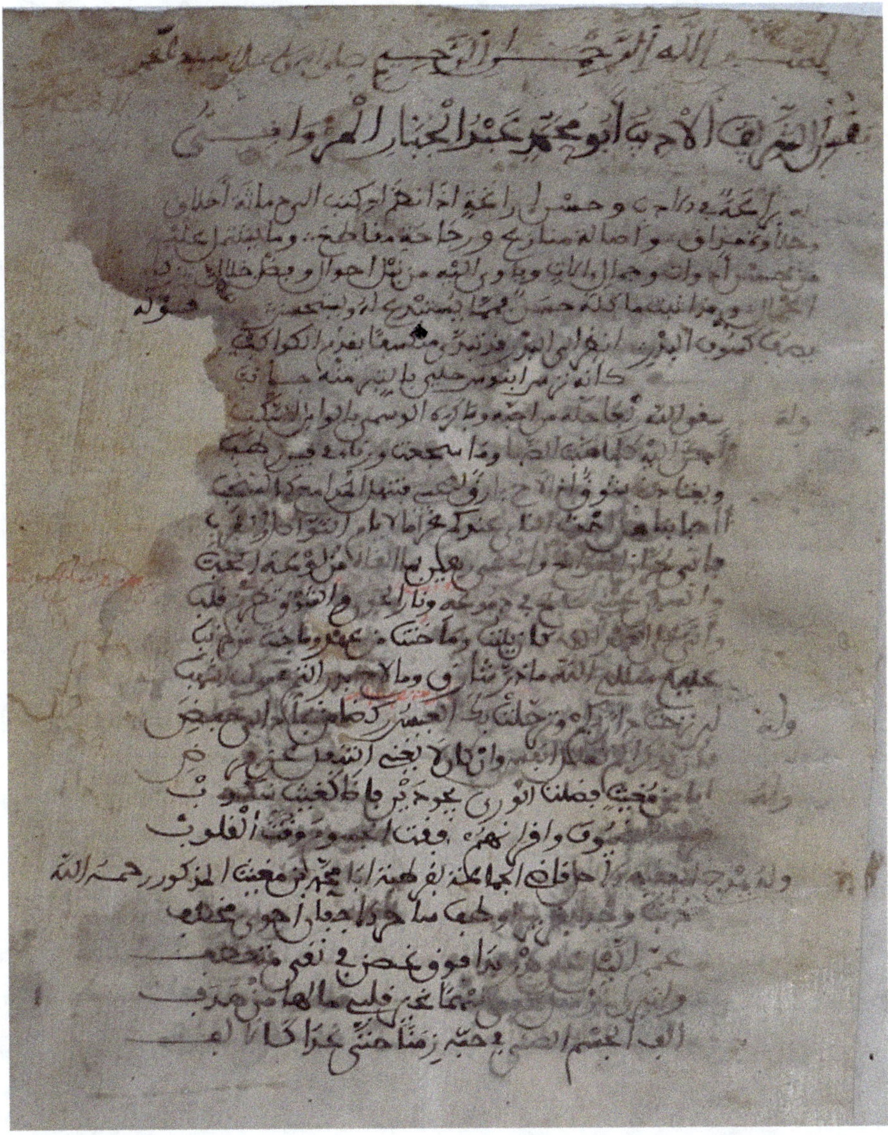

صورة 11.9 لوحة من كتاب جنى الأزاهر النضيرة وسنى الزواهر المنيرة في صلة المطمح والذخيرة مما ولّدته الخواطر من المحاسن في هذه المدّة الأخيرة؛ لأبي الحسن علي ابن الفخار الرّعيني (ت666هـ)؛ محفوظة في خزانة أزاريف

وكتبه لنفسه بخط يده عبيد الله محمد بن أحمد بن يزيد القرشي الناصريّ المرواني بمدينة إشبيلية، أعادها الله إلى الإسلام، وفرغ منه بعقب شهر ربيع الآخر سنة ثلاث وستمائة، وبالله التوفيق، وهو المُعين والمنجد؛ لا رب سواه، نسخته من أصل المؤلف، وقابلتُه به جُهد الاستطاعة. والله الموفق لطاعته لا رب سواه، وفرغت من نسخه بيدي عقب ضحى يوم السبت الموفي لأربعة عشر من شهر ذي القعدة من سنة ثمان وستين وتسعمائة، قاله كاتبه لنفسه ولمن شاء الله أن ينتفع به معه وبعده؛ إبراهيم بن يحيى بن عبد الله بن الحسن التسوسختي بداري بزاوية المرابط الخير سيدي عمر بن عبد الرحمن بن عبد الله بن أبي بكر».

- جنى الأزاهر النضيرة وسنى الزواهر المنيرة في صلة المطمح والذخيرة مما ولّدته الخواطر من المحاسن في هذه المدّة الأخيرة؛ لأبي الحسن علي ابن الفخار الرّعيني(666هـ\1268)[7]، محفوظة في خزانة أزاريف، ضمن مجموع نسخة عتيقة، كتبت بخط مغربي مجوهر، ميزت فيها بعض الأسماء والعناوين بقلم أسود داكن، مجلدة تجليداً أصلياً، مسطرتها 25، خالية من التعقيبة، والسماعات والتملكات، والنسخة مبتورة الأخير.

2 مخطوطات ابن العربي بخزائن سوس:

- رسالة في الكلام على مشكل حديث السُّبحات والحجاب؛ في خزانة خاصة ضمن مجموع، في مجلدة متوسطة الحجم، تجليدها حديث، والنسخة تامة؛ في سبع لوحات، خطها مغربي متأخر، مسطرتها 21 في كل صفحة، ميزّت فيها العناوين وبعض رؤوس المسائل بالحبر الأحمر، خالية من التعقيبة، وعليها تصحيحات في الهامش، مما يدل أنها نسخة مقابلة.
- أحكام القران، نسخة محفوظة بخزانة أزاريف ضواحي مدينة تزنيت، عدد لوحاتها 141، خطها أندلسي، ومسطرتها 25 في كل صفحة، عارية من التعقيبة، ميزت رؤوس المسائل باللون الأسود الداكن، والنسخة تتضمن السفر الثاني من التفسير، تجليدها حديث، خالية من تاريخ النسخ، وذكر ناسخها، والتملكات، جاء في ختامها: «تم السفر الثاني من أحكام القاضي أبي بكر ابن العربي –رضي الله عنه-، ويتلوه السفر الثالث –إن شاء الله- الآية الثانية عشر؛ قوله تعالى: (إِنَّ عِدَّةَ

7 ينظر ترجمته في: برنامج شيوخه بتحقيق إبراهيم شبوح، وصلة الصلة، القسم الرابع، الترجمة: 300، ص:147–146.

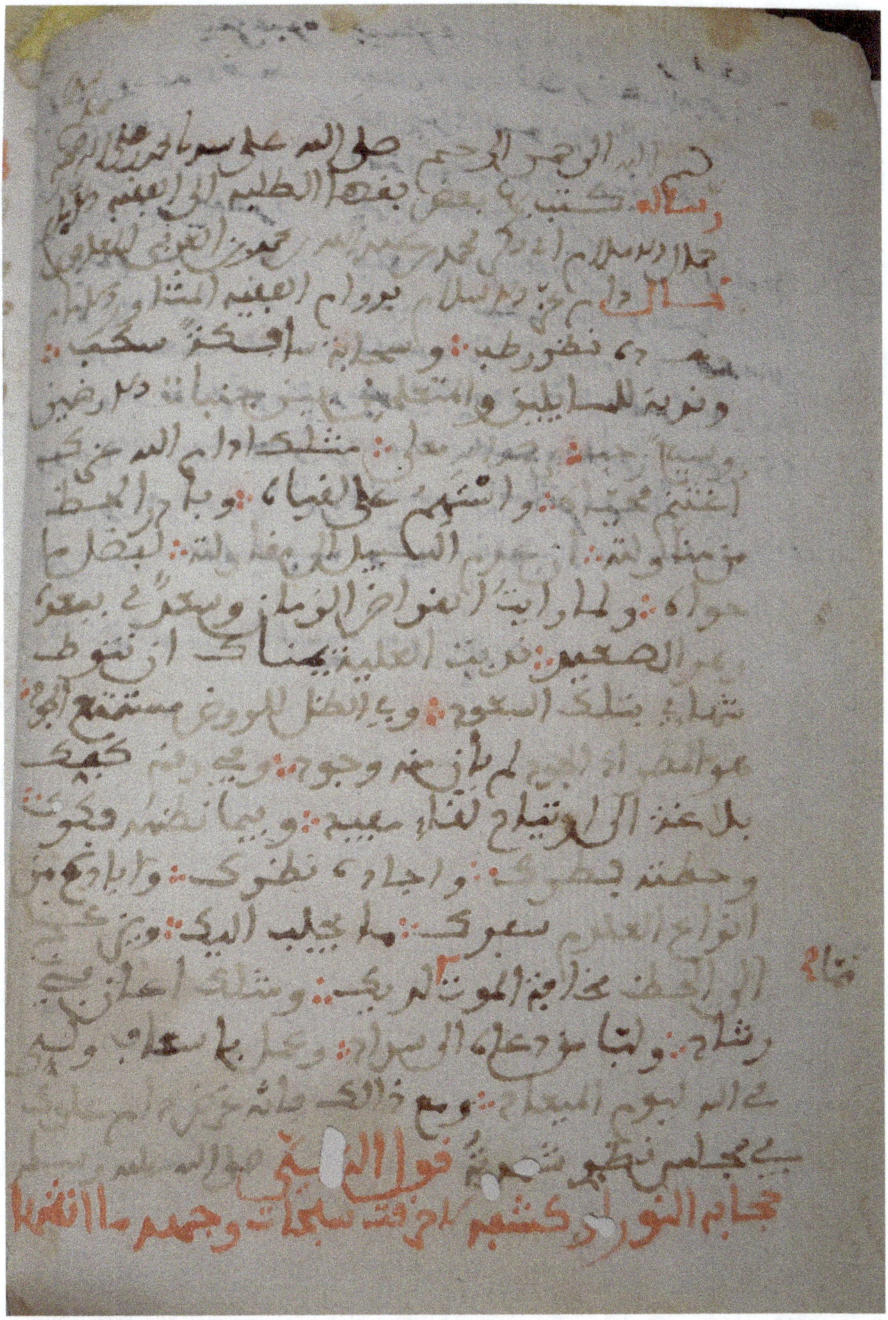

صورة 11.10 لوحة من كتاب رسالة في الكلام على مشكل حديث السُّبحات والحجاب لابن العربي المعافري؛ محفوظ بخزانة خاصة بأكادير

صورة 11.11 لوحة من كتاب أحكام القرآن لابن العربي المعافري؛ محفوظة بخزانة أزاريف

الشُّهُورِ عِنْدَ اللَّهِ اثْنَا عَشَرَ شَهْرًا فِي كِتَابِ اللَّهِ)[التوبة، الآية:36] »، والورقة الأخيرة كتب بخط متأخر.

- الأمد الأقصى في شرح أسماء الله الحسنى، منه نسخة محفوظة بخزانة أزاريف، عدد لوحاتها 110، نسخت بخط أندلسي مليح، عدا 15 لوحة الأولى كتب بخط متأخر، مسطرتها 23 سطراً لكل صفحة، عارية من التعقيبة، ميزت رؤوس المسائل باللون الأسود الداكن، والنسخة مبتورة الأول والأخير.
- واضح السبيل إلى معرفة قانون التأويل وفوائد التنزيل:

أ النسخة الأولى: محفوظة بخزانة خاصة بضواحي تارودانت، مسطرتها 28، خطها مغربي متأخر، جاء في صفحة العنوان بعد البسملة والتصلية: « السفر التاسع من كتاب واضح السبيل إلى معرفة قانون التأويل لابن العربي رحمه الله في فوائد التنزيل ».

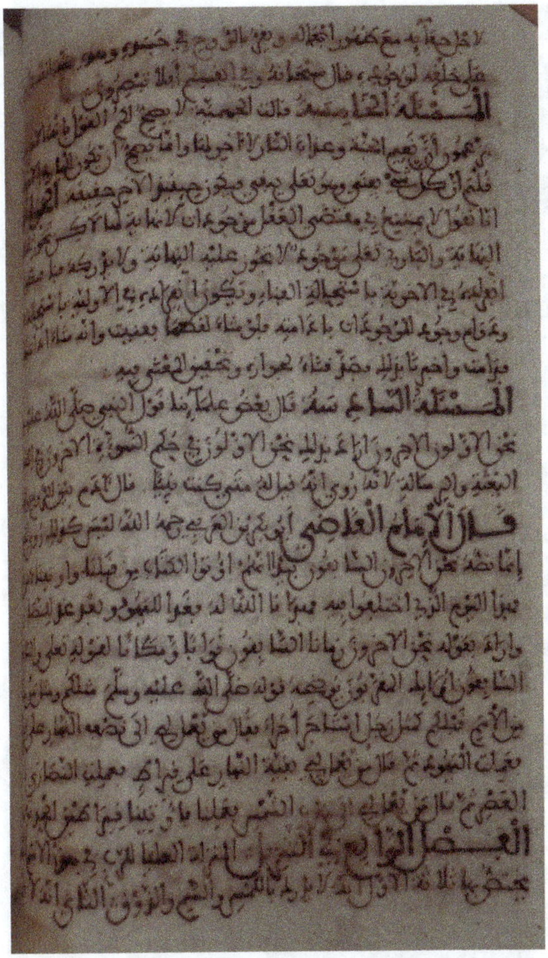

صورة 11.12 لوحة من كتاب الأمد الأقصى لابن العربي؛ محفوظة بخزانة أزاريف

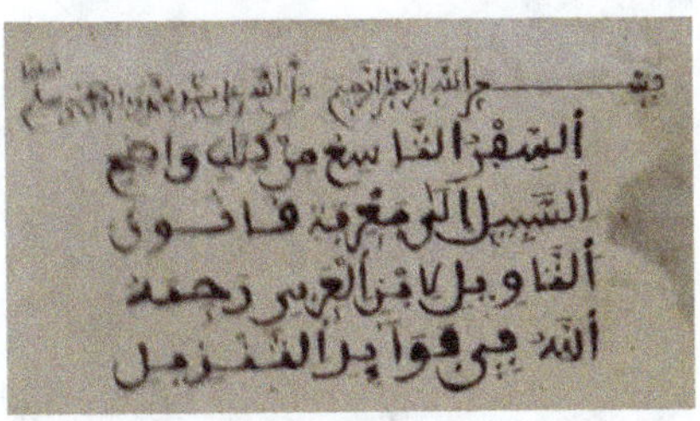

صورة 11.13 لوحة من كتاب واضح السبيل إلى معرفة قانون التأويل وفوائد التنزيل؛ محفوظة بخزانة خاصة بضواحي تارودانت

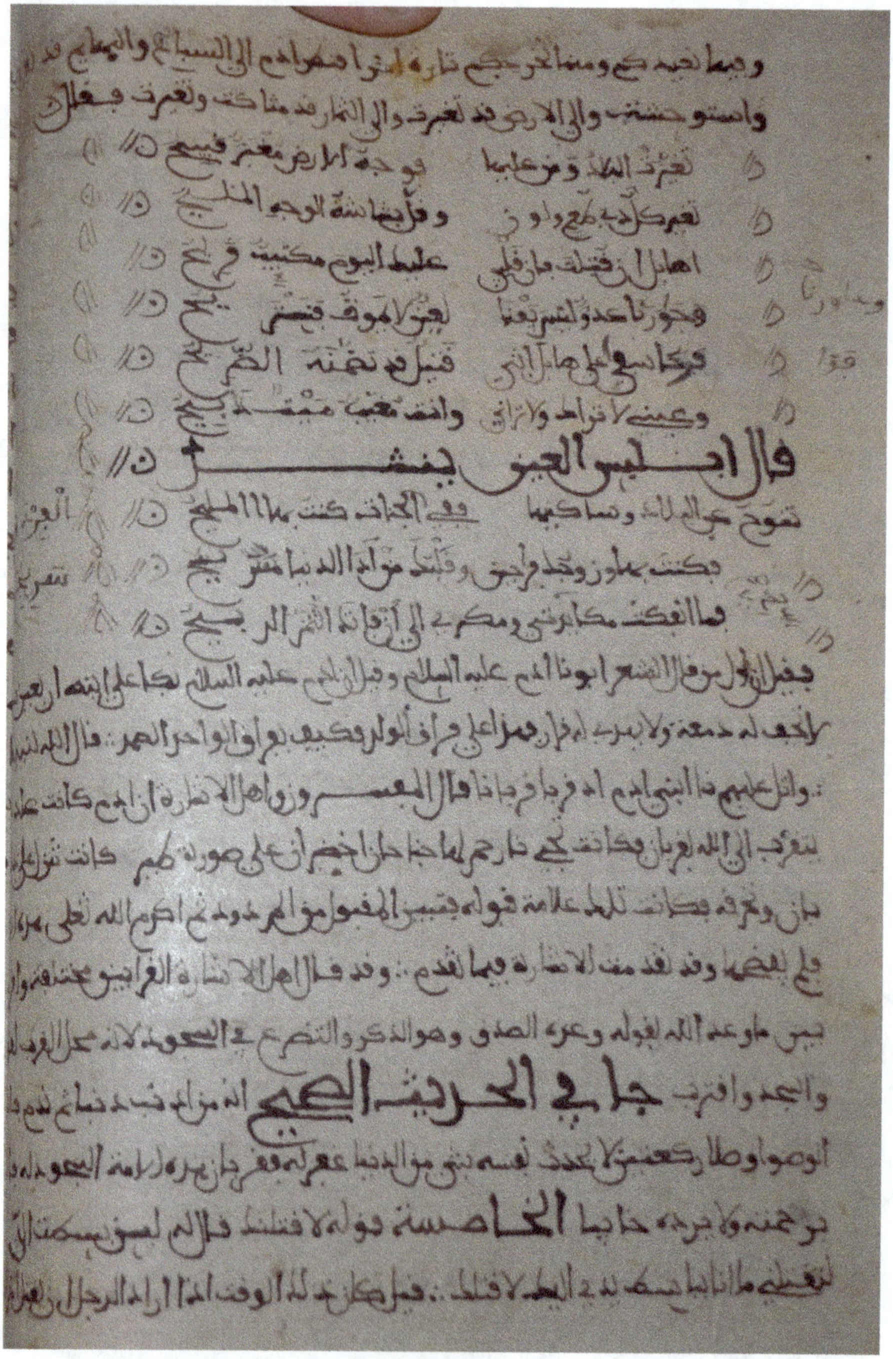

صورة 11.14 لوحة من نسخة أخرى من كتاب واضح السبيل إلى معرفة قانون التأويل وفوائد التنزيل؛ محفوظة بخزانة خاصة بضواحي أكادير

صورة 11.15 لوحة من أول كتاب الناسخ والمنسوخ؛ محفوظة بخزانة الحاج علي بأكلو

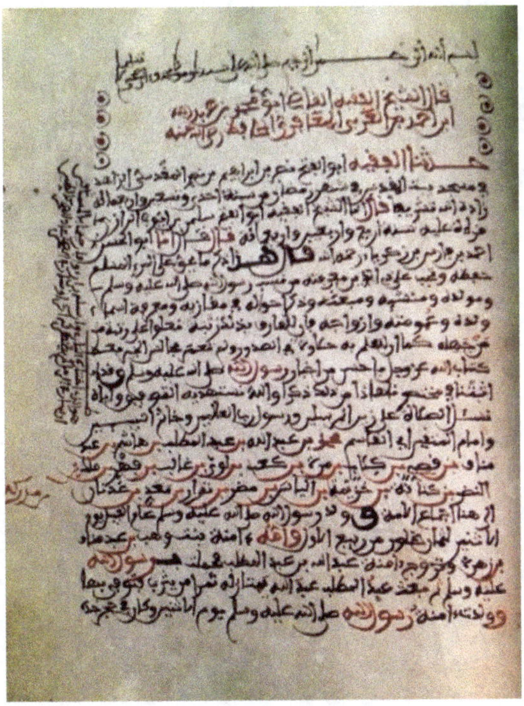

صورة 11.16 لوحة من أول كتاب أوجز السير لخير البشر لابن فارس؛ محفوظة بخزانة تدسي

ب النسخة الثانية: محفوظة بخزانة خاصة بضواحي أكادير، مبتورة الأول والأخير، مكتوبة بخط أندلسي، مسطرتها 23، عدد لوحاتها 127، في أخرها 7 لوحات من تفسير القاضي إسماعيل، والنسخة خالية من التعقيبة، والنسخة تشمل تفسير آيات من قوله تعالى: (يَسْأَلُونَكَ مَاذَا أُحِلَّ لَهُمْ قُلْ أُحِلَّ لَكُمُ الطَّيِّبَاتُ) [الآية: 4، من سورة المائدة]، إلى قوله تعالى: ﴿قَالَ اللَّهُ هَذَا يَوْمُ يَنْفَعُ الصَّادِقِينَ صِدْقُهُمْ﴾ [الآية: 119، سورة المائدة].

تذييل:

ومما ينسب للقاضي أبي بكر ابن العربي في خزائن سوس ولا يصح نسبته إليه:[8]

- الناسخ والمنسوخ؛ وقفت على نسخ منه عديدة بخزائن سوس؛ منها نسخة في خزانة الحاج علي بمنطقة أكلو، ضمن مجموع، عدد لوحاتها 5، مسطرتها 25، وخطها مغربي متأخر، عليها تعقيبة، ميزت فيها بعض العناوين واسم المصنف بالحبر الأحمر.
- مختصر السيرة النبوية؛ (أوجز السير لخير البشر)، لها نسخ عدة بخزائن سوس، منها في: خزانة أزاريف، وخزانة محمد بن العربي الأدوزي اليعقوبي، وخزانة تدسي، وهي رسالة صغيرة، غالبا في ثلاث لوحات، ولا تتعداها، والصواب في نسبتها أنها لأحمد ابن فارس(295هـ\907م)، وقد طبعت عدة مرات، منها بتحقيق الأستاذ هلال ناجي في مجلة المورد العراقية، العدد الرابع، المجلد الثاني، في كانون الأول 1973م.

فهرس المصادر والمراجع

برنامج شيوخ الرعيني. للإمام أبي الحسن الرعيني، حققه: إبراهيم شبوح، مديرية إحياء التراث بوزارة الثقافة السورية - دمشق1381هـ/1962م.

ترتيب المدارك وتقريب المسالك. لأبي الفضل القاضي عياض بن موسى اليحصبي (ت 544هـ)، تحقيق: ابن تاويت الطنجي وسعيد أعراب ومحمد بن شريفة، الطبعة: الأولى، مطبعة فضالة - المحمدية، المغرب.

التكملة لكتاب الصلة. لأبي عبد الله محمد بن عبد الله القُضاعي المعروف بابن الأبار(ت658هـ)، تحقيق: د. بشار معروف، الطبعة الأولى، 2011م، دار الغرب الإسلامي.

جذوة الاقتباس في ذكر من حل من الأعلام مدينة فاس. للعلامة أحمد بن القاضي المكناسي(ت1025هـ)، الطبعة: 1973، دار المنصور للطباعة والوراقة – الرباط.

8 أسباب نسبة بعض العناوين للقاضي ابن العربي متنوعة ومتعددة؛ تحتاج لمقال يتتبعها ويحصيها.

فهرس المنتوري. للعلامة أبو عبد الله محمد بن عبد الملك المنتوري(ت834هـ)، تحقيق: د. محمد بنشريفة، ط/1،2011م-1432هـ، الرابطة المحمدية للعلماء.

كتاب السلك المثنى النِّظام بما للصحابة الكرام على جميعهم الرضوان والسلام من الكرامات والمكرُمات العظام. لأبي محمد حسن بن علي بن عبد الملك الكتامي المعروف بابن القطان (ت.ق 7)، تحقيق: د.حسن العشاب، الطبعة:1437هـ/2016م، منشورات وزارة الأوقاف المغربية.

كتاب صلة الصلة. لأبي جعفر أحمد بن إبراهيم بن الزبير الثقفي الغرناطي(ت708هـ)، تحقيق: د. عبد السلام الهراس – الشيخ سعيد أعراب، طبعة: 1414هـ/1994م، منشورات وزارة الأوقاف المغربية.

مجلة المورد العراقية، العدد الرابع، المجلد الثاني، في كانون الأول 1973م.

مفاخر البربر. لأبي علي صالح بن عبد الحليم الإيلاني، تحقيق: عبد القادر بوباية، الطبعة الثانية: 2008م، دار أبي رقراق.

نظم الدرر. للحسن الرهوني في ترجيز كتاب الإحكام من آي خير خيرة الأنام؛ لأبي محمد حسن بن علي بن عبد الملك الكتامي المعروف بابن القطان (ت.ق 7)، تحقيق: د. خديجة البدوي، منشورات وزارة الأوقاف المغربية.

وصف إفريقيا. للحسن بن محمد الوزان الفاسي المعروف بليون الإفريقي، ترجمة: محمد حجي ومحمد الأخضر، الطبعة الثانية، 1983م، دار الغرب الإسلامي.

Index of Terms

abrogation (*naskh*) 8, 56, 125–126
adab (Belles-lettres) 1–2, 9, 16, 21, 179, 181, 184, 186, 195–196, 199, *adīb* 72, 117, 177, 180, 184, 187, 189, 193, 195–196, 199–200, 215, 222
ahl al-ahwāʾ 62, 64, 65, 67, 69, 79
anthropomorphism (*tajsīm*) 15, 36, 43–44
arbitration (*taḥkīm*) 36, 52–54, 182
ascetism, asceticism (*zuhd*) 25, 92, 209, 214, ascetist (*zāhid*) 31, 95, 105, 225
astrolabe 159
astronomy (*ʿilm al-falak*) 24–25
attribution (*nisba*) 20, 141–142, 207
autorisation de transmettre ou licence de transmission (*ijāza*) 41, 153–154, 157, 162, 165, 207

bibliographical dictionaries (*fahāris, barāmij*) 2, 23–25, 33, 112, 116–117, 128, 158. *See also* Catalogue de transmission (*barāmij, fahāris*)
bidʿa (pl. *bidaʿ*) 61, 64, 66, 69, 77–78, 82–83, 95, *ahl al-bidaʿ* 61–62, 65, 67–68, 82, 85
biographie, biography 4, 7–8, 10, 25, 98, 184, 188–189, 201, biographe 153, 155

caliph 9, 26, 28–29, 31, 56, 71, 91, 222
catalogue de transmission (*barāmij, fahāris*) 160. *See also* bibliographical dictionaries (*fahāris, barāmij*)
commentary (*sharḥ*) 4, 8, 10, 12, 14, 16, 26, 39, 48–50, 52, 55, 180–181, 222, 227
conciliation (*ṣulḥ*) 36, 52, 55

ḍabṭ al-Qurʾān 159
ḍalāl 65–66, 79
database 20–21
debate (*munāẓara*) 36–38, 49, 76, 109–110, 136–137
dīwān 180, 192, 197, 216

East (*Mashriq;* Orient) 1–3, 5, 10–13, 22, 26–31, 36, 39, 56, 71, 83, 85, 90, 92, 97, 100, 103, 109, 121, 146, 150, 152, 157–158, 163, 167, 187, 193, 211, 217
education 2, 13, 25–26, 32, 38, 114

falsafa 3, 11, 71–75, 82, 84, 150, 163
Forum Shopping 36, 54–55, 58

généalogie 146, 157
God's names (*asmāʾ Allāh*) 36, 46–47, 180
grammar (*naḥw*) 24–25, 33, 109, 177–181

ḥadīth 1–4, 13–14, 16, 24–25, 33, 42, 46–50, 54, 69, 80, 83, 85, 99, 101, 122–125, 134–135, 149, 162–163, 203, 206–208, 212, 215–216, 230
ḥarām 119, 125
ḥukm 108, 115–116, 118, 122–123, 125–126, 131, 138–139, 178, 207

ijtihād 46, 110–111, 113, 119–121, 124
ilḥād 15, 65, 74
ʿilla 80, 115, 135, 138, *taʿlīl* 73, 113, 116, 120, 133, 136, 138, 142
isrāʾīliyyāt 58

judge (*qāḍī*) 2, 4, 7, 14–16, 31–32, 37, 41, 51–55, 58, 61–62, 64, 66, 68–70, 76–78, 80, 82–83, 108–120, 122–124, 126–127, 133–136, 140, 146, 157–159, 161–162, 164, 169, 179, 184, 189, 198, 200, 206–208, 211, 215, 221, 232, 238

khabar al-wāḥid 123, 127
khizāna khāṣṣa 230, 232, 234, 238
knowledge (*maʿrifa*) 2–5, 12, 16, 24–33, 36, 41, 43, 45, 47, 49, 54, 56, 78, 91–94, 99, 101–102, 113–114, 120, 134, 136, 137, 177, 180, 218, 234
~ of God (*bi-llāh*) 45

law 1, 9, 14–15, 21, 25, 36, 53–56, 110, 134
Islamic ~ (*fiqh*) 24–25, 33, 85, 101, 117, 121, 123, 135, 149, 152, 165, 177, 180, 192
leadership (*riʿāsa*) 31–32

legal theory or "the fundamentals of the law" or "Islamic legal methodology" (*uṣūl al-fiqh*) 1–2, 8, 14–15, 21, 36, 38, 54, 57, 108–109, 111–113, 115, 117, 119–120, 122, 124–125, 128, 131, 134–135, 137, 140, 148, 169, 180
litiges (*munāzaʿāt*) 162, 164, 168

madhhab 31, 61, 63–64, 67–68, 70, 73–74, 76–78, 80–81, 83–87, 108–109, 115, 124–125, 132, 134, 136–138, 141–143
madrasa (écoles) 91, 95, 114, 131, 143, 146–150, 153–154, 156–157, 196
makrūh 119
manuscript 3–6, 8, 12, 15, 17, 39, 102, 221–222, 232
mathematics (*ḥisāb;* mathématiques) 24–25, 158–159
muqayyad 118
musalsal, tasalsul 16, 203–218
mutawātir 43
muṭlaq 118, 127

naẓar 49, 109, 124, 125, 131–134, 136–139
nifāq 62–63, 68

pèlerinage (*ḥajj*) 91–92, 152, 155, 157, 162–163
poetry 16, 24–25, 33, 39–41, 117, 181
politics (*siyāsa*) 12, 24–25, 33, political power (*sulṭān*) 31, 47, 53, 91, 93, 96, 137
profession of faith (*ʿaqīda*) 43, 71, 77, 161, 168, 225
projection orthogonale 159
public interest (*maṣlaḥa*) 53–54, 109, 115, 122, 138–142

qadar 48–49, 64, 78, 80, 99
qaṣīda 168, 197, 225
Quranic readings (*qirāʾāt*) 169

reform 1, 31, 57, reformist 14
refutation 13, 42, 97, 159
revelation 43–44, 49, 55–56, 119, 124, 158, 203
rhymed prose (*sajʿ*) 9, 195, 200
ruler 6, 26–27, 29, 31–32, 53, 56, 58

sects (*firaq*) 61–66, 68, 70, 73, 77, 80, 84–86, 136
sharīʿa 38, 52–55, 57, 68, 71, 77, 80, 85, 108, 113, 115, 118, 124–125, 127, 139, 177, 180
shirk 62
shūrā 184, 190
speculative theology (*kalām*) 12–13, 25, 36, 42, 62, 66, 71, 76, 78, 85, 101, 113, 117, 146, 148, 150, 159–160, 169, 191
Sufism 2, 13, 36, 45, 82–86, 98–99, 101, 208–209, 225
syllogisme aristotélicien 150

taifa 26–27, 56, 65, 68–70, 72, 74–78, 80, 83–86, 110, 133, 137, 166, 184–185, 188, 191, 199
taʾwīl 64, 70, 74, 76, 77, 80, 82, 114, 123
teacher 15–16, 22, 28, 31, 41, 93–94, 103, 105
theological anathematisation (*takfīr*) 15, 36, 67, 69, 79, 83
traditions (*akhbār*) 16, 46, 58, 70, 149, 195
travel (*riḥla*), travelogue (*riḥla*) 1–2, 5, 8–12, 16, 25–31, 90, 92–93, 95–96, 135, 157–158, 166–169, 186, 188, 190–191
treatise 14, 26, 54, 76, 105, 181, 222

unlawful sex (*zinā*) 36, 42, 50–52, 57

vizir 149, 161, 165, 167, 188, 193, 197
voie des anciens/voie des modernes 150

West (*Maghrib;* Occident) 2–3, 6–7, 14–16, 26–28, 30, 38, 42, 45–47, 61–62, 83, 85–86, 91, 93–96, 98, 100, 103, 109, 111, 121, 137, 141, 146, 150–152, 162, 163, 168, 185, 203–205, 208, 216, 218

zandaqa 61–63, 68–71, 87, *zindiq* 62–63, 68, 70, 80, 84, 86–87

Index of Works

Ibn al-ʿArabī

Aḥkām al-Qurʾān 3, 10, 14–15, 38, 47, 50–52, 55, 121, 137, 178–182, 232
al-Amad al-Aqṣā 3, 11–12, 47–48, 180, 234
ʿĀriḍat al-aḥwadhī sharḥ Jāmiʿ al-Tirmidhī 14, 121, 178
Al-ʿAwāṣim min al-qawāṣim 2, 5, 38, 42, 104, 132
Al-Istishfā min Kitāb al-Shifā 120, 131–135, 140–143

Kitāb al-afʿāl 13
Kitāb al-Mushkilayn 49
Kitāb al- Nāsikh wa l-mansūkh 3, 8, 238

Al-Maḥṣūl fī uṣūl al-fiqh 13, 180
Al-Masālik fī sharḥ Muwaṭṭaʾ Mālik 10
Muljiʾat al-mutafaqqihīn 12
Al-Mutawassiṭ fī l-iʿtiqād 12

Nukat al-Maḥṣūl 13–14, 119–120, 132, 137, 157, 162

Al-Qabas fī sharḥ Muwaṭṭaʾ Mālik b. Anas 8, 10, 15, 38, 48, 49–50, 52, 54, 119, 121, 143, 179
Qānūn al-taʾwīl 2, 5, 7, 10–11, 15, 27–29, 38, 44, 90, 103, 157–158, 169, 177, 180, 234

Risāla fī l-kalām ʿalā mushkil ḥadīṯ al-subuḥāt wa l-ḥijāb 13, 232
Al-Risāla al-ḥākima fī masʾalat al-aymān al-lāzima 13

Sirāj al-Murīdīn 3, 11–12, 30, 96, 104, 132

Tartīb al-Riḥla 1, 4, 6, 8–9, 11, 30, 90, 92, 158

Wāḍiḥ al-sabīl 234
Al-Wuṣūl ilā maʿrifat al-uṣūl 14

Other Authors

Awjaz al-siyar (Ibn Fāris) 238

Al-Bayān al-mughrib (Ibn ʿUdhārā) 222
Al-Burhān (al-Juwaynī) 117, 131

Al-Dhakhīra (Ibn Bassām) 184–189, 192, 195

Al-Fahrasa (Ibn Khayr) 161, 169, 195–196
Al-Faraj baʿd al-shidda (al-Tanūkhī) 197
Fayṣal al-tafriqa (al-Ghazālī) 103

Gharīb al-Muwaṭṭaʾ (al-Ashīrī) 230

Al-Ḥudūd fī l-uṣūl (Ibn Fūrak) 65

Iḥkām ṣanʿat al-kalām (Ibn ʿAbd al-Ghafūr) 191, 195
Al-Iḥyāʾ (al-Ghazālī) 12, 85, 92, 96–103, 105
Al-Irshād (al-Juwaynī) 2, 117, 169
Iṣlāḥ al-ghalaṭ (Ibn al-Fakhkhār) 226–227

Jadhwat al-muqtabis (al-Ḥumaydī) 187
Jāmiʿ al-jalī wa l-khafī fī uṣūl al-dīn (Abū Isḥāq al-Isfarāyīnī) 65
Jāmiʿ al-Tirmidhī 16, 162
Jinā l-azāhir al-naḍīra (Ibn al-Fakhkhār) 232
Khuṭbat al-faṣīḥ (Thaʿlab) 196

Al-Lumaʿ (al-Ashʿarī) 61

Maqālāt al-Islāmiyyin (al-Ashʿarī) 64
Maqāṣid al-Falāsifa (al-Ghazālī) 83, 102–103
Maṭmaḥ al-anfus (Ibn Khāqān) 189
Al-Mughrib fī ḥulā l-Maghrib (Ibn Saʿīd al-Maghribī) 198
Mujarrad maqālāt al-shaykh Abī l-Ḥasan (Ibn Fūrak) 147
Mujmal al-lugha (Ibn Fāris) 197

Qalā'id al-'iqyān (Ibn Khāqān) 189

Al-Ṣaḥīḥ (al-Bukhārī) 163
Saqṭ al-zand (Ibn al-Sīd al-Baṭalyawsī) 180, 196
Sharḥ al-Abyāt al-kindiyya (Faḍl al-Ma'āfirī) 222, 225
Sharḥ al-Jumal (Abū 'Abd Allāh al-Mursī) 227

Al-Tabṣīr fī l-dīn (Abū l-Muẓaffar Al-Asfarāyīnī) 149
Tahāfut al-Falāsifa (al-Ghazālī) 71, 96–97, 103
Al-Takmila (Ibn al-Abbār) 160, 195
Al-Tanbīh (Abū l-Ḥusayn al-Malaṭī) 65
Al- Taqrīb wa l-irshād (al-Bāqillānī) 148, 156
Tartīb al-madārik (al-Qāḍī 'Iyāḍ) 61, 98, 100

Al-Yatīma (al-Tha'ālibī) 197

General Index

Abū Ḥanīfa 31, 81
Abū l-Aṣbagh (ʿĪsā b. Saʿīd, al-Muẓaffar b. al-Manṣūr's vizir) 188
Aḍud al-Dawla 148
Adūzī (al-) 238
Agadir 238
Aglou 238
Aḥmad b. Ḥanbal 50, 115, 125
Alexandria 27, 90–91, 95–96, 98*n*34, 99–100, 105
Alfonso VI de Castille 185
Alī b. Yūsuf b. Tāshafīn 6
Almeria 10, 216
Ascalon 189
Ashʿarī (al-) 30*n*22, 61, 64, 81, 146–149, 152–154, 167, 171
Ashʿarīs 43, 45, 49, 61, 64
Ashīrī (al-) 181, 230
Azārīf 232, 234, 238

Badajoz 184–185, 192
Baghdad 1, 27, 29, 90–94, 96, 105, 178, 198
Bājī (al-, Abū l-Walīd) 21–22, 28–30, 32, 132, 135, 137, 151, 156, 162–164, 166
Bassorah 148
Bāqillānī (al-) 69, 110–111, 124, 126, 134, 147–149, 151, 153–157, 163, 169
Bātinīs, *bāṭiniyya* 15, 41–42, 66, 76–78, 82–83, 85, 97, 99, 101
Byzance 148
Bukhārī (al-) 47, 163

Christians 58

Dabbūsī (al-) 133, 135
Damascus 29, 87
Dānī (al-, Abū ʿAmr) 22
Darrās (al-) 152, 171
Dhakī (al-, Abū ʿAbd Allāh al-Māzarī) 98, 100
Dībājī (al-) 155, 157, 169

Egypt 64, 76–78, 90, 95, 146, 162, 191, 207, 217
Evora 192

Granada 218, 222, 225
Ghazālī (al-) 1–16, 41–42, 45–48, 56, 71, 75, 82, 90, 92–94, 96–105, 110, 114–115, 120, 124, 126, 131, 133, 150, 157, 169, 209

Ḥāḥā 228
Ḥanafīs 138
Ḥanbalīs 43, 110
Hawzanī (al-, Abū Ḥafṣ) 16, 160–162
(Abū l-Qāsim) 160–161, 162, 164, 168
Ḥumaydī (al-, Abū ʿAbd Allāh) 29*n*19, 47, 186–188, 197, 200, 203

Ibāḍīs 86
Ibn al-Abbār 160, 166*n*106, 168*n*114, 169, 195
Ibn ʿAbd al-Barr 28
Ibn ʿAbd al-Ghafūr 184, 191–195, 197, 200
Ibn ʿArabī (Muḥyī l-Dīn) 20–21, 25–26, 31
Ibn al-ʿAṭṭār 226
Ibn Bashkuwāl 29, 116–117, 161–163, 206, 208, 213–214
Ibn Bassām 184–189, 191–192, 194–195
Ibn al-Fakhkhār 214, 226, 232
Ibn al-Faraḍī 187
Ibn Fāris 180, 197, 238
Ibn Ḥabīb al-Ilbīrī 22
Ibn Ḥazm 10, 15, 20–21, 25–32, 43, 46–47, 58, 67–70, 75, 85, 89, 164–168, 177, 186–188, 197, 200
Ibn al-Kattānī 187–188
Ibn Khāqān 184, 189, 192–193
Ibn Khayr 117, 161, 167, 169, 184, 195–198, 205, 208, 211
Ibn al-Qaṭṭān 103, 222
Ibn Rushd al-Ḥafīd (Averroès) 21, 23, 25–26, 31, 109, 156
Ibn Rushd al-Jadd 22–23, 37–38, 45, 51*n*35, 52*n*39, 55, 58, 110, 156–157
Ibn Saʿīd 22, 198
Ibn al-Sīd al-Baṭalyawsī 22–23, 39, 57, 180–181, 193, 196, 200
Ibn Ṭarkhān 165, 187, 197
Ibn Taymiyya 104
Ibn al-Ṭubnī 187, 204

Irak, Iraq 1, 3, 90, 92–93, 113–114, 141, 149, 152, 188, 191
Isfarāyinī (al-, Abū Isḥāq) 65, 124, 147–148
Isfarāyīnī (al-, Abū l-Muẓaffar) 65, 169
Isḥāq b. Yūsuf b. ʿAbd al-Muʾmin (almohad) 222
Ismaʿīlīs, Ismāʿiliyya 41, 85, 97
Iyāḍ (al-Qāḍī) 22–23, 37–38, 41, 45, 58*n*, 61, 68, 98, 100, 116–117, 137*n*, 161–164, 207–208, 211

Jaén 189
Jerusalem 15, 27, 29, 76, 78, 90–92, 94, 101, 137, 169
Jews 42, 52, 56, 58, 72, 78, 229
Juwaynī (al-, Abū l-Maʿālī) 2, 5, 93, 110, 115, 117, 124, 126, 149–150, 155, 157, 169

Kairouan 136, 152–156, 169
Khawārij 61, 66–67, 69, 89
Khorāsān 66, 92, 135
Kufa 196

Maʿarrī (al-, Abū l-ʿAlāʾ) 16, 39–41, 180, 191–194, 196–197, 200
Mahdiyya 87, 100, 169
Mālik 8, 31, 48–49, 52–53, 63, 67, 83, 87, 89, 111, 119, 122–124, 138, 140–143, 152*n*38
Mālikīs, Mālikism 10, 14, 29, 32, 38, 43, 51, 54–57, 63, 67–69, 83, 100, 108–111, 124, 136–138, 140–143, 147, 151–152
Marrakesh 2, 30*n*23, 218, 230
Murjiʾa 61, 89
Mursī (al-, Abū ʿAbd Allāh) 211–214, 227
Murtaḍā (al-, almohad caliph) 222
Mustanṣir (al-) 197
Muʿtamid b. ʿAbbād 167–168, 190–191
Muʿtazilīs, Muʿtazilīsm 36, 61, 67, 69–70, 72–73, 78–81, 85, 89, 126–127, 167, 169

Nāṣir (al-, Cordoba's caliph) 197
Nāṣir al-Dawla (Samanide governor) 148
Nishapur 96, 98–100, 104–105, 148–150
Niẓām al-Mulk 149

Oman 146

Palestine 189
philosophers (Falāsifa) 41, 66, 69, 71–74, 79–80, 83–84, 97, 99

Qadarīs 36, 48–49
Qushayrī (al-, Abū l-Qāsim) 45, 82, 149, 169

Rabat 6, 8, 14, 181
Ruʿaynī (al-) (Almohad chancellery) 156, 213, 222, 232
Rūḥāniyya 63, 84, 86

Ṣadafī (al-) 203
Sanche (king of the Basques) 187
Sancho García 187
Santarem 184–185
Saragosse 188
Saraquṣṭī (Abū ʿAlī). *See* Ṣadafī
Sayr b. Abī Bakr (Sevilla Almoravid governor and Ibn Tāshafīn's cousin) 186
(Muḥammad b. ʿAbd al-Raḥmān) 160–161
(al-, Sulaymān b. Mihrān) 188
Seville 1, 6, 10, 12, 27, 29, 32, 51, 56, 58, 91, 93, 105, 132, 137, 146, 156–157, 161–162, 166–168, 171, 184–186, 188, 191–193, 198, 211–212, 218, 232
Shāfiʿī (al-) 14, 53, 63, 91, 108–111, 113, 115, 126, 134, 138
Shāfiʿīs 50, 53, 63, 70, 109–110, 134, 138, 143
Shahrastānī (al-) 66
Shīʿīs 40, 62, 64–65, 70, 76–78, 85, 110
Simnānī (al-) 28, 156, 166–168
Sufis, Sufism 2, 13, 36, 45, 82–86, 98–99, 101, 104*n*53, 208–209, 225
Sulaymān b. al-Ḥakam (marwanid caliphate) 188
Syria, Syrie 146, 188, 191

Tadnist 228
Tadsī 238
Tanūkhī (al-) 52, 197
Taroudant 234
Thaʿālibi (al-, Abū Manṣūr) 194, 197, 230
Thaʿlab 159, 178, 196
Tibrīzī (al-, Abū Zakariyyāʾ) 2, 156, 196–197
Tirmidhī (al-) 4, 16, 162

Tiznit 232
Ṭūfī (al-) 177
Tughril Bek 149
Ṭurṭūshī (al-) 6, 9–10, 15, 45, 56, 61, 90–92, 94–96, 98–105, 112–113, 117, 137, 169

Yaʿqūb (Abū Yūsuf, Mu'minid prince) 26
Yūsuf b. Tāshafīn 6, 114

Ẓāhirīs, Ẓāhirism 10, 13, 31–32, 56, 69, 85, 132, 143, 146, 164, 166–168

www.ingramcontent.com/pod-product-compliance
Lightning Source LLC
LaVergne TN
LVHW010641300926
849161LV00005B/1015